Textbooks in Telecommunication Engineering

Era of the Internet and Network Science/Engineering

Series Editor

Tarek S. El-Bawab, Ph.D., Boston, MA, USA

Telecommunication and networks have evolved to embrace all aspects of our everyday life, including smart cities and infrastructures, healthcare, banking and businesses, manufacturing, space and aviation, meteorology and climate change, oceans and marine life, Internet of Things, defense, homeland security, education, research, social media, entertainment, and many others. Network applications and services continue to expand, virtually without limits. Therefore, specialized telecommunication and network engineering programs are recognized as a necessity to accelerate the pace of advancement in this field, and to prepare a new generation of engineers for imminent needs in our modern life. These programs need curricula, courses, labs, and textbooks of their own.

The IEEE Communications Society's Telecommunication Engineering Education (TEE) movement, led by Tarek S. El-Bawab- the editor of this Series, resulted in recognition of this field of engineering by the Accreditation Board for Engineering and Technology (ABET) on November 1, 2014. This Springer Series was launched to capitalizes on this milestone. The Series goal is to produce high-quality textbooks to fulfill the education needs of telecommunication and network engineering, and to support the development of specialized undergraduate and graduate curricula in this regard. The Series also supports research in this field and helps prepare its scholars for global challenges that lay ahead. The Series have published innovative textbooks in areas of network science and engineering where textbooks have been rare. It is producing high-quality volumes featuring innovative presentation media, interactive content, and online resources for students and professors.

Book proposals are solicited in all topics of telecommunication and network engineering including, but not limited to: network architecture and protocols; traffic engineering; network design, dimensioning, modeling, measurements, and analytics; network management and softwarization; cybersecurity; synchronization and control; applications of artificial intelligence in telecommunications and networks; applications of data science in telecommunications and networks; network availability, reliability, protection, recovery and restoration; wireless communication systems; cellular technologies and networks (through 5G, 6G, and beyond); satellite and space communications and networks; optical communications and networks; heterogeneous networks; broadband access and free-space optical communications; MSO/cable networks; storage networks; optical interconnects; and data centers; social networks; transmission media and systems; switching and routing (from legacy to today's paradigms); network applications and services; telecom economics and business; telecom regulation and policies; standards and standardization; and laboratories.

Proposals of interest shall be for textbooks that can be used to develop university courses, either in full or in part. They should include recent advances in the field while capturing whatever fundamentals that are necessary for students to understand the bases of the topic and appreciate its evolution trends. Books in this series will provide high-quality illustrations, examples, end-of-chapters' problems/exercises and case studies.

This series is indexed in Scopus. For further information and to submit proposals, please contact the Series Editor, Dr. Tarek S. El-Bawab, telbawab@ieee.org; or Mary James, Executive Editor at Springer, mary.james@springer.com.

Juan Moreno García-Loygorri

Telecommunication Systems for Modern Railways

Springer

Juan Moreno García-Loygorri
Escuela Técnica Superior de Ingeniería de
Sistemas de Telecomunicación
Universidad Politécnica de Madrid
Madrid, Madrid, Spain

ISSN 2524-4345 ISSN 2524-4353 (electronic)
Textbooks in Telecommunication Engineering

ISBN 978-3-031-82249-0 ISBN 978-3-031-82247-6 (eBook)
https://doi.org/10.1007/978-3-031-82247-6

This Springer imprint is published by the registered company Springer Nature Switzerland AG
The registered company address is: Gewerbestrasse 11, 6330 Cham, Switzerland

Note from the Series Editor

This series has its roots in the Telecommunication Engineering Education (TEE) movement in the USA, which I had the honor of leading from 2008 to 2015. The movement was a result of remarkable impacts that telecom networks had already started to have on society, and many more that were looming on the horizon at that time. Major changes in the theoretical foundations of this field of engineering, its design strategies, and its practical applications were unfolding at the same time. The industries we had known as the Telecom, Cable, and Data Communication industries through the twentieth century were going through technical innovations and were reshuffling, repositioning themselves in the market, and converging, in preparation for the global economy of the twenty-first century. The TEE movement was prompted by a personal experience I went through as I moved from industry to academia. Telecom engineering, or network engineering, since the network is centric to the evolution telecom has been through for three to four decades so far, became a distinct multidisciplinary field of study. This field blends ingredients from electrical engineering, computer engineering, computer science, and other disciplines, but may not be taught merely as a branch of any one of them, per se. It became evident that society would be better served by adding specialized undergraduate/graduate courses and programs in telecom/network engineering to our higher education system, especially in the USA, where this was not the norm. The story of the TEE movement has been told and also discussed within broader contexts in the literature (e.g., IEEE Communications Magazine, Vol. 53, No. 11, pp. 35–9, November 2015). The movement had accomplished its main goal of adding telecom/network engineering to ABET's portfolio of recognized engineering disciplincs. ABET is thc Accrcditation Board of Engineering and Technology, Inc., which accredits engineering programs in the USA and more than 40 other countries. A lot of work had to be done to capitalize on that milestone, however. The technical literature of this discipline was rich in scholarly research papers, but not as rich in textbooks designed to support the development of new university courses and curricula. These textbooks are needed to facilitate rigorous academic training for a new generation of network engineers. The idea of this series came up subsequently to help the telecom/networking community overcome this shortcoming. Since its inception, Springer's TTE (Textbooks in Telecommunication Engineering) has been doing this job. The proximity of the acronyms herein—between TEE and TTE—is coincidental, but worth noting! I am proud of the inroads TTE has made in publishing quality textbooks and in fulfilling the goals and objectives the series has had since its beginning. By the end of 2024, we approach the milestone of having 35 books, including 20 published and about 15 in the making (whether in the writing, reviewing, or production phases). More than 50 authors have contributed to this series so far, and they come from about 20 different countries and cultures all over the world. The topics covered by this series so far scan the entire landscape of telecom and network engineering, including wireless, optical, satellite; access and core; devices, systems, and networks; space-aerial-terrestrial integration; queuing theory and traffic engineering; network management; Internet and Internet of Things (IoT); signal processing; applied electromagnetics; and economy/business aspects of telecom. The series includes a cellulargenerations primer and has a Wi-Fi primer project under progress. It covers the most cutting-edge topics with textbooks that are carefully designed to strike the right balance between the need to create new courses thereto today and the fact that progress is imminent in each of these topics, and in the short term. These include books that discuss the roles of edge and cloud computing, blockchains, and artificial intelligence and machine learning in network engineering. Special topics such as underwater communications, railways telecom, networked navigation systems, semantic communications, video analytics, and broadband to combat the digital divide are also covered by TTE. I have tried my best to make this series a top de facto global home for quality telecom/network engineering textbooks, and I am hoping that many in our community can start to consider it as such now. The fruits of the hard work put by many people into fulfilling this goal are now handy to pick by our community. I want to thank my wife Shahira and my daughter Nadine for their support of my work in this series over nearly ten years. I have put a lot of time into this task, and they have understood how this work is important to me. I owe a big deal to all the authors of this series. Words cannot describe my appreciation of their contributions and talent. I would like to also express my thanks to Springer in general, and to Mary James in particular. Mary has been my main partner throughout this journey, and I couldn't have thought of a better partner than her. Finally, I thank all my colleagues in the field of telecom/network engineering and our readership for their continued support of the mission and the endeavor of this great project.

Boston, MA, USA
December 2024

Tarek S. El-Bawab

Preface

The focus of this book is to help understand telecommunications systems in railways. In this sense, the title is not misleading but, behind it, the reader will find a large diversity of topics. The reason for this is that railways exhibit considerable diversity.

Chapter 1 is an introduction to railways, its parts, and environment, highlighting its condition of "system of systems." In Chap. 2, we explain signaling systems, which are one of the most characteristic systems in railways and an essential foundation of safety. In Chap. 3, we explain noncritical services, like CCTV, passenger information, radiotelephony, etc. Both critical and noncritical services are railway specific and are explained in detail, covering wayside and train-related aspects. Typically, rolling stock issues are left out from academic and educational texts, but here we cover them in a detailed manner.

Chapter 4 briefly introduces network technologies used in railways, although they are not railway specific (like, for example, MPLS). Cybersecurity, which is a major trend in railways is covered as well. Chapter 5 is about radio communication and wireless systems. Some of these systems are defined for railways like Global System for Mobile Communications – Railway (GSM-R), which is now close to the end of its life, but still in use worldwide. Mobile communications systems in railways are very complex and are explained in detail, paying attention to physical-layer aspects among others.

Chapters 4 and 6 are about networks, but this time onboard trains. We explain all the technologies behind train control and other communication services, from legacy industrial buses to modern IP-based networks.

Chapter 7 explains some general topics about the maintenance of railway communications systems, tackling several topics like Reliability, Availability, Maintainability and Safety (RAMS) or obsolescence, which are hard to manage successfully.

Finally, Chap. 8 is a look into the near future of railway communications systems. The author has participated in many research projects that have defined the future of some of these systems, and it is always good to understand what will be next. Moreover, having a glimpse of these future technologies is useful to understand the limitations and bottlenecks that current technology faces.

All these chapters have problems that allow students to evaluate their understanding. On the other hand, many references are provided for those readers who may want to delve deeper into the topic.

Madrid, Spain — Juan Moreno García-Loygorri

Acknowledgments

Writing a book would have been impossible without the help of many people supporting me. Some of my colleagues and friends like Julián Martín Jarillo, Carlos Hernández, and Antonio Girón read some chapters and provided many suggestions to improve the manuscript. Indeed, Jose Manuel Riera and Guan Ke read the whole manuscript with great attention to detail. All of them helped to fix mistakes, but if there are any left, it is my fault and not theirs.

I'd like to thank my other colleagues, with whom I am grateful every day to be working, as I learn so much from them each day, like Juan Pablo García Nieto, Mario Rodríguez, Javier Gonzalo, Antonio Pérez Yuste, César Briso, Marion Berbineau, Antonio Melgarejo, and Danping He.

I express my gratitude to Tarek S. El-Bawab, Textbooks in Telecommunication Engineering series editor for his help from the very beginning of this venture. Kaitlyn Bonomo, Mary James, and Pradheepa Vijay, from Springer, helped me throughout the complete process. My gratitude is for all of them.

Finally, I want to write a big thank you to my family: my wife Araceli and our kids, Araceli and Antón. In these two years, writing this book took a lot of time away from being with you, and I'm sorry for that. I really hope you don't end up hating railways because of this. Your support means everything to me.

Competing Interests The author has no competing interests to declare that are relevant to the content of this manuscript.

Contents

1 Introduction to Railways

This chapter is not about communications systems or technologies. Instead, we focus on the foundations of railways as a transportation system. So, here we introduce railways up to a level that allows readers to understand properly all the other chapters of the book (which are about communications for railways), even if the readers know nothing about this topic.

In case the reader is a railway professional or an enthusiast with at least an average understanding of how railways work, rolling stock, signaling, etc., this chapter can be skipped and the second chapter can be commenced.

This book is not about railways history, but delving into it is something we absolutely recommend. Many times, history explains the present. There are many excellent books about this topic in English, like some by Christian Wolmar such as *Blood, Iron, and Gold: How the Railways Transformed the World* [1] or *The Great Railway Revolution: The Epic Story of the American Railroad* [2]. In the further reading section of this chapter, the reader can find much more useful references.

The structure of this chapter is the following: first of all, we explain what railways are and then the differences between railway lines, addressing the diversity of this field; then we introduce the most important agents and organization involved in railways to understand better the way railways are organized; finally, in several independent sections, we explain specific subsystems within railways, such as rolling stock, energy and traction, signaling, stations, and traffic control, and we finish this chapter introducing some maintenance generalities. It is noteworthy that almost all these topics are covered in a specific chapter focused on the intersection between communications and the topic itself (communications onboard a train; communications for signaling, etc.).

Since this chapter provides an overview of railways, the additional reading list at the end of the chapter comprises books, papers, reports, and websites to allow interested readers to know more about some of the topics that we have briefly introduced in this chapter. Notice that in many cases, these references do not agree with our point of view on some topics, but are worth reading anyway. Go read them and form your own opinion!

1.1 Concept of Railways

What we call "railways" is a more diverse thing that may appear at first sight. High-speed trains have little in common with tramways: the whole user experience, operational aspects, etc., are very different, but both run on rails. Between high-speed trains and tramways, we have many other types of railway lines and services, like regional trains, intercity, commuter trains (usually connecting the city center with the outskirts), and subways. On the other hand, we have other transportation systems that are not, strictly speaking, railways like magnetic levitation vehicles (called MAGLEV) or more recent technologies such as the highly publicized hyperloop, which basically consists of a capsule that goes through a vacuum-air tube, with reduced friction.

Before we go into further details, we have to understand what the word "railways" means. The Cambridge Dictionary provides two meanings for this word (the same for its US counterpart, "railroad"):

1. "The metal tracks on which trains run," that is, the track itself
2. "The system of tracks, stations, trains, etc.," that is, the transportation system.

Even though the first meaning is more accurate, throughout this book we will use more frequently the second one. More precisely, we will identify a transportation system as

Supplementary Information The online version contains supplementary material available at https://doi.org/10.1007/978-3-031-82247-6_1.

J. Moreno García-Loygorri, *Telecommunication Systems for Modern Railways*, Textbooks in Telecommunication Engineering,
https://doi.org/10.1007/978-3-031-82247-6_1

"railways" if and only if there is a guided contact (that is, guidance plus contact, which leaves out magnetic levitation vehicles) between the track and the wheel. This definition leaves out both magnetic levitation and hyperloop-based vehicles, but not rubber-tyred or rack railways. On the other hand, guidance is also a key concept here because it has a strong influence in the railways' environment. Please see Fig. 1.1 to understand wheel-track interaction. The reader can see how the wheel flange guides the train. This means that train cabins have no steering wheels because the routing function of the vehicles is performed by the infrastructure, not by the train.

The most general scenario is a metallic wheel and a metallic track.[1] Having a metallic wheel over a metallic track implies a reduced rolling resistance, which means higher efficiency for the movement of the train. However, it also means a reduced adherence, which can lead to inadequate traction and, more importantly, implies that trains need longer braking distances than lorries or automobiles. These limitations must be considered when designing railways lines, for example, avoiding steep ramps.

Digression: MAGLEV

Even though it is not strict-sense railways (there is no wheel-rail contact at all), the magnetic levitation vehicle is worth bringing up, or at least very briefly. It comes from an old idea from the 1930s, but it was put into service for the very first time in 2003 (30 km long, to connect Shanghai's city center and the Pudong International Airport, in China; see Fig. 1.2). It has never been a very popular technology, with very few real-world implementations, but it has some remarkable features, like an impressive peak speed of 430 km/h, 260 km/h average speed, and a low-noise profile. However, its CAPEX per kilometer is twice that of high-speed railways (HSR), which is a major drawback.

There are five MAGLEV lines in the world, three of which are in China, one in Korea, and one in Japan. The Shanghai Pudong one is the only one that is in the high-speed range, as the other three run at 120 km/h or less [3].

Fig. 1.1 Wheelset (wheel + axle) running over the tracks. The conical shape of the wheels guides the wheelset

[1]Actually, the track can be made of many different materials (even stone). For example, Paris Metro has several rubber-tyred lines (i.e., the fully automated meteor line), and the track is made of concrete.

Digression: Hyperloop

Definitely, hyperloop is not railways as it consists of small capsules traveling through low-pressure tubes. The speed target is even higher than magnetic levitation vehicles (indeed it is also based on this levitation). By the time this book was written, there were no real-world hyperloop systems, neither operational nor under construction, but since it was first proposed, it has attracted more attention from academia, media, and even some transport authorities. Rather than a real-world transportation system, perhaps it is a good example of how news and discussion topics are managed today.

1.1.1 High-Speed Railways

Strictly speaking, there is no worldwide definition for both "high-speed railways (HSR)" or "high-speed trains (HST)." Only some intuitions which we disclose in this section of the book, are discussed. A common way to explain the HSR

Fig. 1.2 Shanghai's magnetic levitation train [4]

challenge is to make an analogy with supersonic travel in airplanes [5]: different vehicle design, higher safety standards, better passenger experience, etc. This means that the best available tracks, best trains, maintenance, facilities, etc. are needed, which means higher costs but also a higher usage level of the infrastructure and the trains.

For sure, HSR is a very popular technology all around the world, and several high-speed lines (HSL) have been or are being constructed.

It is generally assumed that a HSL must allow peak speeds higher than 250 km/h, but this is a tricky definition because some lines allow very high speeds sometimes only but have very poor average speeds (which are the most important for transportation). Anyway, the very first HSL were Tokyo-Osaka (1964) in Japan and Paris-Lyon (1981) in France. However, when this book was written (2023), the world leader in HSL length was China (whose first HSL was in 2008, for the Olympic Games), distantly followed by Spain (started in 1992). In all these years, many countries have built several km of tracks for high-speed trains, and more are under construction or being planned.

More precisely, the non-exhaustive list of the main requirements for a HSL are the following (notice that there are some exceptions for each one of the list and that numbers are not written in stone, so they may vary across different countries):

- The separation between the two tracks (that is, not the rails) needs to be greater.
- Wider tunnels (minimal cross section of 75–90 m^2), to minimize shockwaves due to crossing trains, or even, separate tunnels for each track (strongly recommended if the tunnel is larger than 6 km).
- Less curves (minimal radius of 7000 m for 350 km/h).
- Less slopes (up to 1.25%). This requirement usually implies more viaducts and tunnels in the design of HSL. Actually, in many cases, there are higher slopes than this (even up to 4%).
- Stations are more separated from each other, to allow trains to achieve high speeds. HST needs several km to both achieve the maximum speed and stop the train completely.
- Removal of level crossings (see Fig. 1.3).

Fig. 1.3 Level crossing in the USA. Notice that there are signals at both ends of the crossing but not barriers [6]

- AC electrification at 25 kV (typically at 50 Hz, but this may differ in some countries like Japan or USA), which allows more distant substations and thinner wires.
- Lower axle load than in conventional railways (up to 17.5 Tm/axis in HSR). This is a very important parameter when designing the infrastructure and the rolling stock.
- Having a distributed traction scheme onboard the train against the old "locomotive and wagons" allows higher speeds. Some authors say that for speeds higher than 230 km/h, distributed traction is much needed, but there are centralized traction trains running at 330 km/h, like the Talgo series 102–112.
- Rolling stock aerodynamics must be considered because it is very relevant at high speeds.
- If the train sped is high (<200 km/h), there must be in-cabin signaling. We explain this in Chap. 2 of this book.

It is always troublesome to specify the benefits of a transportation mode which varies from one country to another,[2] but in general terms, for a country, the main advantages of HSR are:

- Higher occupation, that is, more passengers (indeed, the proper metric here is passenger · km, which considers how many km every passenger travels).
- Territorial cohesion. Having large cities connected to each other on a fast and safe way, both for national and perhaps international routes. In Europe, for example, there are many international services, such as London-Paris-Brussels and Barcelona-Lyon.
- Industrial development.
- Reduction of gas emissions (trains are much greener than cars) [5].
- Reduction of accidents (people move to trains, which are safer than road traffic).
- Time savings, which means higher economic productivity.

Notice that some of these advantages are extensive compared to other railway services rather than HSR, but are more noticeable here. Among the disadvantages, perhaps the most significant one, is the high costs of constructing the infrastructure. Rolling stock can also be quite expensive, but not to the same extent.

As a fun fact, the current speed record in railways 574 km/h was set in 2008 in France, on a regular high-speed track using a commercial train, but with many modifications and improvements to allow the train to go faster than its design speed. Among these changes were [7] removal of some trailer cars of the train, tightening of the catenary, higher voltage (31 kV instead of 25 kV), higher cant, larger wheels, etc. All these things do not diminish the credit that this impressive record has, but the usual maximum speed that HST achieves in commercial service is almost never higher than 350 km/h.

[2]For example, in the 1960s both Germany and France tried to improve their railway networks to achieve speeds higher than 200 km/h, to remove passengers from road traffic. They won that war but lost against airlines. Only HSR can beat airlines in terms of time travel and ticket cost.

1.2 Gauge

Another important concept in railways is the gauge or "track gauge." There are many of them around the world (see Fig. 1.4), but the most common one is the so-called standard one or UIC (Union Internationale des Chemins de Fer, in English, International Union of Railways) gauge, which is 1435 mm. This distance is measured between the internal faces of the head of both tracks. Actually, this is measured 10 mm below the tread (being the tread the band where the wheel-track interaction is). However, some countries have built their railway networks (or a large part of it) on different gauges, like Spain (1668 mm) or Russia (1520 mm). It is important to understand that having different gauges limits how far trains can go. See Fig. 1.4 for more details on the different track gauges around the world.

It is possible to change the gauge of a train to allow it into a line with a different gauge. This is not very common, but in some scenarios, it provides railway operators a more efficient use of the infrastructures. A very good example of this are the Spanish track gauge changers, strategically placed in some locations of the railway network, allowing some trains to use both the high-speed lines (UIC gauge) and the conventional lines (Spanish gauge). The needed wheelset is far more complex (and therefore, expensive), but the operational savings of having less trains offsets the costs [8].

1.2.1 Loading Gauge

Another important aspect of railways is the loading gauge. This is the definition of the maximum height and width for trains and other railway vehicles. The objective of this definition is to make sure that vehicles can pass through the infrastructure (tunnels, stations, viaducts, etc.), without interfering with them. There are many different gauges, like the static and the dynamic (which depends also on how the train behaves when in motion, that is, especially in curves). See Fig. 1.5 for an extreme example of loading gauge of a London Underground tunnel and the associated rolling stock.

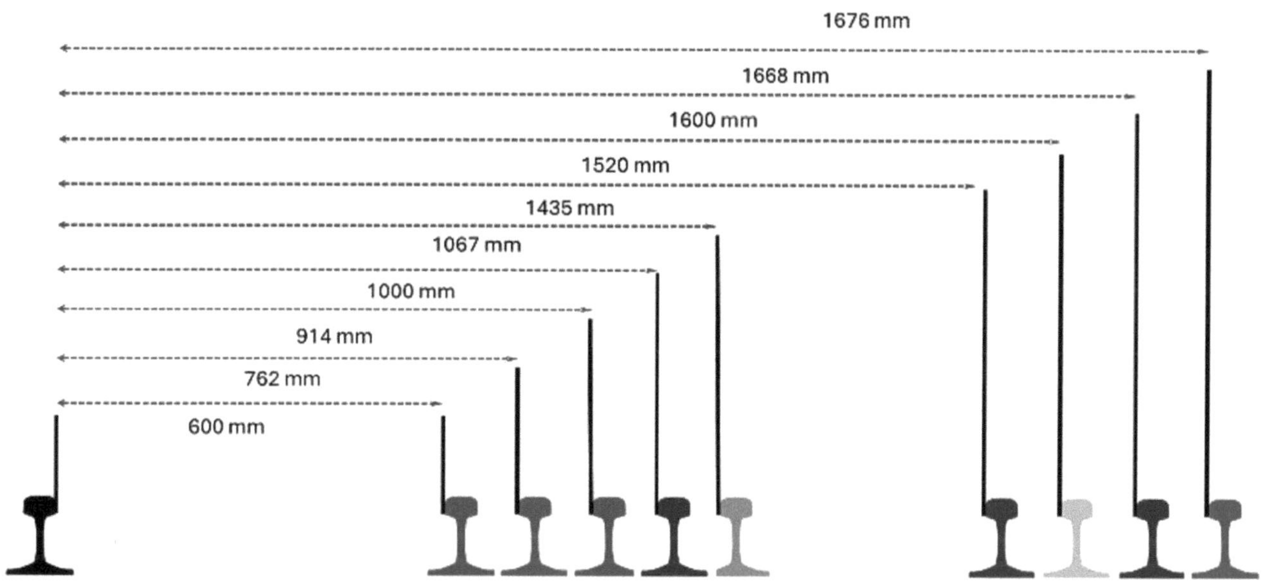

Fig. 1.4 Most common track gauges around the world

Fig. 1.5 London Underground train at the entrance of a tunnel. The clearance space between train and tunnel is remarkably short in this case [9]

1.3 Railway Services

As we introduced before, there are several types of railway services. There is no standard way to classify them, so we follow different criteria:

Speed Some trains, like high-speed trains, reach 350 km/h, yet at times, in some others, such as tramways or metro trains, their average speed equals a pedestrian's. This depends heavily on its circumstances, for example, tramways run in urban areas, surrounded by other types of traffic, like pedestrians, cars, bikes, etc. Its speed profile does not exceed 50–60 km/h in the best case, and there are many stops on its route, so the average speed is much lower. Metro trains have a platform of their own, usually in tunnels, and the maximum speed is usually 80–100 km/h. Commuter trains connect the outskirts of the cities with the city center and sometimes achieve higher speeds than metro trains (around 140–150 km/h). Regional trains or intercity services connect cities located at distances below 200–250 km. Their speed is usually higher than commuter trains but still not close to high-speed trains. Finally, high-speed trains (HST) run on railway lines designed and built on purpose for these trains and can run up to 350 km/h.

Distance Between Stations This is a very important factor because it has a strong impact on many aspects of railways, including the rolling stock design. Metro lines and tramways have many stations very close to each other (around 1 km, sometimes even less, especially in tramways). This has a strong implication in the rolling stock, which needs high acceleration and braking capabilities, not very high speeds (less than 100–110 km/h), and the aerodynamics are not very relevant. On the other hand, high-speed trains do not need to accelerate very fast, but the design of the vehicles must consider the aerodynamic profile of the train, to save energy and make less noise. Regenerative braking increases the efficiency of those services with more stops on its route.

Headway The time interval between trains is usually called headway. Headway is a key quality performance indicator for passengers and an engineering challenge as well (as we will see in Chap. 2 when we discuss signaling and train control). Metro and sometimes even commuter trains run around 1–2 min from each other, but regional and HST usually are separated around 20–30 min. This parameter has also a strong impact on the number of trains needed to operate a line and the capacity (measured in passengers per hour) of the line itself. Headway must always consider the safe distance between trains, that is, the separation between them to avoid collisions or dangerous situations.

Freight and Passenger Trains We have talked about passenger trains only, but there are also freight trains. These trains have a very different function, which is to carry goods efficiently. Although it is not the typical freight train, in Fig. 1.6, a HST freight train, the TGV la Poste is shown (Table 1.1).

Fig. 1.6 TGV la Poste. This freight train, operated by the SNCF and la Poste in France between 1984 and 2015, was the fastest freight train in the world, able to run up to 270 km/h [10]

Table 1.1 Relevant figures for the main railway services

	HSL	Urban	Regional	Freight
Max. number of passengers per train unit	400	1000	120	0
Number of staff per train	4	1	1–2	1
Trains per hour per direction	15	30	1–2	1
Maximum train speed (km/h)	250–350	80–120	160	120
Average line length (km)	150–600	10–40	150–600	–
Average distance between stations (km)	100–200	0.5–1.5	10	–
Headway (s)	85–240	85–130	60–120	120–240

Adapted from [11]
All of them are approximate, with possible exceptions out of the provided values

1.4 Railway Organization: "System of Systems," Agents, and Standardization

A very common way to define railways is saying that it is a "system of systems." This is an ensemble of several different parts like trains, tracks, signaling systems, communications system, energy, stations, depots, ticketing and billing systems, etc. What makes this approach interesting is that it allows us to focus on the interfaces between systems to address each one of them.

For example, the way trains are powered is a key aspect of the operation of the line. In the dawn of railways, all trains were steam powered (even the London Underground ones), but nowadays, this technology is limited to some tourist trains and have been replaced by diesel trains and, more recently, electric trains. Diesel trains have a combustion engine and are powered by diesel fuel, whereas electric trains get their energy from the infrastructure. Therefore, electric trains are significantly different from diesel trains because a path must be set for the electric current to arrive at the engines of the train. This path is usually formed by several parts, like the pantograph onboard the train, a catenary or overhead line, electric substations, transformers, etc.

Intelligent transportation systems (ITS) is a recent term to address some transportation modes, like modern buses, electric vehicles, and railways. Under the umbrella of "connected vehicles," many technologies are helping transportation to be more efficient and safer, such as artificial intelligence (AI), edge and cloud computing, communications, Internet of Things (IoT), and many more.

1.4.1 Railway Agents

Railway agents are those that play a significant role in this transportation system. It is important to understand all of them or, at least, the most relevant ones.

Perhaps the most important roles are the infrastructure manager (IM), who is in charge of the maintenance of the existing lines, construction of new ones, setting up of the tariffs to be paid by operators, traffic management, and several other functions related to the infrastructure (tracks, catenary, stations, tunnels, communications systems, etc.). On the other hand, railway operators (usually called "railway companies") are the entities in charge of the train operation and selling the tickets for passengers. Freight trains are operated by these companies or by different ones, but the role is the same: putting trains to work.

There are many ways to organize railways: vertical integration, holding, and vertical separation.

Vertical integration: It is the same company for both IM and railway operation (trains). No competitors are allowed. This approach is followed by many countries like China or Russia.

Holding: Different companies for IM and trains, but both belong to the same entity (usually, national or regional government or other public sector agencies), and train operators run a monopoly with no competing companies. That is, there is a certain management separation between these two entities. This scenario is usually a transient one between the first model and the third one, but sometimes it can last for years, as in many European countries like Spain, Italy, or France until they allowed some competitors for the historic operator.

Vertical separation: The third model is the IM completely separated to the operators and has at least two competing operators. As we said before, this is now the model for some European countries and for the USA. However, some countries like the United Kingdom have many different operators, but they do not compete against each other because they have an exclusive concession over one line. When we talk about maintenance at the end of this chapter, we address the impact of this third model in the economics of railways and many recent developments.

Other railway agents are maintenance companies, which help railway operators and IM to have all railway assets in good condition. Sometimes the IM and operators do all the maintenance, but in some cases, they rely on third parties to do some specific tasks. For example, small operators (less than 20 to 30 trains) usually hire train manufacturers to maintain the rolling stock.

Another important agent in railways is manufacturers, like train manufacturers, which are the companies that build trains. As car makers, usually they don't build the whole train by itself, but integrate third parties' solutions, such as wheels, bogies, brakes, seats, and many other parts that we will see very soon. Together with train makers, there are signaling companies (in charge of the development of signaling systems), energy companies, communications, construction, and so on.

In many countries, there is an independent agency in charge of all safety issues. Accident investigation, vehicle authorization, railway license management, and many other tasks are among their duties.

Regarding the strategy (that is, planning new lines and extensions of the existing ones), transportation departments (or ministries), local and regional governments, usually rule all these tasks. In general terms (but with very relevant exceptions), railway infrastructures are public but not trains, which are sometimes privately owned. The reason for this, apart from the historical ones, is that building a railway line is very expensive, with a large impact on several aspects of everyday's life.

Even in those countries where railway operation is a liberalized market, that anyone cannot simply operate a train on the tracks. Railway operators must comply with all regulations, safety rules, etc.[3] This compliance of safety requirements is a complex process which usually takes many years to have all trains certified, staff trained, and everything ready to go. Indeed, there are two alternatives for this competition:

Competence in the market: There are many competing operators in the same region or line. This model is present in Spain and Italy, for example.

Competence for the market: There is only one operator for a particular line or region. This is the model for the UK railways.

In the UK model, railway operators operate under a franchise agreement which typically lasts 7–10 years. This means that sometimes the rolling stock is owned by a third party (called "Rolling Stock Companies" or ROSCOs) which leases trains to the operators. This model, which has many pros and cons, has enforced relevant research on maintenance optimization which has made railways more efficient (as we will see in Chaps. 5 and 8).

Apart from manufacturers and maintenance companies, there is still room from other companies which usually help the former and also operators and IM: the service companies, for installations and other similar tasks which do not need high-tech profiles but skilled workers aware of the railway world.

Engineering companies are a relevant agent in railways as they do all the studies and projects related to both the building of the infrastructure and large modifications of the existing elements.

Research centers and universities have an essential role in the technologies that are making railways greener and better every day. For example, the European Shift2Rail program and, more recently, its successor Europe's Rail are both good examples of how innovation can improve this sector when putting these research entities to work together. Since around 2012–2015, we have been living in an offspring of railway-related research activities, in several fields, like rolling stock, energy, materials, digitalization, and so on.

Another agent involved in railways are law enforcement agencies, like security staff, police, civil defense, firemen, etc., which need to be integrated in the daily workflow of railway operation to do their duties properly (prevent crime, protect passengers, etc.). We will see some details of this in Chap. 2.

Finally, we left the most important one at the end of the list: passengers. Their relevance not only comes from the fact that they make railways possible by using them every day but also because of the way they interact with some of the other agents, requesting improvements, providing feedback, etc.

1.4.2 Railway Standardization Bodies and Institutions

Why is standardization so important in railways? It is a key element when it comes to improve safety, security, and passenger comfort, while reducing cost and improving the overall quality and competitiveness of the railway sector [12].

Standardization in railways is a hot topic today, yet historically, it has received minimal attention. For example, not only between neighboring countries but also within a country or region, the trains were not interoperable, that is, they were not able to run on different lines. But first, what does interoperability mean?

Interoperability is about a train being able to go through different lines, in different places without changing trains. For example, if a train can go from London to Koln, crossing the United Kingdom, France, Belgium, and Germany, several aspects must be considered, like the voltage of the line, signaling system, gauge, track width, and some others. Perhaps the train must be able to work with different gauges, signaling systems, or voltages, but the optimal solution is to have only one on each case: an interoperable technology. In the next chapter, we will go into further details when we talk about Global System for Mobile Communications - Railway (GSM-R) and European Rail Traffic Management System

[3] This compliance of safety requirements is a complex process which usually takes many years to have all trains certified, staff trained, and everything ready to go.

(ERTMS). For the moment, it is enough if we understand why we need interoperability in railways.

The first entity that must be mentioned when talking about railway standardization is UIC (Union Internationale des Chemins de Fer, in English, International Union of Railways) is an association that "promotes rail transport globally with the objective of responding effectively to current and future challenges relating to mobility and sustainable development." Indeed, it is also very active in some standards, like GSM-R and Future Railway Mobile Communication System (FRMCS), which are both communications systems for railways and are covered in other chapters of this book. It has its headquarters in Paris (France).

UIC is also important in terms of notation standardization of locomotives and trains. This system allows to convey specific information about the vehicle, like country, gauge, type of train, or locomotive (freight/passenger/…) plus the number of the train within the series.

Besides the UIC, there are other regional agencies like the "Union of African Railways," founded in Addis Ababa, Ethiopia, in 1922, or the "Association of American Railroads" (for freight railways in Canada, Mexico, and the USA) and many others around the world.

Another important body is the UITP (Union Internationale des Transports Publics, in English, International Association of Public Transport), which is more related to urban transport. Indeed, it has bus companies among its almost 2000 members. In the United Kingdom, the Rail Safety and Standards Board (RSSB) is in charge of monitoring and enforcing safety in railways. There are many other agencies around the world with a similar mission as RSSB, independently from IM, government, and railway operators.

The European Union Agency for Railways (ERA) helps standardization and safety enforcement in European railways, for example, in ERTMS-related activities, among others.

UNIFE (Union des Industries Ferroviaires Européennes) is an association of the railway industry in Europe, which promotes railway standards and also has a very active role advocating its members' interests. It has developed the International Railway Industry Standard (IRIS) family of standards for requirements in several railway-related aspects, like audits and some others.

CEN/CENELEC/ETSI: These three bodies oversee European standardization. There are subtle differences between them, but in general terms, CEN (European Committee for Standardization, in French: Comité Européen de Normalisation) in charge to develop European Standards (EN), not only for railways but for many other fields as well. It is in charge of fostering the European Single Market as well. For example, a norm that will be mentioned several times when we address onboard train communications is EN 50155. This norm must be followed by European countries, but many other regions also do the same. CENELEC ("Comité Européen de Normalisation Électrotechnique," in English "European Committee for Electrotechnical Standardization") is focused on electrical engineering[4] topics. In many cases, CEN/CENELEC work together. ETSI (European Telecommunications Standards Institute) is a certification organization focused on IT and communications standardization, such as Global System for Mobile Communications (GSM), Terrestrial Trunked Radio (TETRA), Long-Term Evolution (LTE), or Digital Enhanced Cordless Telecommunications (DECT).

In China, the most important standardization body is China National Institute of Standardization (CNIS). It is responsible for the development of standards in several fields, not only limited to railways. Besides, the "China Academy of Railway Sciences" (CARS) is also involved in research and standardization tasks in railways.

Finally, there are many private organizations across the world that are important agents in the railways sector either on a regional or an international area. For example, in the USA, American Railway Engineering and Maintenance-of-Way Association (AREMA) fosters education for railways professionals and publishes "best practices" documents which are very helpful for its members. A similar example in the United Kingdom is the Institution of Railway Signal Engineers (IRSE), which is for those related to both railway signaling and telecommunications.

To sum up, a recommendation for newcomers to railways could be to get to know which organizations and agents are relevant in this field, as well as to understand the main applicable standards.

1.5 Railway Infrastructures

1.5.1 Train Stations

We are familiar with train stations or railway stations[5] because that is where passengers get on and off the trains. There are one or more tracks, one or more platforms, ticket machines, and other facilities like waiting rooms, etc. These buildings are frequently structured in several layers, perhaps one for the tracks, another one for the main lobby, etc. In some other cases, there are different railway lines that are served by the same station (this is very common in subways and commuter trains).

However, there are some considerations that must be considered, like its location along the railway line has a strong influence on the performance of the line; there are terminal

[4]Please note that it is widely accepted that the term "electrical engineering" (EE) addresses electrical and also electronics and telecommunications engineering.

[5]We leave the term "depot" which is much used in the USA for railway stations, to identify the rolling stock maintenance facilities.

(or "terminus") stations (the line ends and begins there) and through stations, where trains pass through, do not ending or starting there. Platforms can be isolated (platform is placed between both tracks for the two directions) or non-isolated (two separate platforms outside the track). In some low-density lines, there are some small stations where there is almost no staff and, in some special cases, trains only stop on demand. The height of the platform is a critical parameter to have a suitable (and safe) onboarding of the train. Indeed, it has a strong impact on the rolling stock design (train doors must open at the same level as the platform), but this is something difficult to achieve if there are different platform heights and different trains in the same line. However, only few countries, like Japan, standardize this height.

A very important thing in train stations is accessibility. Routes inside the station must be carefully planned to avoid architectural barriers for impaired people, trolleys, etc.

Moreover, there are some stations that have very high throughputs of passengers every day (like, for example, Shinjuku Station in Tokyo (more than three million people every day), a huge station with more than 200 exits.

Regarding communications engineering, stations are special places for many reasons. Here we provide only a summary, but in the other chapters of this book, we go into further details.

1. Sometimes many trains meet at the same time at a station, if it is large enough, as central stations in large cities are.
2. There are many facilities and systems that may require connectivity in train stations, like CCTV cameras, lifts, escalators, intercoms, etc.
3. Railway and security staff require access to communications services when in stations.
4. Whenever possible, communications equipment must be placed in technical rooms in stations. It is better than having them somewhere in the line.
5. In automated lines, with no driver onboard the train, sometimes platform screen doors (PSD) are placed at the edge of the platform. These doors are opened only when the train doors do the same and provide extra safety for passengers, who will not fail to the tracks by accident. These doors must be remotely controlled by the signaling system and require connectivity.
6. Train stations require modern Public Announcement and Passenger Information Systems (PAPIS) not only for impaired people but for all the passengers. In Chap. 3, we see more details about these systems.

1.5.2 Track, Tunnels, and Viaducts

The layout is a generic term to address the infrastructure, that is, where trains run. It is formed by three parts: the main line (where commercial services are provided); branch or secondary lines (from depots or workshops to main line), and stations (see Fig. 1.7).

For telecommunications, the layout is very important due to its impact on the physical layer. For example, on a regional or HSL, there are viaducts, tunnels, trenches (the tracks lie several meters below ground level, with slopes at both sides of the tracks), etc. In subways, it is very frequent to alternate different tunnel cross sections (rectangular, circular, etc.) and outdoor segments, whereas tramways typically operate within urban areas. In Chap. 5, we deal with radio aspects of communications, and we see the impact of all these scenarios, which is considerable.

The track is composed of a pair of steel rails (see Fig. 1.8) and has many important functions:

1. Support the loads generated by the train. These loads can be either vertical, longitudinal, or lateral.
2. Guide the train (remember that trains have no steering wheel).
3. Allow track circuits to work properly (as can be seen in the Signaling section of this chapter).
4. In electric trains only, close the circuit that begins in substations, allowing the electric current to flow back to the ground.

There are different types of rail profiles, but the most common are Vignole for light trains and tramways and the UIC for almost all the others, including HSR. Regarding the trackbed (what lies below the rails), there are also two different options: ballast track (that employs crushed stones; see Fig. 1.10) and ballast less (or slab track, as seen in Fig. 1.9), where the trackbed is composed by concrete. In HSR, it is very common to have slab track, but in old lines, ballast is still dominant. The mission of the trackbed is twofold: to allow drainage of the track and avoid vegetation to grow up and damage the infrastructure, and second, to distribute and transmit forces to the terrain. Ballast track demands more maintenance workforce but is cheaper to construct (Fig. 1.10).

1.5.3 Powering the Trains: Catenary and Substations

When the engines of the train are electric, a power source is needed (if trains are not electric, there is no need to have any catenary or substations). This power sources are called electric substations. These substations can be either alternate current (AC) or direct current (DC). DC is more common in train systems that demand less power, like subways (typically, 600 VDC, 750 VDC, and 1500 VDC), tramways (600 VDC and 750 VDC; with some cases at lower ones, such as 550 VDC and 300 VDC), and commuter trains (1500 and 3000 VDC). On the other hand, when it comes to AC, we find HSL which are usually powered at 25 kV (50 Hz) or 15 kV

Fig. 1.7 Platform screen doors in the Elizabeth Line in the London Underground [13]

(16.6 Hz). Regional lines are sometimes AC based, but others remain at DC. For new lines to be constructed, the de facto standard is to go to 25 kV at 50 Hz for HSL and 1500 VDC for subways, leaving for example 600 VDC for legacy lines only.

This electric power must get to the train engine to move the train and energize all the onboard systems. To do this, there are several methods, like the catenary or overhead line (OHL), which is the most common (see Figs. 1.11 and 1.12), and the so-called third rail (or live rail, which is a rigid conductor at one side of the train, allowing the electric power to get into the train as well). For the catenary, there are different types, but all of them require a pantograph onboard the train to retrieve the electric current. Once the current is on the train, it faces some transformations to be consumed by the engines and other onboard systems, which may have different electric requirements. Finally, as we saw in the track section, the ground return circuit is completed by the wheels of the train and, finally, the tracks. Therefore, the loop is closed.

Catenary is far from being a simple element. Here are a few examples to demonstrate this:

1. It must be mechanically compensated to not have dependence of temperature in its position. As usual, there are exceptions: rigid catenaries, which are mostly used only in some metro lines, do not require mechanical compensation.
2. It is not a straight wire, but a zigzag one, to avoid excessive wear of the pantograph collector (which is the part of the pantograph that is in touch with the catenary). The wire must have a specific tightness and stiffness, to ensure good electric contact and minimize energy losses and voltage decay. The wire is electrically insulated from other parts of the system (masts, etc.).
3. In AC lines which have three phases, there must be a neutral zone with no voltage to separate the section for each different phase. In these neutral zones, the train must never stop.

Fig. 1.8 Single-track trenched railway in Spain

Fig. 1.9 Slab track

Fig. 1.10 Ballast track

Fig. 1.11 Overhead line mast in Spain

Fig. 1.12 Catenary overhead line [14]

4. Imperfections in the catenary-pantograph contact may lead to sparks or even electric arcs that can severely harm both the OHL and the pantograph. This is more common in DC lines with low voltages than in AC but can happen there as well.

It is out of the scope of this book to explain all the technical details of power electronics that energizes electric trains, but in the "further reading" section of this chapter, we provide some useful and interesting lectures.

1.6 Signaling

Signaling is a key topic to understand both railways and its evolution. The term comes from the signals that are placed to the sides of the track, bearing the most important information in railways: if the train can move forward or not. Not all the safety in railways belongs to signaling, but a large part of it. Communications are essential for signaling and many times, signaling professionals are highly skilled in communications (and vice versa). However, this book is not about railway signaling, so in this section (and in Chap. 2 as well), we introduce only the key concepts that are needed to understand properly signaling and its impact on communications systems.

1.6.1 The History of Signaling Is a Story of Learnings

Back in the nineteenth century, signaling was only about signals, but today with trains running at 350 km/h and subway trains with a headway as low as 60 s, signaling is much more complex than that. For the moment, it is enough if we consider that signaling is the set of systems and technologies that allow the safe movement of the train. However, this definition is incomplete because it leaves aside efficiency. If the capacity of a railway line or the regularity of the trains is higher, we can say that we have better efficiency. These two aspects are very important in modern railways, but it is important to note that neither of them is safety related.

Therefore, signaling is mostly for safety and, next for efficiency. In the remaining of the section, we introduce the key signaling concepts interleaved with its history. This history ends with the last large innovation in railways: in-cabin signaling, which is explained in Chap. 2 together with the communications systems that make it possible.

In the early days of railways, there were no protection systems apart from the expertise of train drivers and other railway staff. Unfortunately, accidents started occurring, so technology came to the rescue. The first signals (remember: to let drivers know if they can progress or not) were flags. Later, they were replaced by discs, directly controlled by the railway staff. In 1841, the first semaphore is installed in the London-Croydon line. This signal is nothing more than a mechanical arm as depicted in Fig. 1.13, which works in the following way:

- Arm makes an angle of 90 degrees with the pole (arm down): means stop.
- 45 degrees means "caution" (or sometimes, "clear").
- 0 degrees (arm and pole are aligned) means clear (or in some cases, "signal off").

Railway Slang

It is much more accurate, and the reader will get much more credit saying "signal" instead of "semaphore." A semaphore is something that changes periodically and railway signals change based on events (a train getting into a particular track, for example). We have used the term "semaphore" here for the very last time in this book. From now on, there will be signals only.

Fig. 1.13 Mechanical railway signals [15]

Please notice that if the arm fails somehow, it will go down, meaning "clear track." This is not a safe scenario. The opposite to this is the basis of robust systems and is called "fail-safe," which means that in case of a failure, safety is not compromised. Always design systems with this idea in mind: "failures, breakdowns, software bugs, etc. will occur, but safety must never get compromised."

Manual switches are another key technology from the same era. Today, there are still manual mechanical switches. Most of the early days railway lines had very low densities, and the speed of the vehicles were also low. Fortunately, railways were a huge success from the very beginning, so the demand skyrocketed, and safety requirements went higher as well. For example, switches and signals became more concentrated on a single point in stations or near them.

Shortly after, the world's first formal railway legislation appeared: it was the "Regulation of Railways Act" [16], which was a law passed by the British Parliament in 1889. This law was passed after a deadly accident that happened shortly before (the Armagh Accident, which happened on June 1889 and resulted in 80 deaths and more than 200 injuries). Strictly speaking, this is something very common in transportation systems: accidents with fatalities very often lead to changes in regulation. These lessons learnt from accidents are used to improve the safety of the service.

1.6.2 A Key Element: The Interlocking

From this act, interlockings became mandatory, which are another key element in railways and are still widely used. Indeed, interlockings are the essential pillar of safety in railways, because they prevent the authorization of train routes that are potentially conflictive for two or more trains (in the simplest scenario, having two incoming trains at the same time in the same track). That is, interlockings "lock" a route for a train, forbidding it to other trains.

The very first interlockings were mechanical, soldering together some levers of the switches and also individual signals for a complete itinerary. They were invented by P.A. Vignier, a French railwayman, and were installed for the first time in the Batignolles station in 1855 [17]. Safety improved a lot with interlockings because they allowed for the avoidance of many human errors.

1.6.3 Track Circuits

Another key aspect for signaling is how to locate and detect trains, that is, the detection of the occupation of the tracks. In the early days of railways, this was done manually, but eventually, both the need of capacity and some accidents enforced a new technology for train detection: the track circuit (TC). It is widely used around the world even today, for more than 150 years. It was invented in 1872 by William Robinson; the first installation was in 1873 in Pennsylvania (USA).

Track circuits must be installed in the whole railway line and work as follows:

1. Installation: A transmitter and a receiver must be wired to the rails. The rails will act as the conductive wire for the electric current.
2. The transmitter will send a known signal (a certain voltage) which will arrive to the receiver if:
   (a) The rail is not broken (open circuit).
   (b) The transmitter and the receiver do not fail.
   (c) There is no train in between the transmitter and the receiver.

If a train is between the transmitter and the receiver, due to the very low impedance of the wheelset (wheels + axles), the electric current will flow through the wheelset instead of arriving to the receiver. That is, the train shunts the electric circuit. Therefore, the receiver will notice this and report an occupation. Therefore, if a certain part of the track (a track circuit, i.e., between a transmitter and its associated receiver) is flagged as "occupied," no other train can get inside that very track circuit. This procedure is depicted in Figs. 1.15 and 1.16.

It is very important to notice some important things about track circuits:

Trains must shunt the circuit: To have this detection system working properly, the train must shunt the electric circuit (remember that train wheels are metallic). This means that its impedance must be very low (less than 1 Ω)

Fail-safe: In case something fails (a broken rail or a faulty transmitter or receiver), the system will flag this scenario as an occupation ("no signal" in the receiver means "track circuit" occupied).

Broken rails: TC can be used to detect broken rails (in case an occupied TC is reported, and no train is actually there). Not all broken rails imply an open circuit in the TC receiver, but some of them can be easily detected this way.

This system is limited by the maximum distance where the TC receiver can detect the incoming signal sent by the TC transmitter. This distance is related by the transmitted power, the attenuation due to the rails, the distance, and the sensitivity of the receiver. There are other influential parameters like rail rustiness, trackbed (ballast or concrete), presence of water, etc., that can affect the maximum TC length. Therefore, usually a margin is given from the technical specifications of the TC system. In HSL, the typical maximum length for a TC is 3 km.

Receivers: On a TC, there is only one transmitter, but it can have more than one receiver. This is especially useful if the track section is not straight and contains one or more switches (which leads to more than 1 ending; all endings must be ended with a receiver, otherwise that section will not be protected). Having more than 1 TC receiver shortens significantly the length of the TC. Indeed, TCs were first used only in stations with more than one track. Later, they were also installed between stations.

If there is more than one train, the track circuit will show occupation, but there is no way for the receiver to notice that there is more than one train. To be more precise, we should have said "wheelset" (two wheels plus the axle between them) rather than train, because one shunting wheelset is enough to be detected, and there is no way to notice if there is more than one between the transmitter and the receiver.

In Figs. 1.16 and 1.17, we show how do TC work when a train is in the TC and when there is no train on it.

1.6.4 Blocks

Please also notice that estimating the train location with TCs is not precise in the modern sense. Once a TC is reported as occupied, we positively know that a train is somewhere between the transmitter and the receiver (that section is *blocked*), but we do not have further precision. This is something that severely limits the performance of the railway line (in terms of capacity or number of trains per unit of time). We see more details about this in Chap. 2 when we introduce transmission-based signaling (TBS) systems, which can improve performance without compromising safety at all.

In Fig. 1.14, the reader can see a schematic of a single-track railway line, divided in several TC. Every TC is protected by a signal, and we call block to each one of the sections on which the line is divided. Depending on the status of each TC, the signal has a different color. Notice that there are signals for both directions because on a single-track line, there are trains running in both directions.

To ensure safety in circulations between stations, blocks started to be widely used. An interstation was blocked if a train was somewhere between the stations. Therefore, no train was allowed to depart a station once the previous one released the block (that is, the previous train arrived to the next station). To implement blocks, several technologies were used, like physical tokens which were given to the train driver (or other staff onboard the train), and they gave it back to the chief of station as soon as they arrived. Another technology that increased both safety and capacity was first the telegraph and later the telephone. When a train was about to depart a station, it was reported to the next station by telegraph/telephone. Telephonic block is still available in some countries, as a backup system or as the main train detection system.

1.6.5 Modern Signaling Concepts

Signaling technology continued improving by the introduction of the following:

1. Entrance signals in stations: Protect the entrance in stations. They are placed several hundred of meters before the station.
2. Advanced signals: Give information of the status of the next signal, giving more time of reaction to the driver.
3. Automatic block for single-track lines: Allows the removal of the station master.

At the same time, improvements in mechanical engineering allowed faster trains, which would have been useless unless signaling technology permitted them as well. This is something usually forgotten when planning railway infrastructures: wayside elements and rolling stock must be considered together.

Eventually, mechanical interlockings were replaced by the novel electric ones (based on safety relays) and later by electronic interlockings (based on microprocessors). Redundant systems were frequently used in both power supplies and other components.

All these improvements led to safer railways and to a better transportation system, capable of carrying more passengers. However, there were still some problems to solve, like interference between TCs. This was solved modulating the signal at a different frequency for each neighboring TC. In jointless TCs (like regular TCs, but without a gap separating two of them), this frequency planning is mandatory.

Another big issue was the noisy environment, and the solution for this was digitalization of the signal. Regarding TC high price, for low density lines, a suitable solution for the train detection problem is the axle counter. This device counts the number of wheels that enter a certain point (where

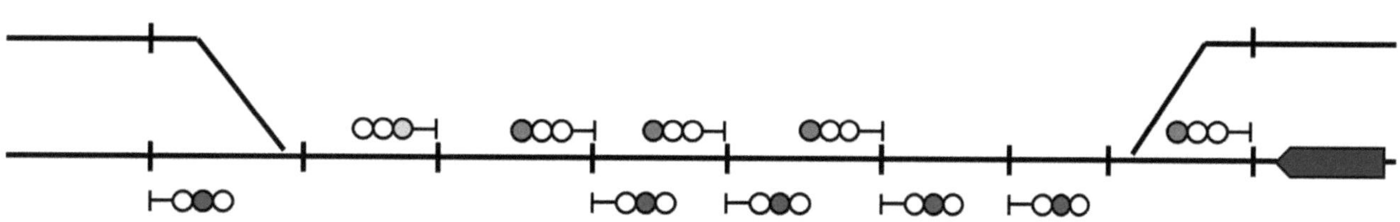

Fig. 1.14 Single-track line, with all the signals in the route

the axle counter is placed) and the number of wheels that leave that place.

Finally, operational control centers (OCC), a major innovation, were introduced to centralize the operation of one or more railway lines, managing together all its signals, TCs, switches, and other elements. As we see in Chap. 2, these OCCs have evolved largely and now can monitor the whole line, including station systems (elevators, escalators, etc.), energy, trains, and even control some of these systems (Figs. 1.15 and 1.16).

The next step in signaling technology was onboarding all the information to the train cabin. We explain this in Chap. 2.

1.6.6 Detectors

Strictly speaking, detectors do not belong to signaling, but they are very relevant for railway safety, so we introduce them in this section. Detectors are much used in HSL and not very much on the other railway lines, but there are remarkable exceptions. They are placed in some points of the track, and its purpose is very specific: to trigger alarms under certain conditions. They are of many types like:

Lateral wind detectors: Lateral wind can be a serious treat for HST, especially in viaducts and at the end of trenches. These wind blows can derail the train, and the detectors are used to slow down trains (or stop them in the worst case). These detectors are complete weather stations who send data periodically to the OCC, where the decisions are made. Wind detectors are placed in sensitive places (viaducts, tunnel entrances, etc., and in some other places where the weather models demand them).

Hot bearing detectors: They detect overheated axle boxes and brakes of the trains. They are placed in strategic locations where train brakes and axle boxes are more prone to heat and, in the worst case, break.

Foreign object debris (FOD) detectors: Once the train gets a certain speed, it is impossible for the train driver to react and stop the train safely if something is in the track that could compromise safety. These detectors are very useful to avoid automobiles falling from an overpass to the railway track and are usually complemented with CCTV cameras that help avoiding false positives.

Vertical impact detectors: Tracks can be severely damaged by some particular trains. The idea behind these detectors is to measure the vertical load that every train puts on the

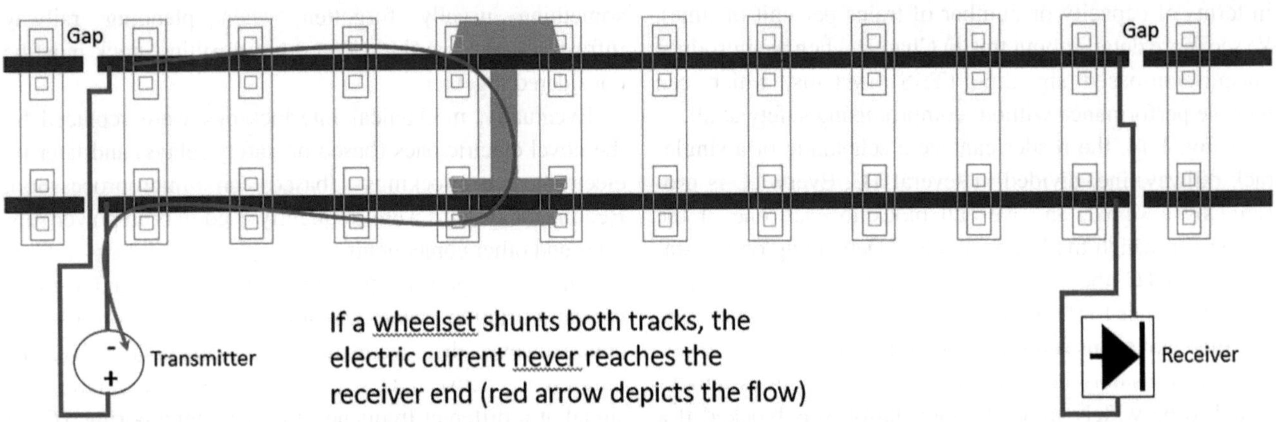

Fig. 1.15 Track circuit. The wheelset shunts the tracks and no current reaches the receiver. The train is detected

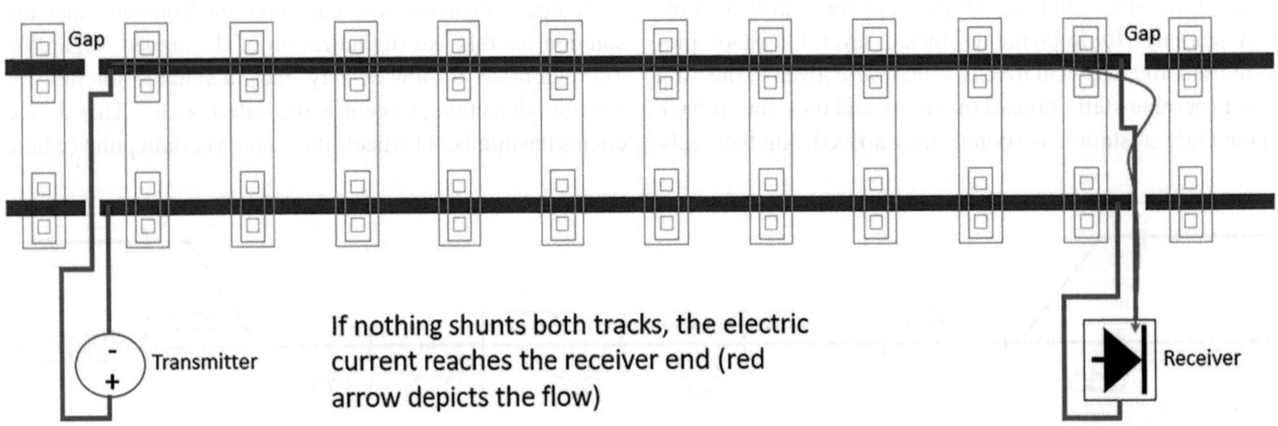

Fig. 1.16 Track circuit. The electric current flows through the tracks and gets to the receiver. No train is detected

track, to detect faulty wheels (flat wheels are especially troublesome), excessive train weights, and other causes that may harm the track.

Earthquake detectors: In some regions, earthquakes can be harmful, and the railway track is especially sensitive to their effects. The idea for these detectors is to gather the information from the seism and help the OCC to decide whether to stop the trains or not. Earthquake detectors can be either external to the railways' system detectors or not (that is, the detector control center can receive alarms triggered by its own earthquake detectors, from third parties or both). Perhaps the most relevant example for this is Japan, quickly safely stopping all its trains few minutes after the 2011 Great East Japan Earthquake.

Level crossing obstacle detection: Its mission is to make sure that tracks are clear in level crossings. Consider that level crossings can be protected by two barriers, one barrier only, or have no protection at all. The technology for detection can be radar based, optical (like Lidars), or CCTV related (using computer vision or not). Unfortunately, at the time of writing this book, there are still many fatalities in level crossings around the world.

Many others, like flange lubrication detectors (FLD), prevent excessive wear on both wheels and tracks, and many more. When we discuss maintenance in Chap. 7, we introduce another detection systems used for rolling stock maintenance.

All these detectors placed in the track are connected to a control center (sometimes dedicated for the detectors only) to process the information collected by them. In case an alarm must be triggered (perhaps to stop the train or to slow it down), this information is transmitted through the network to the OCC.

1.7 Rolling Stock

Trains are, by far, the most noticeable element of railways. There are uncountable websites and forums devoted to rolling stock where railway enthusiasts around the world share their thoughts, pictures, and other posts. In this section, we introduce the basics of trains, paying some extra attention to its integration with wayside elements; in Chap. 6, we go deeper on onboard systems and networks.

1.7.1 Types of Trains

First, let's delve into rolling stock taxonomy. There are many ways to classify trains, but the most popular one is by the traction system:

Steam trains: The oldest ones: the classical postcard of the nineteenth century survived until the 1970s–1980s almost everywhere. Now they remain only as tourist trains. They had external combustion engines, and the propellant was typically coal.

Diesel trains: Trains powered by diesel engines (or, more generally, internal combustion engines). The propellant is consumed on an internal chamber, therefore generating energy.

Electric trains: The first electric train was manufactured in 1879 by Siemens in Germany, and today it is, by far, the most popular type of train. Strictly speaking, all HST are electric trains, which do not mean that some diesel-powered ones can't achieve very high speeds (the Spanish company Talgo currently holds the world record for a diesel-powered train: 256 km/h in 2002). Nowadays, all metro trains are electric[6] because the emissions of diesel or steam trains are not compatible with underground tunnels.

Hybrid trains: Have two different traction systems onboard, being able to switch between them, like diesel and electric, for example.

Minority ones: Trains propelled by jets, rockets, etc. They are nothing more than prototypes (very interesting ones).

Staying on this subject, another important distinction is between centralized and distributed traction. In centralized traction, the train is formed by locomotives (which have engines) and trailer cars (which do not have engines). Locomotives can be at one end of the train or at both, but there are exceptions where the locomotives are in the middle of the consist. On the other hand, in distributed traction, the traction is distributed among many cars in the train consist. This approach is getting more and more common, and many modern trains are built this way. This has a strong impact in the overall design of the train, and in the maximum slope, the train can navigate [17]. In Tables 1.2 and 1.3, many details about some trains can be found.

In some cases, it is the legislation which takes the decision. The Eurotunnel, which is the infrastructure that lies behind the English Channel, connecting France and the United Kingdom, had trains with centralized traction because it was not allowed to have thick electric wires (for traction) in passenger cars. However, this has changed, and since 2015, Eurostar e320 trains, manufactured by Siemens, have distributed traction. Eurostar train exemplifies interoperability nightmares which are explained in Chap. 2 as an introduction for ERTMS.

[6]The London Underground opened in 1862 and its trains were all steam-powered units. Eventually were replaced by electric trains, but there are still remains of that age: for example, the fake houses at 23 and 24 Leinster Gardens, which concealed the part of the track where trains released their steam.

Table 1.2 Main characteristics of some representative high-speed trains

Parameter	Siemens Velaro E	TGV Duplex	Shinkansen N700	Talgo 350 (S112)	Acela Express
Length (m)	200	200	204.70	200	202.91
Number of cars	8	10	8	14	8
Passenger capacity	404	516	546	365	304
Weight (M/R) (Tm)	425	380	357.5	322	565
Power (kW)	8800	8800	8540	8000	9200
Voltage (V)	25 kV 50 Hz	25 kV 50 Hz 1.5 kV DC	25 kV 60 Hz	25 kV 50 Hz	25 kV 60 Hz 12 kV 60 Hz 12 kV 25 Hz
Max speed (test/commercial operation) (km/h)	403/301	574/320	–/270	330/300	266/240
Country	Germany	France	Japan	Spain	France/USA

Table 1.3 Main characteristics of some representative metro trains

Parameter	Alstom Metropolis	Hitachi Rail Driverless Metro	Bombardier Movia (S7)	CAF Metro train	Rotem
Length (m)	86	109.80	117.35	109.31	166
Number of cars	5	2–6	7	6	8
Passenger capacity	960	1094	865	1270	1800
Weight (M/R) (Tm)	157	194	268.1	205	335
Power (kW)	3200	3168	–	3040	3600
Voltage (VDC)	1500	750 and 1500	750	1500	1500
Max speed (test/commercial operation) (km/h)	80	110	100	110	135
Country	France	Italy	UK	Spain	South Korea

1.7.2 General Description of How Electric Trains Work

In this section, we explain at a very high level how an electric train works. First, the train must retrieve the electric current from the catenary (or third rail) using a pantograph. Then we find the circuit breaker, to protect train equipment from excessive electric currents. Then we have the converters, some for traction and others for the auxiliary systems (air conditioning, doors, etc.). Traction converters are for supplying electric energy to the engines and for electric braking. The electric current goes back to ground through the train wheels. In case the train is braking, this means that the current goes backward. This is called regenerative braking, and notice that two conditions must be fulfilled:

1. The braking train must be able to turn around the current.
2. Another train must be near to receive this current. Otherwise, this energy is dissipated onboard the braking train (in large resistors, and we call this rheostatic braking). This requirement only applies to DC lines (like subways) and not to AC lines (like HSL).

Regenerative braking is an essential feature in modern trains which allows operators to save considerable energy (around 20%, depending on the circumstances) in their daily operation. Apart from the electric braking (regenerative or rheostatic), there is also the pneumatic braking, which is not electric at all.

1.7.3 Main Characteristics of Modern Trains

To end this section, in Tables 1.2 and 1.3, we provide a summary of some of the most representative HST, trams, and subway trains.

Tramways are considered as very standard vehicles, but they can have many particularities, so it is hard to compare them reasonably. For example, the percentage of low-pass floor, the number of cars, if the vehicle must run on batteries for a while (typically, some parts of the city center), the minimum curve radius, the maximum speed (as in suburban areas stops are more separated and trains may run faster), and many others are sources of potential variations. Furthermore, there are several generations of all the tram models provided in Table 1.4 since they have been commercialized for decades. Consequently, Table 1.4 must be handled with care.

Table 1.4 Main characteristics of some representative tramways

Parameter	Siemens Combino	Alstom Citadis	CAF Urbos	Flexity	Stadler Tango
Length (m)	19–54	20–40	18–56	20–40	25–44
Passenger capacity	100–250	140–300	129–327	135–275	150–265
Power (kW)	Up to 8 × 100	Up to 720	–	Up to 640	440–600
Voltage (VDC)	750	750	600 and 750	600 and 750	600 and 750
Max speed (km/h)	70	70	70	70–100	80–100
Country	Germany	France	Spain	Germany	Switzerland

1.8 Operation and Traffic Management

Railways are a scale-driven industry. The denser the network is across the territory, the better [5]. However, railway operation is hard to explain because it tends to be site specific, that is, largely adapted to the circumstances of the place. This means that generalizations are almost impossible, and providing here details of a particular country would be useless. Assuming that it is the core of the railway industry, in this section, we try to sketch some general aspects to allow readers to understand easily the operational aspects of their own countries with less effort.

In a nutshell, railway operation (which includes traffic management as well) is about keeping trains running on schedule. To do this, several things must be considered. First of all, the specifications of the line, like gauge (UIC, metric, etc.), voltage (3000 VDC 1500 VDC, 750 VDC, 600 VDC, 25 kV AC, etc.), or non-electrified line; signaling system, speed profile of the line, maximum length of the train, maximum slope, platform height, maximum axle load (that the tracks can bear), minimum radius of a curve, and minimum speed in ramps: this is very important in freight trains, etc.

The problem here is those are many parameters, and it is hard for the rolling stock to be compatible with all of them. This is the main driver for interoperability. A Polish train may have the same gauge as a German one, but perhaps it is not ready to be powered with the same voltage or, more likely, the signaling system would be different. A train running from one country to the other needs to be compatible with both, and this is why there are dual-voltage trains: to work in two different railway lines (or, sometimes, even more than two).

Therefore, an operator must have a very good knowledge of the characteristics of the lines; their trains will traverse to adapt the rolling stock to the infrastructure. But the job does not end here, because very often it must attach to a service level agreement (SLA) where operators are requested to provide several if not all the following features.

A minimum capacity: For example, 30,000 passengers per hour and direction for a metro line. This is something that will impact for sure the number of trains in the line and the number of cars per train unit.

Headway between trains: This is strongly dependent on the signaling system of the line (not all of them allow very low headways), but it is impacted by the dwell time (the time a train stops at a station). Dwell time is also influenced by the number of people that gets on/off a train in that station and in the number of train doors (less doors mean larger queues and more time). However, headway is the main driver for dense lines only (that is, many metro lines, some commuter lines, and some specific HSL). Metro trains usually have three to four doors per car and side of the train to allow many people to go on/off of the train. Metro lines have frequent stops, and dwell time is critical. But HST trains only have one door per car and side, because stops in HSL are less frequent and dwell time is not critical.

Timetable (and schedule): The time where the train is expected to depart a station and arrive at the next one.

To measure its performance, railway operators use several parameters:

- Number of trains at peak hour and, perhaps, number of cars, because they can have a line with 20 trains formed by 3 cars (that is, 60 cars) or 15 six-car trains (90 cars). In the first scenario, more drivers are needed, but they have less transport capacity.
- Passengers: This is not very useful because it accounts the same for a passenger who travels 1 km only and also for another one who travels 600 km.
- Passenger km: The sum of the km that every passenger travels. This key performance indicator (KPI) is one of the most important in railways.
- Train km: The same as passenger km but for trains.
- Punctuality rate.
- Rolling stock heterogeneity: Having many different trains is a complicated task for an operator, because both the maintenance and the operation (has impact on training, for example) become harder[7] if there are many types of

[7] The typical example of this is Ryanair, the Irish airline, which only has Boeing 737 airplanes on its fleet. This way, training is simpler (all pilots are well trained to fly every plane) and maintenance is more efficient (better prices for parts, more skilled staff, no need to have stocked parts for small fleets, etc.).

trains. An extreme example of heterogeneity is FGC in Spain, which operates in five different gauges: Standard (1435 mm), meter (1000 mm), Iberian (1668 mm), 600 mm, plus a rack railway.
- Total incomes and incomes per passenger.

The same for infrastructure managers' performance:

- Length of the network.
- Number of stations.
- Number of train services on its network, or, better, usage of the network. This is how many trains they have out of the maximum they could have.
- Density: This could mean track length divided by the surface of the region or track length by population. The idea here is to provide an indicator of how dense a network is. For example, in large cities, it is very common to give a percentage of people that lives 500 m or closer to a metro station. The denser the network, the higher the percentage.

And, for both infrastructure managers and operators, the most important KPI is the safety record, which is the number of incidents (and accidents). This record is usually given by the passenger km number or, typically, train km.

For freight trains, the picture is very similar, but operators must consider the characteristics of the freight stations, like classifications yards, dry ports, containers stations, etc., depending on the goods to be carried. Freight trains are the most efficient transportation mode in terms of energy per transportation unit, and it is even more efficient when a large amount of goods share origin and destination (because railways is not a "door-to-door" mode). Moreover, to compute the maximum load for a train, freight train operators must account for the maximum number of wagons and maximum force at the coupling (as well as maximum slope and other parameters mentioned above for passenger operators).

Depending on the railway organization (see Sect. 1.4.1) and the specific conditions for any region or country, there are some other aspects to consider, like:

Railway fee: If vertical organization is not the railway model, operators pay a fee to the IM. This fee can be different across countries, but it is usually the main source of incomes for IM. Typically, this fee accounts for both maintenance and energy costs, plus the railway passenger service (capacity) [5].

Loss-making routes are operated under the public service obligation (PSO) paradigm. In these unprofitable routes, operators do not make enough money to pay for both operation and maintenance, so the government subsidizes them (under certain conditions).

Decision-making on stopping at a station or not: This decision is usually made considering both the incomes (train tickets) and the costs. The costs are related to the energy needed for another stop, the extra time for the complete route, and the wasted seats (which can mean less passenger km). Typically, other aspects have impact on this decision, like socioeconomic reasons.

1.9 Maintenance

Railway maintenance is nothing more and nothing less than preserving the condition of all railway assets. This section could end here, but before we address this topic in depth in Chap. 7, we would like to give readers some intuitions about maintenance.

1.9.1 Types of Maintenance

First, we need to learn about the types of maintenance:

Corrective maintenance: Fix only what ceases to work properly. This one is the most obvious and common around the world. However, it is not very efficient as all failures can impact the service and, perhaps, safety as well.

Preventive maintenance: Perform periodical checks to see that everything is fine. If not, replace the damaged parts. This approach is costly as it takes many resources, and many times no failure is prevented as everything is OK. Moreover, it is time-consuming and, depending on the strategy, it can be costly as well. However, it is largely used in industrial environments as it provides a balance between costs and risks.

Predictive maintenance: Try to determine in advance when an asset will stop working and replace or fix it just before. This type of maintenance is optimal (both guarantees safety and minimizes costs with no service disruptions) but requires an accurate model for the assets to maintain which can predict the condition of the asset. Many times, this is impossible to achieve due to the great complexity of the railway systems. However, in some scenarios where the behavior of the assets is predictable (because it is linear with time or depends on one or two variables only), this strategy is worth following. Hopefully in the future, helped by large machine learning (ML) or deep learning (DL) models, this type of maintenance would be more extended.

Legal maintenance: This is the maintenance which must be done by legal imperative. In railways, there are many regulations and applicable laws, and this maintenance must fulfill all the requirements that come from them.

Condition-based maintenance (CBM): This type of maintenance is based on continuously sensing the assets to maintain to react in case it is needed. No models to predict its condition are needed, because it is assumed that it can be directly measured. The main idea here is to measure one or more variables (train axle boxes' temperatures, for example) and trigger an alarm if this measurement is above a certain threshold. This threshold can be set based either on previous knowledge from the manufacturer or, maybe, data from similar failures of this system. Current technology (as of 2023) is ready to make CBM work but requires large expenditures to place sensors, connect them to the network, transmit the data, collect them, etc. We will go further on this topic in Chap. 7.

1.9.2 Advanced Maintenance Strategies

No need to say that maintenance, alongside with operation, takes a large share of railway companies' budget. Therefore, there is a lot of pressure on them to be as much efficient as they could be, but without compromising safety. A good tool to perform an efficient maintenance and operation of a railway line is life cycle cost (LCC). It consists of considering all the costs of an asset, whatever they come from, including raw material cost, manufacturing, maintenance, operation, and asset retirement costs (addressing recycling issues). This is a good solution to avoid failing in some fallacies, like buying a cheap communications system than along its whole life may require heavy maintenance, skyrocketing costs, or buying a cheap train that, even if it does not require much maintenance, perhaps is very heavy and requires much more energy than other trains.

There are many guidelines to estimate the whole LCC of an asset, but it requires both future predictions (when starting the project) and good accounting (in the lifetime of it).

1.9.3 Obsolescence

Another important aspect to tackle in railways is obsolescence. Obsolescence happens when something ceases working because it is outdated. This impacts two areas: technical obsolescence and cybersecurity.

Technical obsolescence: A component or subsystem cannot be repaired because the manufacturer does not support it. Railway companies must ensure that their assets will have support for a certain period (typically, until they are upgraded) avoiding service disruptions. This means paying attention to component level (for example, CPUs and other ICs can be outdated by the manufacturer without being noticed by either the system integrator or the railway company).

Cyber security: Obsolete parts can become a security threat if they cannot get security patches and other updates. Old PCs with legacy Windows versions dealing with critical tasks are a typical scenario for this.

For all this, obsolescence must be managed properly, forecasting all the potential issues before they impact railway service. There are many ways to address obsolescence, but all of them are based on common sense and many times not very easy to achieve, for example, not working with providers who refuse to commit to large obsolescence periods, having a suitable obsolescence plan, etc.

1.10 RAMS

RAMS stands for reliability, availability, maintainability, and safety, which are the four pillars of industrial performance. Indeed, these four parameters are essential in railways to account for the overall quality of service. RAMS is a very deep world, so here we only introduce it very briefly, but it is really at the heart of railways, with a remarkable impact on design, operation, maintenance, etc. Indeed, RAMS is something that railway engineers must keep in mind from the beginning of the design phase to achieve their objectives (very demanding, as we will see in this section). Fault-tolerant systems (that is, redundant) are a good example of this. Otherwise, patched designs, with on-the-go changes, usually fail and damage severely the railway service. Let's go with it:

1.10.1 Reliability

Reliability is the ability to work without failures. It is usually expressed as a function of time (failures per year, for example, in some wayside systems) or distance (failures per 100,000 km or per million km, in rolling stock). This is very important because failures may lead to service disruptions and may take time to be repaired (see maintainability). It is also important to notice that handing over an asset to maintenance sometimes means putting a train in the depot, stopping all railway traffic in the line or waiting to the end of the day. All these alternatives are costly and can have a large budgetary or reputational impact on both IM and operators. Therefore, railway systems are usually designed to be not only fail-safe but also very robust and reliable, which means higher acquisition costs.

1.10.2 Availability

Availability is the ability for a system to keep working. For example, track circuits are usually required to have at least 99.999% of availability. If the reader does the maths, this is an outage of slightly more than 5 min per year. A very common architecture in interlockings is the "2 out of 2," which means having two separate systems, and when both outcomes are unanimous, it means that everything is OK. Another architecture (used for example in onboard systems) is "2 out of 3" which has three redundant systems and a voting system: the overall outcome is the one that have at least 2 out of 3. This one has a higher availability and reliability. A common approach in railways is to average the time of unavailability over the global time of operation, but another one is to consider only the availability at a particular time of the day. The latter is very common in train operators, which need a certain number of trains at peak hour in the morning and want to maximize the availability at that time.

1.10.3 Maintainability

Maintainability is the ability to be easily (that is, timely) maintained. This is measured in time units (in railways typically, hours). This KPI has a special importance because failures in railway assets must be fixed fast to avoid delays and service disruptions. A typical KPI for maintainability is mean time to repair (MTTR). It comprises the total time for corrective maintenance divided by the number of actions (in each time period). As in the other RAMS aspects, this MTTR is usually a requirement for the designers and can be largely impacted by its external conditions (for example, its location onboard the train: it is not the same to place an RF transceiver in the train roof or in a locker in the passengers' area. In the first case, electric traction must be switched off, staff must climb to the top of the train, and then repair the issue. In the second case, it is only needed to open the locker). In this sense, it is important to distinguish "repair" and "recovery":

Mean time to recovery: Time starts when the failure is discovered.

Mean time to repair: Clock starts ticking when maintenance staff can access the malfunctioning system. This could mean that the train is stopped in the depot, for example, or the track is out of service to be fixed.

1.10.4 Safety

Safety is the protection from harm, more precisely, being free from accidents and malfunctions that could harm people or other assets. The fail-safe concept we introduced in the signaling section of this chapter is a good example of how safety is managed in railways. However, it is important to notice that no human system can be completely safe, no matter how hard we try. Indeed, trying to pursue the "100% safety" paradigm could be tricky, because this would increase costs significantly, impact ticket prices, remove passengers from trains, and, ultimately, have more fatalities (only if they move to less safe transportation modes, like private cars). We go deeper on safety in Chap. 2 when we discuss transmission-based signaling.

Reliability and availability are closely related, because if something fails at a high rate, both its reliability and availability will be degraded. Maintainability, however, is not closely related to the others and depends largely on how accessible a system for the maintenance staff is.

In recent years, some authors have started talking about RAMSS (with two "S") to acknowledge the importance of security (including cyber) as well. In Chap. 4, we provide more details about cybersecurity in railways.

Outline of the Book

Throughout this book, we will explain many details of telecommunications systems for railways. The organization of the book will be top bottom, that is, we will start explaining the use cases and applications (the top of the stack, as seen in Fig. 1.17); then we will go progressively down until we finish with some topics about the infrastructure (technically, not the bottom of the stack, which is the physical layer, but the networks).

Therefore, Chap. 2 covers one of the most relevant systems in railways: signaling, which is one of the bases of safety. In Chap. 3, we explain non-safety-related applications, like CCTV, radiotelephony, etc. All these use cases and applications are very important to understand because they are the main input for all the requirements for communication layers below (network, physical, etc.).

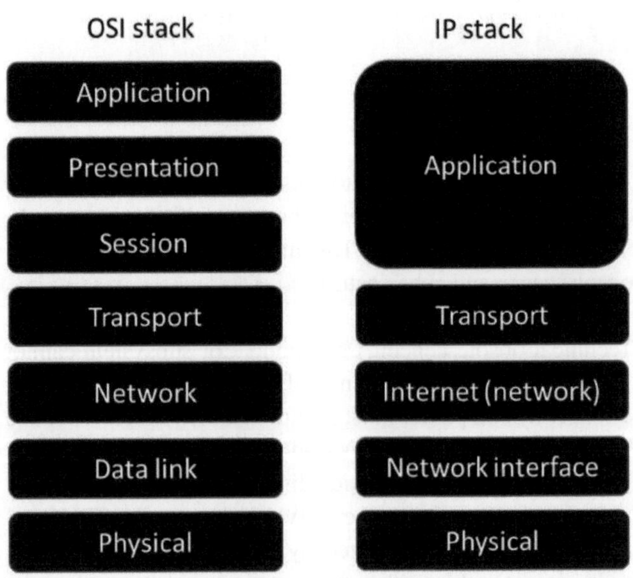

Fig. 1.17 OSI stack and IP stack

Chapter 4 is about the network layer, and we explain several technologies (including some legacy ones) and topologies. Onboard communications systems (that is, the internal networks within trains) will do the same in Chap. 6.

Consequently, Chap. 5 is about radio communications systems, mostly train-to-ground systems like GSM-R, and we explain some physical layer aspects which have an outstanding importance in vehicular scenarios (like railways), and it is very often left out.

Finally, two short chapters: Chap. 7 is perhaps more an annex to all the previous than a chapter itself, because it covers maintenance aspects. The purpose behind this is to let students and readers understand that any railway activities can be successful without considering the maintenance aspects, or, more generally, the whole life cycle of technology and operations. In Chap. 8, we try to look into the future, explaining some of the upcoming technologies, use cases, and challenges for the communications systems in railways. We don't forecast the future; what we do is to describe some (past) projects that will star the new versions of some railway-related standards for railways. Luckily, this chapter will be outdated sooner or later, but its mission is to help understand both the upcoming future and also the state of the art, looking at both its challenges and bottlenecks.

All chapters include short questions and problems that will allow students to reinforce what they have learned. Further reading references are recommended in case the reader wants to learn more about the topic.

Summary of the Chapter

This chapter is all about introducing readers to railways. Therefore, we started with some definitions, learning about high-speed rail (a concept that is much more than going fast, because this requirement has implications at many levels). Then we learned about gauge, differences between railway lines, like regional or urban rail, plus some "weird" transportation systems which are not strict-sense railways, like MAGLEV and hyperloop.

In engineering, standardization is very important, so we paid some attention to this, and we introduced the main agents involved in railways around the world. There are many, but the most important ones are infrastructure managers and operators. This is responsible of the wayside and the rolling stock (i.e., trains), respectively.

We introduced many wayside elements, like the substations, catenary, signaling systems (which will be explained in more detail in Chap. 2), detectors, the track itself and many more. Then we had a subsection to explain some general aspects of the rolling stock, introducing the different types of trains, plus some onboard systems.

The chapter ended with a very short introduction to two core tasks in railways: operation and maintenance. Regarding maintenance (which is covered in Chap. 7), we explained the key parameters: reliability, availability, maintainability, and safety.

Exercises

1. Choose any railway line near your home and find its main parameters, such as length, number of train services, maximum speed, minimum speed, average speed, number of stations, average/maximum/minimum distance between stations, and voltage (if any).
2. On a metro line, discuss the impact of long stops (larger than 1 min) in the headway of the line. What about only one stop with more than 3 min?
3. Tramways usually do not have an onboard signaling system. Discuss why they do not have such a system and what should be done to let them have an efficient one.
4. Power calculations. Consider any of the HST from Table 1.2 and compare it with your favorite automobile. Calculate the ratio between power and number of passengers.
5. Why railway crossings are usually not present in HSL?
6. We have explained how a single-track stretch implements the track circuit detection technology. But what about a line stretch with more than two ends (like a junction)? Draw a line with a junction and place the track circuit to detect any train on that stretch.
7. Track circuits can be used to detect some broken rails, but not of all types. Considering what we have learned from this technology, draw a sketch of a broken rail not detectable by track circuits.
8. The system under study was not working for 2 h last year. Please calculate the availability of this system.

Further Reading

We had no room to explain the engineering details of railway construction, which are really interesting. There are many books on this topic, like this one about design, construction, and operation of railways [18]. Tunnels are engineering marvels, and this book about early British railway tunnels provides historical context and highlights the work of some rockstar engineers of that time like Brunel and some others [19]. Stations is another topic that may interest general readers: there is a recent book about some London stations [20], a paper about historic trends on station design [21], a guide for station design released by Network Rail [22], and an UIC report [23]. There are many useful and well-presented documents in the UIC website as well, covering many different railway aspects. For high-speed rail, there is a nice brochure by the UIC [24], a paper published a long ago in Scientific American [25] on vehicular dynamics and other mechanical aspect, economics [26], and many more references. In chapter 12 of [18], there is a lot of information about specific requirements for HSL and HST.

Let's go for a polemic issue: railway privatization in the United Kingdom. A pioneering initiative which implied the division of a large public enterprise, British Rail, into more than 100 small companies. Without entering the discussion, it had both good and bad consequences, but there are many things to learn about. This report [27] from the House of Commons summarizes many of them (even though it is focused on discussing the future of rail, it does explain the recent past as well). Franchises, ROSCOs, COVID19 impact on railways, and many things are addressed (with data!). Here is a more academic paper on the privatization ok UK rail [28]. Even if some points may be arguable, these two references show the great complexity behind railway policy and operation.

Japanese railways' technology is in some aspects the most advanced of the world. This can be seen when learning about some examples, like this one about the train control system [29], an old one about technologic advances that made possible Japanese train bullets in the 1960s and 1970s [30], and other innovative developments in rolling stock to improve it [31].

Finally, speed records and track gauges are two top things to know about in railways. For the speed records, the best resource is Wikipedia and in Visual Capitalist, there are some nice infographics about the trains involved in these speed records.

2 Transmission-Based Signaling

The structure of this chapter is as follows: first, we explain non- transmission-based signaling (TBS) systems, that is, distance to go and some other signaling systems that do not rely on radio communications to work. Then we go into the core of the chapter: European Rail Traffic Management System (ERTMS) and communications-based train control (CBTC), which are best two examples of TBS and the dominant ones in the market for both high-speed line (HSL) and main line (for ERTMS) and subways (CBTC). Then we explain the different grades of automation of a railway line which are enabled by TBS. Finally, we provide an insight of operational control centers (OCC) to explain how those lines equipped with TBS on its tracks and trains are operated. As usual, there are two more sections at the end of the book: Exercises and Further Reading.

2.1 Concept of Transmission-Based Signaling

In the previous chapter, we introduced the basics of signaling: switches, crossings, signals, track circuits, and interlockings. In this chapter, we start where we left off. The main focus is the transmission-based signaling (TBS) systems, which are nothing more and nothing less than what its name indicates: signaling systems, that is, for train protection and control, whose data is carried by a (radio) communications system. The term "radio" is placed between brackets because, in radio communications, only in one part of the end-to-end communications system, data goes through the air.

For sure what is needed is a communications system between wayside (that is, ground) and the train. In this chapter, we cover only the signaling system itself: how it works and its requirements to the communications systems.

Conventional signaling relies almost only on track circuits (TC) and lateral signals. In-cabin signaling is about making all that safety-relevant information available to the driver onboard the train. This is the main difference between the old-fashioned (conventional) signaling systems and the modern ones: in the latter, the information for the driver is not only in the lineside signals but onboard the train as well. This is called "in-cabin signaling" or "cab signaling," whereas the other is called "lineside signaling" or "conventional signaling." There are two reasons that trigger the evolution toward modern systems:

1. Studies have shown that when speeds are higher than 200 km/h, it is almost impossible to see the signals on the track. Therefore, a reliable and safe system cannot trust only what the driver sees or think they see.
2. The need for efficiency. Even if the speed is not that high (for example, in metro trains), it is a very good policy to have the essential route information onboard the train. This information includes:
   (a) Speed profile (including limits)
   (b) Status of upcoming signals
   (c) Other useful information about the track

   There is almost no limitation for the information that can be provided, as we explain later. Once again, the most important words in signaling: safety and efficiency. Therefore, with this new in-cabin signaling concept, we should stop thinking about wayside signals and start visualizing the complex control desk of a train (see Fig. 2.1).

These systems transmit safety data (and, sometimes, non-safety data as well) from the ground systems to the onboard systems. This information is depicted on a screen (or something else) to let the driver decide and act.

Supplementary Information The online version contains supplementary material available at https://doi.org/10.1007/978-3-031-82247-6_2.

J. Moreno García-Loygorri, *Telecommunication Systems for Modern Railways*, Textbooks in Telecommunication Engineering,
https://doi.org/10.1007/978-3-031-82247-6_2

Fig. 2.1 Control desk of a commuter train. Notice that several displays are used to provide information to the driver. (Courtesy of "Escuela Técnica Profesional de Mantenimiento de RENFE". Printed with permission)

2.2 Automatic Train Control (ATC)

ATC is the system that controls safe train movements. It has a mandatory part, called automatic train protection (ATP) plus two optional ones (see Fig. 2.2): automatic train operation (ATO) and automatic train supervision (ATS).

ATO (automatic train operation) performs speed regulation functions, stops the train at the stations, as well as other functions related to the operation of the train but not to the protection of the train. That is, ATO is not safety related.

ATS (automatic train supervision): It is about train monitoring and keeps headways at the desired values (that is, regulates the flow of trains, slowing or speeding them up if necessary).

These two (ATO and ATS) are not mandatory, but it is very uncommon to have an ATP system without an associated ATS. ATO is very often left only for metro trains, but we have more details on this in this chapter.

ATP is about safety, that is, protection of the train. In the next section, we go into this.

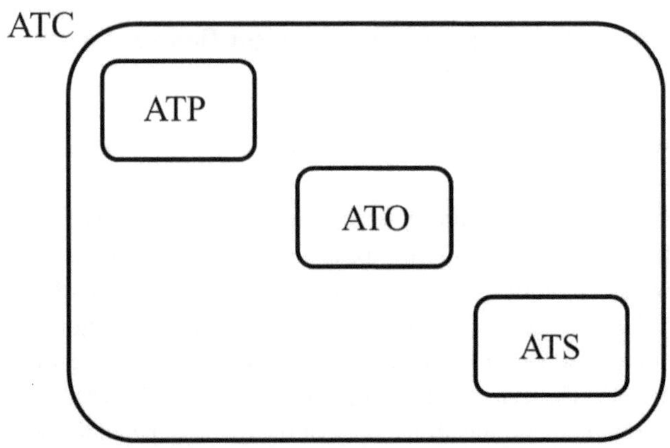

Fig. 2.2 ATC is formed by ATP plus ATO and ATS subsystems

2.2.1 ATP Functions

Automatic train protection (ATP) is a system to perform all safety-critical functions, which are to be implemented by the signaling system.

Train location: It is the determination of the location of each train. It optionally includes a secondary train location determination system, perhaps as a fallback method or any other scenario (more on this in the "degraded scenarios" section).

Speed determination: Train-speed plus train-direction determination are ATP functions. Given that all measurements have errors or inaccuracies, all the related processing must consider that errors. This is all safety-related calculations coming from wherever they come must have safety margins to skip fatal errors. This many seem something obvious, but train-speed measurements rely heavily on wheel diameters, which are not a static parameter but a variable one. The reason for this is the wheel-track contact that progressively erodes the wheel, decreasing its diameter. We provide more details on speed determination in the "movement authority" section in this chapter. Another important aspect is the redundancy: both in terms of number of equal devices and of different devices running in parallel (i.e., performing the same function).

Train separation: ATP systems guarantee that trains are separated such a distance to allow safe braking (that is, without collisions or other unsafe situations). Speed determination and safe train separation can be seen as two parts of the same function, because one influences largely the other. It is needed to determine in advance the braking characteristics of the involved trains. The closer the trains run, the more important is the safe train separation.

Overspeed protection and brake assurance: ATP system controls the speed of the train to avoid overspeed.

Rollback protection: ATP system prevents trains going backward for any reason. Indeed, sometimes a train goes past the stop and must move backward a little. This maneuver is quite delicate to perform, because it can easily imply a safety threat with the upcoming train.

End-of-track detection: Based on the safe-braking model that the ATP system has for the train, this function allows the safe movement of the train near the end of the track. There are many possible end of tracks, and they shall be addressed carefully. For example, the typical bumping post (stopblock in the USA), another railway line, a work zone, etc. The safe braking model is a mathematical tool that represents the performance of the train brakes. It must consider worst-case scenarios, potential failures, degraded modes, etc.

Consist protection and safe coupling and decoupling of train units: Many times, trains can be coupled together with decoupled units to form different trains (more about types of trains in Chap. 6). Depending on the operator logistics and convenience, this can be done at the end of the day, from time to time or even in normal operation. This procedure has several implications for onboard networks which are explained in Chap. 6. For the ATP system, it is needed to detect the coupling/uncoupling of the train. Determination of the train length is especially critical for the safety, but this is not the only aspect to consider.

Door opening control: Train doors must open only when the train is stopped at a station or any other authorized location.

Departure interlocking: Train shall only depart the station when it is safe to be done and certain requirements are met (doors closed, green lights, etc.).

Emergency braking: In case it is needed, the ATP is responsible to trigger the brake. This function is twofold: the watchdog to determine to trigger or not the emergency brake (based on the brake model of the train) and sending the command to the brake system. Notice that the simplest ATP systems were only about emergency braking to protect the train. For example, Spanish ASFA system works this way: avoiding overspeed at certain points of the track and emergency braking if necessary. There are many similar systems all around the world.

Route interlocking: This is very similar to the classical route-locking function performed by conventional interlockings, which we briefly explained in Chap. 1.

There are some others than can be consulted in the CBTC standard [32]. Notice that we have explained ATP functions considering CBTC as a reference. This may seem an awkward decision, but the reason is that CBTC systems are the most advanced signaling systems, and the set of performed functions is larger than in other technologies. However, this means that there are other ATP technologies that do not perform all the functions we have explained in this section. ERTMS can be addressed on a similar way, and the many related functions are defined there as well.

Before we continue explaining the other ATC subsystems (ATO and ATS), we will look on safety threats, how to account them, and how to manage them.

2.2.2 SIL Levels

Once again, in this section, we only introduce the foundations of safety management, to allow readers to understand some of the following sections in this chapter . Our recommendation for interested readers is to start studying the following standards, which are the basic corpus for safety management (among other topics, like RAMS, which was introduced in chapter):

- EN 50126 [33, 34] railway applications: The Specification and Demonstration of Reliability, Availability, Maintainability and Safety (RAMS) Generic RAMS Process: Generic RAMS process to optimize railways

companies' and railway suppliers' reliability, availability, maintainability, and safety. See Chap. 1 for more details.
- EN 50128 [35] Railway applications: Communication, signaling, and processing systems. Software for railway control and protection systems: in this standard, safety levels are defined and structures software development, maintenance, validation, and documentation, that is, the complete software life cycle. In this section, we discuss mostly safety levels, with some comments on software.
- EN 50129 [36] Railway applications: Communication, signaling, and processing systems. Safety-related electronic systems for signaling: that is, the same as EN 50128 but for the hardware. It addresses development, maintenance, quality control, etc. We focus on this standard for the rest of the section.

These standards are sometimes vague, sometimes very precise but are the basic reference for this section, and are essential to understand how modern railways are engineered, managed, and operated.

Consider that safety engineering is a very complex discipline which covers many fields, like analysis of threats, estimation of the probability for potential failures, accounting the human influence, and so on.

Notice that having 100% safety, that is, a risk level of ground zero is not only impossible but is even not recommendable at all. It is impossible because there will always be unpredictable threats, and it can take a huge sum of money and time to do it. It is not recommendable because such a system would be very expensive;[1] some people will stop using it and move to unsafe options. Therefore, risk must be kept at a manageable level. This is a very important concept. Moreover, both deciding which risk is manageable and measuring it are very difficult indeed.

The key idea here is that the more critical a system is, the safer it must be. This means that the tolerable hazard rate (THR) must be low. THR is defined as "the maximal frequency of occurrence that is acceptable for the hazard in a given scenario" [33]. According to the three standards mentioned before [33–35], there is a range of THR for each safety level (safety integrity level, SIL). In Table 2.1, we provide a summary of these THR ranges.

The ideal situation would be to have everything SIL-4 certified, but this is unpractical for several reasons, like cost and complexity. Therefore, a careful analysis must be performed to determine the SIL for every system and function. The usual approach is the following:

- SIL 4: ATP systems, track circuits, and interlockings
- SIL 3: Train brakes and perhaps door control units (both PSD and train doors)

Table 2.1 THR vs SIL

SIL	THR per year
1	$10^{-6} \leq THR \leq 10^{-5}$
2	$10^{-7} \leq THR \leq 10^{-6}$
3	$10^{-8} \leq THR \leq 10^{-7}$
4	$10^{-9} \leq THR \leq 10^{-8}$

- SIL 2: Traffic control systems and ATO, monitoring of the axle box temperature, etc.
- SIL 1: Speed monitoring and displays, maintenance and diagnosis systems
- SIL 0: Everything else, which means no safety

As the reader can see, this list is far from being exhaustive, but the philosophy is to require the highest SIL for safety-critical systems, whose failures will probably mean a safety threat. Notice as well that the THR for higher SILs are very low numbers.

Machine Learning and Safety

Nowadays, where AI is nearly ubiquitous, a relevant question arises: Will machine learning (ML) improve railway signaling and safety? While it's difficult to give a short answer, the response is likely "no." Or more precisely, "not with the way safety is currently managed in the railway industry.".

What do we mean by "as safety is addressed nowadays"? To answer this question, we must refer to the standard EN 50128, which governs SIL determination. The key point is that software validation follows a clear, structured process that ensures traceability from high-level requirements down to low-level code. This code is typically modular and decomposed into atomic functions, following the best practices established over decades of industry experience and hard work.

But ML works in a very different way. The function that relates inputs and outputs of the system is "learned" from a large dataset, instead of decomposed into low-level code. This is the dataset is the requirement specification. Therefore, this approach has some problems. First, it raises questions about the representativeness of the dataset to account all possible scenarios for the safety-critical software. An incomplete dataset, with limited coverage, may lead to poor results. Another issue is validation. In data-based algorithms, it is harder to trace requirements into code, so more limited black-box tests should be done.

Therefore, please note that the answer is "not for the moment," but it is not a "no, never." Hopefully, railway industry finds a way to overcome these difficulties and allow certification of AI-based safety systems, as the automotive industry is doing.

[1] Needs several redundancies, long testing time, and complexity will skyrocket, making such a system unmanageable and not very practical.

We are finishing this section here instead of getting deeper on this topic, which is very complex and very relevant. Before moving to ATO and ATS functions, just a remark: communications systems do not have an associated SIL level, because they do not perform safety functions. Other systems do. In Chap. 5, we explain this in more detail when we discuss GSM-R. But this does not mean no need of RAM requirements or certifications. Indeed, these requirements are quite remarkable, as we see in Chap. 5, especially reliability and availability.

2.2.3 ATO Functions

Automatic train operation (ATO) functions are, as we have said, optional. The transport authority states which of them must be provided by the system and which not. Notice two important things: ATO is not safety related, and the decision on which ATO functions are to be implemented can be based on several aspects, being more likely operational convenience, passenger comfort, etc.

The following are the most relevant ATO functions:

Automatic speed regulation: This one is the most relevant one and the more complex as well. The basic idea is to set the speed profile for the train, freeing the driver to manually control it. This is more efficient that the manual driving and allows a faster movement. Moreover, a better regulation of the headway and inter-train distances is achieved as well. Indeed, this speed profile does not rely only on basic speed calculations, but also on the brake model of the train, passenger comfort, and performance of the traction system of the train. For example, the acceleration rate of the train must be accounted in the model. Passenger comfort is related to accelerations, deceleration, and rate of change of them both (called "jerk"). For example, in Europe, the standard for comfort measurement and evaluation onboard trains is EN 12299 [43].

Door control: Train doors are an essential system for passenger's safety. Therefore, they must be opened only in some specific locations (typically, stations).

Safe train departure: Once passengers have completed the onboarding and offboarding of the train, the train must depart from the station, that is, to resume the journey safely.

Train stops at the station: Trains must stop at some specific points in the platform, to avoid having parts of the train out of the platform or unnecessarily delaying passengers (as passengers expect the train on certain locations).

Automatic turn of the train: This function is a man oeuvre very useful for railway lines with a very short headway (mostly metro lines). When a train gets to the last station of the line, open doors, passengers leave the train, closes doors, and the train changes itself to the other track automatically, that is, requesting a route to the interlocking which must change the switch to allow this turn. This man oeuvre can save time at the end of the line and help to have a short headway. The overall headway will be severely degraded if only one station becomes a bottleneck and trains get stuck for more time there than other stations. That's how automatic turn can help.

2.2.4 ATS Functions

As in ATO, recall that ATS is an optional subsystem of the ATC, but in fact is mandatory to allow operators an efficient operation. On the other hand, among ATC systems, there are very simple ones and state-of-the-art ones like CBTC or ERTMS (which are both explained in this chapter). This means that some of the following functions are not available in many ATC implementations, but the most advanced ones provide the complete list plus some others.

ATS user interface (UI): The UI shall show all the ATS information with low latency, to enable the safe provision of all the other ATS functions.

Train identification and train tracking: Vehicles have a static identifier which does not change, and there is also the train number, an operational identifier which may change every day or even within the same day. This train number allows operators and infrastructure manager (IM) to efficiently manage train operations, because the static identifier is more complex than train numbers. Typically, train numbers are correlative integers (1, 2, 3, etc.) or, like airliners, have a larger number ("05078") or many others. Train static IDs typically follow UIC naming rules [37], but some operators have their own ones. In fact, these rules are maintained by the ERA.

Train routing: Updating the route of the train implies merging trains at junctions, turn back trains, etc.

Automatic train regulation: This includes monitoring trains to maintain headways, keep schedules, optimize energy, manage junctions, etc.

Station-stop functions: This includes stopping a train at the next station, skipping a station (in case it is crowded, for example), holding a train at a station, control door inhibition, etc.

Restricting train operations: This includes stopping a train in route, temporary speed restrictions, block/unblock a switch, or a signal, track, etc., or defining work zones. This only applies if the ATS can provide these functions safely,

Interfacing with passenger information systems (PIS): We see more details on PIS in Chap. 3, because it is a noncritical service. The idea here is that the best source for the train location data is the signaling system (best in terms of reliability and availability). So, integrating the PIS (which relies heavily on the determination of the train location) with the ATS can be a good idea.

Fault reporting: Either in the trains or in the wayside. These faults shall be reported and displayed to be managed by the operator or the IM.

2.3 Train Protection Systems

However, these systems not only inform the driver but also protect the train. For example, stopping the train if any of the following circumstances happen:

1. The driver runs a red light. This means a safety risk, so, in this case, an emergency break in immediately launched.
2. The driver exceeds the maximum speed. Sometimes a timeout is triggered to allow the driver to slow down the train. If this timeout expires, an emergency stop is triggered. The timeout allows avoiding unnecessary stops, because each time an emergency brake is launched, the train will stop completely, and it may take some time to get in motion again. Once triggered, the emergency brake stops completely the train. That is, it is irretrievable. On the other side, the service brake is retrievable.
3. The driver does not acknowledge a request from the train. This system is called "dead-man's switch" and could me a pedal, a handle or a button, which must be actuated by the driver. If the driver does not react, the system will stop the train. Therefore, it prevents accidents when the driver has lost consciousness, is dead or any other cause that means a loss of control from their side.

Please notice that this emergency stop needs the onboard signaling system to command the brake system, that is, to be integrated with the brakes that will stop the train. We will discuss these integrations later in this chapter.

2.3.1 Types of Train Protection Systems

Train protection systems can be of two types: punctual or continuous.

Punctual: The system only protects at some points, like signals, stations, rail switches, "dangerous" curves, or any other point that trains are more susceptible to achieve speeds higher than the recommended ones.

Continuous: Protection is along the complete track, or at least, main line. Depots, auxiliary lines, cul-de-sac, and others are usually not protected this way.

Punctual systems are cheaper than continuous because they require less hardware and are much less complex. However, providing only protection at some points means that the level of safety is significantly lower. The most common approach is to implement continuous systems in dense lines, with short headways or if the speed is high. Otherwise, sparse lines with non-fast trains usually are only punctually protected.

Therefore, a punctual train protection system is based on protecting the train at specific locations. The first system of this type was the "Train stop" developed by Siemens in 1928. It is a mechanical system which stops the train if it disregards a red light or any other advice (like a speed limitation). In Fig. 2.3, a balise of a punctual train protection system is depicted. The information must be read by the train and the driver must react to that command.

We see more details of continuous train protection systems in both ERTMS and CBTC sections.

2.3.2 Movement Authority

In this section, we explain four different methods to compute the movement authority on a railway line: speed codes, distance-to-go, virtual track circuits, and moving block. Please notice not only the specifics about each one of them but also the impact on other systems (for example, the need of a bidirectional communications system or not).

Before we begin, let us introduce "civil speed limit," a new concept, which is the maximum speed authorized for each section of the line. For the sake of simplicity, in this text, we assume that a section is a track circuit (TC), but this is not necessarily true. See Fig. 2.4 for a summarized depiction of the movement authority.

2.3.2.1 Speed Codes

In speed codes, the maximum speed is the same for the whole TC, and the wayside system transmits this maximum speed for the end of the current TC and the next one as well (see Fig. 1.5). The location accuracy is at TC level (that is, it is only possible to know on which TCs the train is, but not where it is within the TC). The onboard systems, with the brake model, then set the speed curve.[2] If the speed limit is lower than the current speed of the train, the onboard system will trigger the emergency brake.

[2]Extrapolates a speed curve based on both the punctual speed limits sent by the ground systems and the brake curve of the train. Notice that the brake profile of the train depends heavily on the load it has. That is the reason why modern trains sense the load to adapt the brake force to it.

Fig. 2.3 Track balise for a punctual train protection system (The Spanish ASFA). These balises are located in two different locations: few hundred meters in advance to the location to be protected and also few meters in advance to the same location

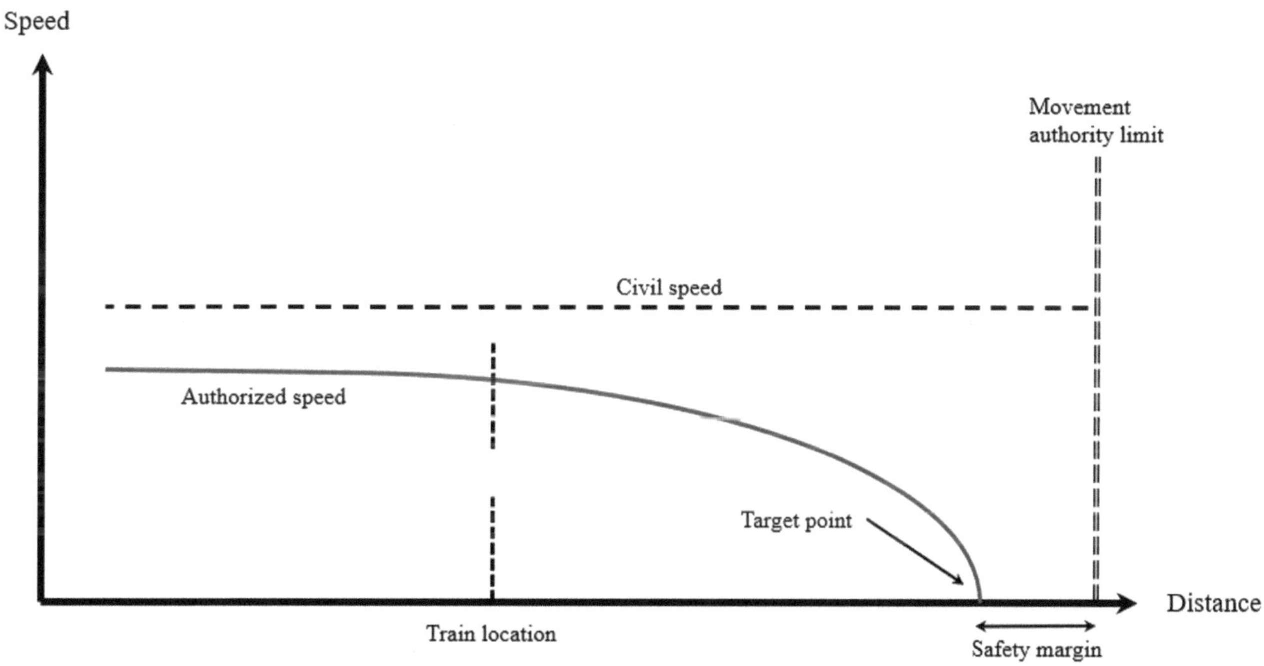

Fig. 2.4 Movement authority according to IEEE 1474.3-2008 [38]

Notice four important things:

1. There is no knowledge of anything past the next TC, so the speed control is done by the onboard signaling system with this degree of uncertainty. However, it is assumed that if the speed for the current TC is 60 km/h and for the next one is 60 km/h as well, for the next one, it will not be 0 km/h (unless this last TC is long enough to guarantee a "smooth" braking, that is, a non-emergency one).
2. The same information is transmitted to the train when it is on a particular TC.
3. Only one train can be in one TC at the same time.
4. Moreover, usually, an empty TC is left between two trains. This is due to the uncertainty of the location of both trains inside each TC (they can be one at the end of its TC and the other one at the very beginning of the next TC, that is, very close to each other). Having an empty TC between them sets a minimum safe distance.

The information is received by the train using loop readers or other type of antenna receivers (the information is only ground-to-train), and the volume of data is very small (only the speed profile for a particular TC, which can be quantized, and the next one as well). On the wayside, the transmitters are often wire loops fastened to the tracks or placed somewhere near them. This is a low-technology approach, but safe and simple. Trains are more distant than in distance-to-go systems, as we explain in the next section.

There is no need to have many different speeds. For example, on a metro system, the speed can be defined in steps like 0, 25, 50, and 75 km/h. This needs only 2 bits (0 km/h can be "00," "25," "10," etc.) plus error control and detection bits and some others. This is not optimal in terms of speed control but removes a lot of complexity from the whole system and makes simpler the onboard processors.

A depiction of the performance of this method is provided in Fig. 2.5.

2.3.2.2 Distance-to-Go

Distance-to-go (DTG) systems are one step ahead in both complexity and performance. The novelty here is that the train knows the location of the next train better (but limited to the TC level of accuracy), that is, each train knows on which TC is located the precedent one.

Moreover, the TC before the precedent train can be occupied and must not be left unoccupied. This TC is used for stopping the train, because at the very end of the TC, the train must be completely stopped. Please see Fig. 1.6 for more details.

As usual, we provide some important things about DTG systems: A safe margin is left to avoid trains trespassing into the TC occupied by the precedent train. These margins are very common in safety systems of any kind, because they protect from measurement errors, delays, and other issues that may happen and lead into an unsafe situation. DTG systems allow a significant reduction of the distance between train which implies shorter headways. DTGs also permits higher speeds because the train starts to brake based on the distance to the end of the movement authority (that is, most of the times, the beginning of the TC occupied by the precedent train plus a safety distance) and not based on the maximum speed defined for each TC. There are cases where the limit of the movement authority does not come from the precedent train, but any other limitation, like a switch, the end of the line, or some others. The transmission is unidirectional as in speed codes and is usually performed in the same way.

The data to be transmitted can be either safety critical (signal status, TC occupied by the precedent train, temporary speed limitations, etc.) or not critical.

Both ATP and ATO functions can be performed under this framework. We see this in detail in the "grades of automation" section of this chapter, but for the moment it is enough to understand two things:

1. DTG system allows the calculation of the maximum speed of the train at any moment and location. In case the human driver is in charge, one must not exceed this limit anytime.
2. ATO is for the automatic speed control of the train (based on ATP maximum speed, train characteristics, and some others).

Once again, the specific details depend heavily on the manufacturer. It is very common to have a map of the line stored onboard the train to save information in the transmission and provide extra safety.

An example of a DTG system is European Train Control System (ETCS) level 1, which is explained later in this chapter in the ERTMS' section.

A depiction of the performance of this method is provided in Fig. 2.6.

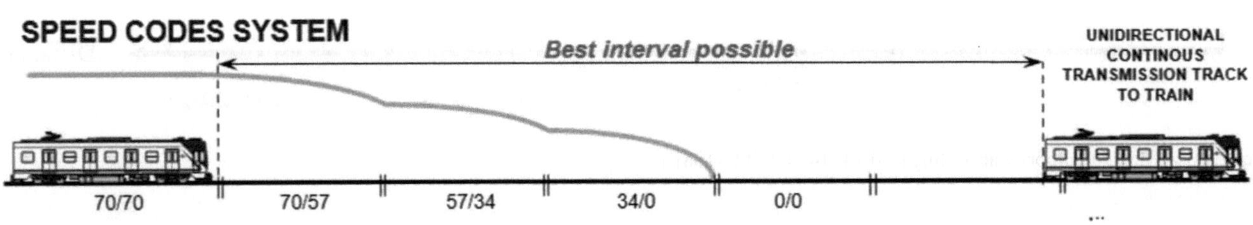

Fig. 2.5 speed codes

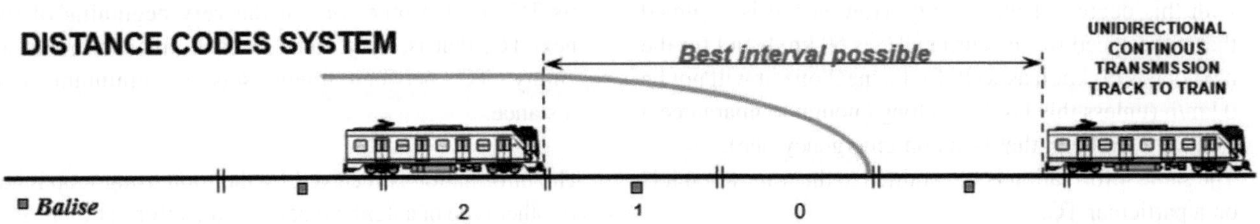

Fig. 2.6 Distance-to-go

2.3.2.3 Virtual Track Circuits

The next method for the determination of the movement authority is "virtual track circuits" (VTC). It is based on a more precise determination of the location of the train. In both speed codes and DTG systems, the knowledge of the location of the train is limited to the TC level. In VTC, using balises placed on the track and onboard sensors to measure the train speed, the train can know better its location.

However, the train location is now limited to a shorter scale than the TC, but the "virtual TC." Virtual TC (or VTC) are virtual divisions of a physical track circuit (see Fig. 1.7), and the location of the train is not related to the TC but to the VTC. This is done using the balises.

Balises usually only have static information about its own location plus the track number and other useful information, so no there is no need to have them online.

This approach has three important implications:

1. The train needs computation resources onboard to calculate the location of the train. This calculation is done by reading a track balise with information about its location and then, estimating the travelled distance from the balise. The distance is computed by integrating the measured speed.
2. Given that this approach is very sensitive to errors, balises must be placed frequently in the track.
3. Safety margins must be accounted to counteract these errors (speed, distance, etc.).

Each train transmits its location to ground systems, which use all of them to compute the movement authority for all CBTC trains.

This method has strong implications for both onboard and ground systems. The main advantage comes from the high-accuracy location determination, which allows shorter headways and higher safety. The disadvantages are extra costs and more complexity. For example, balises must be regularly checked to avoid service disruptions for this reason.

Figure 2.7 shows an example of an authority movement of a VTC system. In the next subsection, we explain "moving block" systems, which are the state of the art of signaling technologies as of today.

2.3.2.4 Moving Block

Lastly, moving block is the evolution of DTG and VTC systems. TCs are no longer needed because the train location estimation is done using only balises (plus distance measurement using speed meters, as in VTC). The difference with VTC is that in moving block systems, the location estimation is not discrete (VTC 1, VTC 1&2, VTC 2, etc.) but it is a pure estimation of the location in the track.

This location is more precise because it does not include the whole VTC, but the sole location estimation, plus some safety margins for both the head and the tail of the train. See Fig. 2.8 for more details.

This method is the most advanced one which is currently in service today. As the name suggests, the block is now in motion with the train. This is the section of the line which is blocked by the train is around the train, without the fragmentation that comes from TCs.

Some remarks on this are:

– Implementing moving block systems allows railway companies to remove many TCs, but not all of them. For

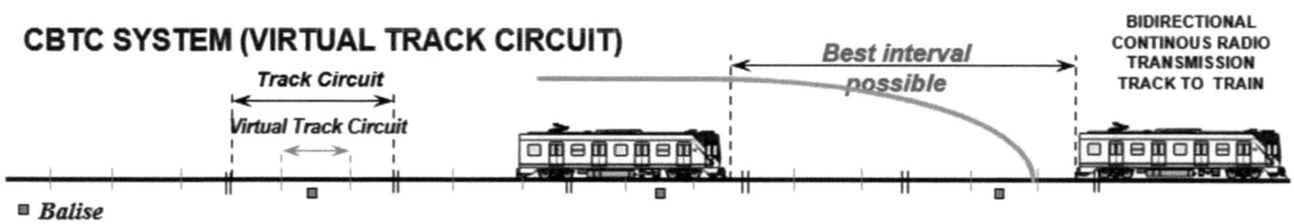

Fig. 2.7 Virtual track circuits

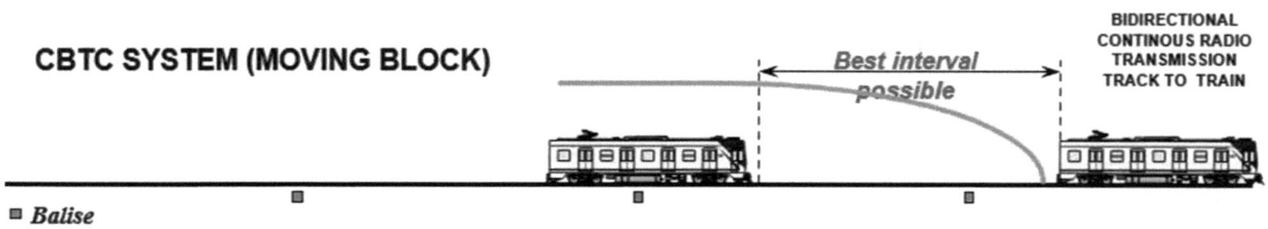

Fig. 2.8 Moving block

example, at line-entrance points, most moving block systems require a TC.

- The "moving block" of each train itself is at least the length of the train, plus some extra margin. This margin is proportional to the speed of the train plus and estimation of the impact of errors in the signals to be measured (speed, typically).
- There are many CBTC-moving block systems currently in service across the world, and it is remarkably impressive how close two train can be. Details on CBTC are provided later in this chapter.
- However, as some authors point out, headways, and therefore transportation capacities in DTG systems, are very similar than CBTC ones [39], but costs figures are much lower. This decision should be made carefully and considering many aspects (see Table 2.2 for some guidance on this).
- Finally, ETCS level 3 is a pure moving block system. As of today (2023), it has not been implemented anywhere, and only some limited tests have been carried out. In this chapter's further reading, we discuss this.

Figure 2.9 provides a comparison of all four methods.

2.4 Grades of Automation

This section is about urban trains only and does not apply to any other railway system.

Once the signaling data is onboard the train, there are several potential benefits, like improving safety and automating railway operations. Regarding extra safety, as we have discussed, having the route information and signal "colors" is safer than the old-fashioned lineside signaling.

On the other hand, with this information and integrating more systems, more functions can be automated. These systems can be either ground or onboard: for example, traction and brake systems of the train to automate the driving (this is called ATO, and we see more details about it later); or opening and closing platform doors when the train opens its doors.

In this section, we introduce train automation, which has several grades. We call them Grades of Automation (GoA) and are ranked from 0 to 4:

Grade of Automation 0 (no automation). This is also the baseline GoA, and all the others are built from here.

In this scenario, there is no automation at all, so driving is completely manual. Driver has all the responsibility for

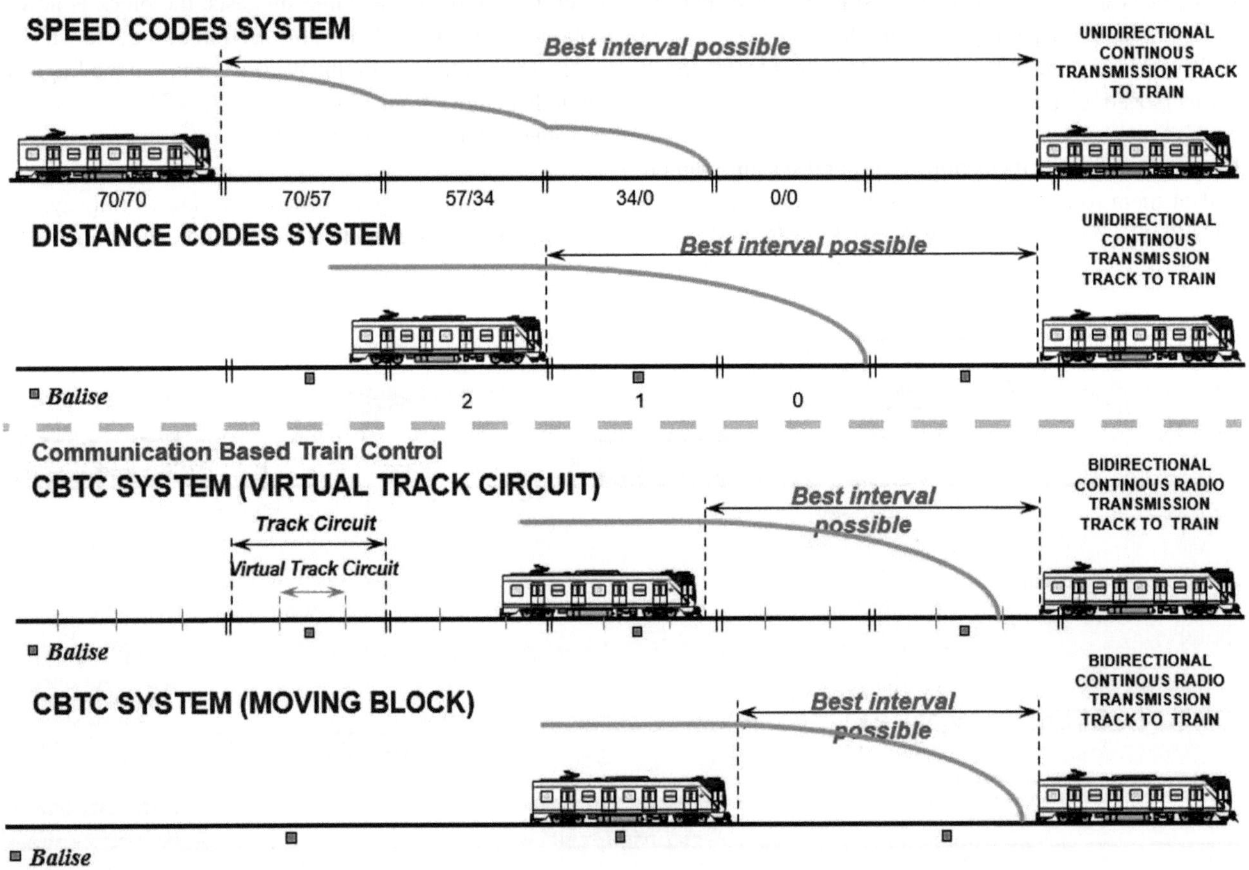

Fig. 2.9 Summary of movement authority methods: speed codes, DTG, VTC, and moving block

75

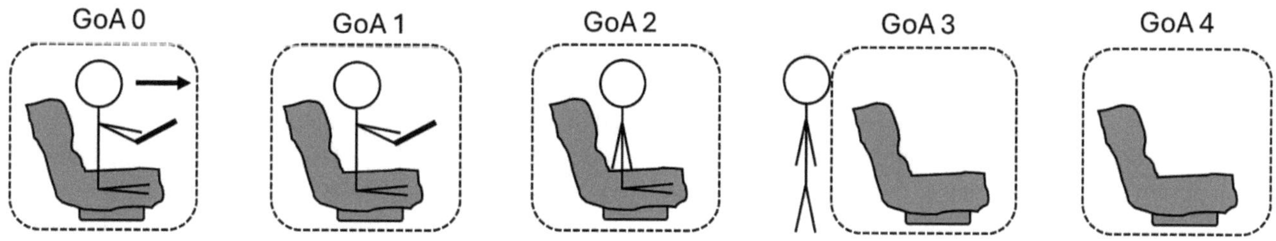

Fig. 2.10 Grades of automation depending on the role of the staff onboard

both safety and efficiency of the operation. Therefore, train driver can drive the train to an unsafe state with no system able to stop the train. There is no protection against any kind of errors. The slang term for this mode is "on-sight."

Movement authority is provided by either lineside signaling, communication with the operational control center (OCC), or static operation rules.

Notice that some protections implemented in the train can apply here, like hardwired speed limitations[3] or the dead-man switch we introduced in Chap. 1. These systems are sometimes apart from the signaling system and may protect the train if they are not bypassed. Figure 2.10 provides a visual summary of all grades of automation defined by the standard EN 62290.

2.4.1 GoA1

Automatic train protection (ATP). This is still a manual operation mode, because the train is protected against overspeed, running a red light, keeping the distance to the previous train and some others. In these cases, the ATP system will stop the train (emergency mode). This is the ATP output in case of emergency is the emergency brake. When the detected threat is overspeed, some systems trigger a warning signal for the driver and wait few seconds before the emergency brake is triggered. This allows the driver to slow down the train and avoid the ATP emergency brake.

Train driving is manually done by the driver. They are also in charge of door opening/closing and watching the condition of the tracks. Movement authority is provided by the wayside system and sent to the onboard signaling system.

Notice that these ATP functions can be punctual or continuous. Where the protection is only punctual, it is limited to some specific points only, being all the others completely unprotected.

In many cases, the "dead-man switch" we introduced before is also a function provided by ATP. In some others not, because this function can be provided only in trains which do not have any signaling system at all.

Protection here is restricted to some specific threats only. This means less safety and less efficiency.

2.4.2 GoA2

Automatic train operation (ATO). In this scenario, trains are more automated because the driving (in the sense of speed profile, that is, braking and accelerating) is performed by the system. Therefore, among driver duties are opening and closing doors, communicating with the OCC and reacting to emergency situations (which includes monitoring the track ahead just in case). Door opening/closing is a critical task, expensive, hard to automate, and could also lead to safety threats if somebody gets trapped within the door sections. On the other hand, emergency situations can be due to a failure or to unexpected situations. This mode of operation has many advantages:

1. More efficient operation because the system can consider in calculations ramps, regenerative braking, and other energy-saving features.
2. More uniform braking curve, which has impact on passenger comfort.
3. More repeatable speed profiles. This means higher regularity (8–10% [39]).
4. Train stops at the same place in the platform every time. This has also implications on train punctuality because it allows a faster onboarding/offboarding of the train.

GoA2 can be considered a semiautomatic mode of operation, because some tasks are still manual, whereas some others are automated.

3 Some trains have a handle to limit speed up to a certain value. This is useful for maneuvers in depots, like coupling trains (for example, up to 3 km/h to avoid damages in the couplers) or limiting speed in depots up to 20 km/h. These systems sometimes must be selected by the driver and sometimes not.

2.4.3 GoA3

GoA3 is called driverless train operation (DTO). In this mode, the system does all the driving and opens/closes doors as well. Humans are required onboard just in case an emergency arises or to provide customer service onboard. Therefore, onboard staff is not required at the driver's cabin and are usually out of it, assisting passengers. That is, in normal operation, humans are not for driving. This is why in this mode, onboard staff are called attendants instead of drivers.

GoA3 is a remarkable step over GoA2 because many systems are redundant and because many situations that the driver used to solve now must be done by the system. Robust broadband train-to-ground communications systems are recommended here but are not as essential as in GoA4. Staff onboard can handle with almost every emergency situation, but voice communications (that is, not broadband general purpose) are required for the communication with the OCC. Data for the TBS is not included in this count.

Given that the staff is no longer needed in the driver's cabin, some functions previously assumed by the driver are now automated. For example, derailment detection can be easily done by a trained driver, but on an automated train requires sensors and actuators.

Notice that this implies extra budget and much more testing than GoA2 as well. On the nontechnical side of the problem, GoA3 is often a solution to avoid dismissing train drivers when a line gets upgraded from GoA2 (they are transformed into onboard staff). On the other hand, OCC staff must be trained to operate DTO/GoA3 lines. We see more details about this later in this chapter. The removal of human errors is the main advantage of GoA3. However, it has a remarkable impact on costs, so the decision for such an upgrade must be done carefully. Finally, as usual in non-interoperable railway operation, there are no clear rules for what is required in GoA3 or 4, and the final decision is made by the operator based on the safety analysis.

2.4.4 GoA4

GoA4 implies unmanned train operation (UTO). This is the top-rank grade because it implies the complete removal of humans to operate the train, including emergency mode. There are some scenarios where a driver should go to the train, but the system is designed to make this as unlikely as possible. This means that both reliability and availability must be excellent, with multiple redundant systems.

As explained in GoA3, automatic hazard detection systems are mandatory in GoA4, like derailments, smoke, fire, etc. These situations can happen in both the train and the wayside, so detections system apply in both. However, detection of obstacles in the track is often performed using wayside sensors, like video analytics, laser detectors, or directly installing platform screen doors (PSD) that remove or almost remove the probability of having objects or people in the track. PSD can be beneficial also for heating ventilation air conditioning (HVAC), allowing more comfortable platforms and decreasing energy expenditures.

GoA4 requires a reliable train-to-ground communications systems for both safety and non-safety data. Watching onboard CCTV cameras from the OCC is needed, because there is no staff onboard. Bidirectional voice communications between train intercoms and the OCC are needed as well.

Strictly speaking, these detection systems are not mandatory, as operators decide where to stop addressing risks and where they begin to export them. For example, an UTO train with no derailment detection system can be a nightmare if the train derails, but the possibilities for such a scenario can be very low if both trains and tracks are maintained very carefully, speed at some points is limited, etc. Notice that PSD are in the same situation: they are much recommended but are not mandatory in GoA3/4 lines.

Notice that GoA levels are related. That is, ATP requires ATO, and DTO is a generalization of ATO (like UTO). It is also assumed that the train protection is continuous, but this is not always true, as there are still metro lines with punctual signaling systems. Therefore, ATP is limited to these spots only. A summary of all the GoAs is shown in Table 2.2.

ATO functions are not safety related (only ATP functions are). ATO regulates train speed, but the maximum speed is specified by ATP.

There are many functions for each one of the GoAs that we have not explained here, like avoiding wrong-side door opening or automatic turns at the end of the line (both of them ATO functions); ensuring the safe train departure. This is usually done by human drivers up to GoA2 and by the

Table 2.2 Summary of GoA-related functions

GoA	Onboard staff provides	System provides	Requirements
0	Driving	Nothing	–
1	Driving	Train protection	–
2	Specific functions, like door opening, safe train departure, and degraded scenarios	Operational functions, like driving, and stopping at stations	–
3	Degraded modes, passenger assistant, and interfacing with OCC	All functions but some degraded modes and passenger assistance	Recommended T2G system
4	No need	Everything	Mandatory redundant T2G system plus CCTV

machine in GoA3 and 4; and train condition supervision is performed by humans up to GoA3.

GoA3 and GoA4 modes usually have no train cabin, but for degraded modes, trains have a small train deck that allows operational staff to take over the train.

Regarding emergencies, there are some scenarios where onboard staff are not in the strict sense required and could be replaced with technology, but are extremely helpful: for example, when evacuating a train, we all people prefer to follow orders from flesh-and-bone humans rather than from speakers and screens.

Some operators fully automate the operation in the line, but not in depots or maintenance yards. However, depending on the number of trains involved, this procedure can imply a large workforce, so it is often not recommended.

There are intermediate GoAs like GoA2.5, which is between GoA2 and 3, meaning "more than GoA2, but driver is still required in the cabin, unlike GoA3."

The reference for this is the EN 62290 standard "Railway applications—Urban guided transport management and command/control systems" [40] and applies only for urban trains. Mainline and HSL are (as of today) automated only up to the equivalent to GoA1, that is, train protection, but the driving is manually done by a human at all times (within the limits set by the signaling system). As we explain in Chap. 8, future trains will be more automated, so this GoA will rise for HST among others. On the other hand, the reader may notice that automation of railway lines is not about technology only, because it involves people and organizations. Therefore, it has three legs:

1. Technology: All the hardware, software, tests, etc., that are involved on a transportation system.
2. People: Involving people in a new technology or a new way of working is often challenging due to the resistance that can arise in some cases, from unions or other collectives.
3. Processes: Corporate culture is a relevant force and must not be underestimated. Resistance to change could lead to less employee engagement, which could harm the service after all.

The signaling technology more often used to increase the GoA of a line is CBTC. However, it is not mandatory, and there are many examples around the world of automated lines without CBTC.[4] CBTC is explained in the next section of this chapter.

[4]For example, Lille Metro Line 1 in 1982 had no CBTC and was the first metro line in the world to be fully automated. Some versions say that the first one was Port Liner in Kobe (Japan) 1 year earlier. We take no sides in this war, because neither of them needed a CBTC system to be fully automated.

2.5 CBTC (Communications-Based Train Control)

CBTC stands for communications-based train control. This train control information is delivered by a (radio) communications system. Therefore, CBTC is pure TBS.

Higher GoAs and CBTC systems usually go hand in hand. As for other requirements that we envisaged when we described some particularities of the GoAs, CBTC is not mandatory but is very close to be a market standard in urban trains. However, apart from some basic concepts, it is hard to describe because it does not follow a strict-sense standard (like ERTMS) and, more importantly, CBTC systems are not interoperable between each other. This is a key concept in signaling, and we come back to it when we discuss ERTMS in the next section.

Indeed, CBTC is not interoperable because there is no need to have it this way. Interoperability means higher engineering costs.

Moreover, this means that it is hard to explain "standard CBTC" because it does not exist. In this section, we introduce the basics (supposedly common to all manufacturers and implementations of CBTC) but not a real CBTC system. All CBTC implementations have subtle differences.

2.5.1 CBTC Concept

There is an IEEE standard for CBTC. Indeed, a family of standards formed by four parts:

1. IEEE Std 1474.1 IEEE Standard for Communications-Based Train Control (CBTC) Performance and Functional Requirements [32]. This norm was released in 2004 and is based on a previous one of 1999. Something very relevant about standard is the list of mandatory ATP (and ATO & ATS if apply) functions must provide a CBTC system. For example, a CBTC system unable to locate trains will not work.
2. IEEE Std 1474-2-2003 1474.2TM IEEE Standard for User Interface Requirements in Communications-Based Train Control (CBTC) Systems [41].
3. IEEE Std 1474.3-2008 IEEE Recommended Practice for Communications-Based Train Control (CBTC) System Design and Functional Allocations [38].
4. IEEE Std 1474.4-2011 IEEE Recommended Practice for Functional Testing of a Communications-Based Train Control (CBTC) System [42].

All IEEE 1474 parts of the standard are worth reading, so we encourage interested readers to download and study them. However, remember that CBTC is not interoperable, and there is no "standard CBTC," but several interpretations of the standard made by several manufacturers.

When we explained the methods to compute the movement authority, we explained three basic bottlenecks for conventional methods:

1. Train location independent from track circuits. The accuracy of the location determination shall be high. It makes no sense to have more uncertainty than using track circuits.
2. Processing of critical data both in the wayside and onboard the train.
3. Train-to-ground communications. This system must be bidirectional, continuous and ultra reliable.

Solving them allows train operators to achieve reduced headways and higher capacities with the same safety levels or even higher. This is very important. CBTC aims is to overcome all these difficulties [32] for several following transportation systems, like tramways, subways, commuter trains, and automatic people movers (APM).

That is, CBTC is not for HSL, regional, or long-distance trains. The equivalent system for them is ERTMS, which is explained later in this chapter.

The formal definition of CBTC according to the standard [32] is: "*A CBTC system is a continuous, automatic train control system utilizing high-resolution train location determination, independent of track circuits; continuous, high-capacity, bidirectional train-to-wayside data communications; and train-borne and wayside processors capable of implementing automatic train protection (ATP) functions, as well as optional automatic train operation (ATO) and automatic train supervision (ATS) functions.*"

Now we will explain this definition term by term with its implications:

1. *Continuous, and automatic train control system*: The protection that CBTC provides includes (at least) the line. That is, once a CBTC-equipped train is registered in the system, it is protected all the time. Depots, maintenance yards, and other auxiliary tracks are optional.
2. *Utilizing high-resolution train location determination, independent of track circuits*: The estimation of the location of the train must be done in a more precise way than track circuits can do. Remember that the accuracy with this method is severely limited to the length of the track circuit itself, with no information inside it.
3. *Continuous [train-to-wayside data communications]:* If the CBTC protection must be continuous, the T2G system that supports it must be continuous as well.
4. *High capacity [train-to-wayside data communications]:* In my interpretation, here "high capacity" does not imply high data rates (which are not needed in CBTC implementations) but capable of managing many trains. This requirement is not straightforward but is very relevant. We come back to it later in this section.
5. *Bidirectional train-to-wayside data communications:* This refers to both train-to-wayside and wayside-to-train. The information to be sent is:
   (a) The precise location and measured speed of the train, which is transmitted by the train itself to the wayside system (uplink data)
   (b) The movement authority, which is sent by the wayside system to the onboard ones (downlink data)
   Both can be transmitted anytime, so the coverage must be as thorough as possible.
6. *Train-borne and wayside processors capable of implementing automatic train protection (ATP) functions*:
   (a) Onboard processors will receive balise lectures and speed data from the train sensors and will estimate both the location and the speed of the train.
   (b) Wayside processors will collect the information sent by all the trains, compute the movement authorities, and send one to each train.
7. *Optional automatic train operation (ATO) and automatic train supervision (ATS) functions*

Actually, there is a third type of data which is non-safety data originated either onboard the train or in the ground systems. This data is related to ATO and ATS functions and about the condition of the CBTC system (that is, telemetry, logs, etc.).

2.5.2 CBTC Functional Architecture

The idea for this book is to provide a general overview of many systems and technologies and not highlight some vendor-specific technologies. Therefore, and once again: in this section, we explain the general CBTC architecture, as it is depicted in the standard [32]. We encourage the reader to visit the references we point out at the end of this chapter in case they are interested in specific details for real-world implementations.

The general CBTC functional architecture is shown in Fig. 2.11. CBTC system is divided into five layers of systems:

1. The ATS: This is the automatic train supervision, which includes interfacing with users (in this case, operators or railway authorities that may require to establish a new route, set temporary speed limitations, etc.). Specific details about the information to be displayed are described in the second part of IEEE 1474 standard [41]. Both mandatory data and optional data are considered.

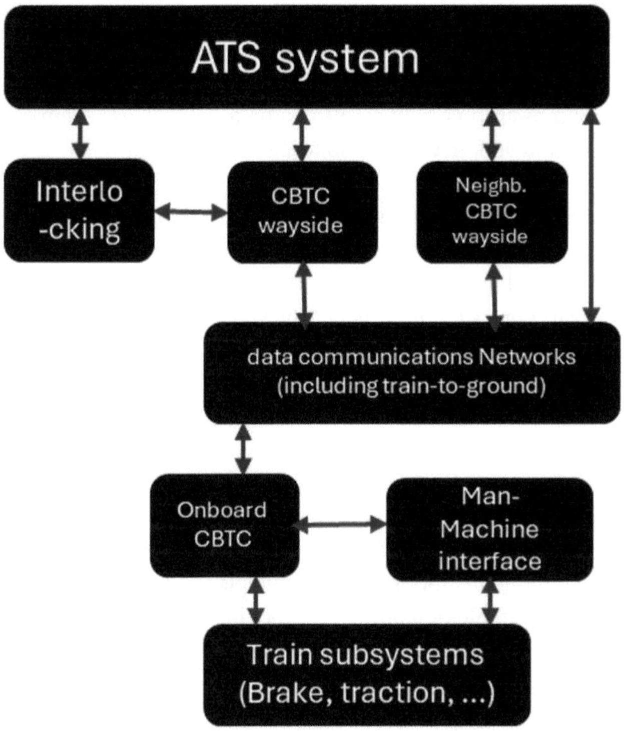

Fig. 2.11 General CBTC architecture. (Adapted from [32])

2. Wayside equipment: It includes interlockings, all CBTC wayside equipment, and neighboring CBTC systems. Interlockings have the same duties as in conventional signaling (handling switches, signals, TCs, routes, etc., but this does not mean that a CBTC upgrade does not require an upgrade of the interlocking as well. In some cases, it is necessary), but in CBTC, they also interface with CBTC-specific equipment which is in charge of interfacing with the (CBTC subsystem of the) train.
3. The communications network: This network includes both the fixed wayside part, the train-to-ground, and the onboard network. It is used to communicate all elements. In Chap. 4, we explain network technologies, and in Chap. 5, we provide details about train-to-ground communications systems.

   Notice that it is not mandatory to connect each element with TCP/IP protocols. Some CBTC subsystems may be connected by legacy technologies (i.e., Profibus [43] instead of IP networks). The standard is not against these solutions, but the trend is to have all-IP systems.

   The standard is tech-agnostic. As we see in Chap. 5, there are common places for train-to-ground in CBTC (like IEEE 802.11), but this is not mandatory. Do not mix up de facto standards with strict-sense standards (like ERTMS over GSM-R).

   The communications network is not critical for safety. That is, in case it is not available, the train will stop and then, if available, move to a degraded mode where the journey may restart.

   In driverless modes, typically, the safe condition for the train is to advance up to the next station (unless a derailment occurs or other incident that hinders the movement of the train). If a driver or other onboard staff is available, the typical safe scenario is the train stopped.

   Finally, the network does not perform any CBTC function.
4. Onboard CBTC subsystem. This is in charge of all ATP and ATO functions onboard the train, as we detailed in section {ATP} and {ATO}. Kindly observe that this subsystem requires:
   (a) Certain processing capabilities to acquire data from sensors, process movement authorities, compute braking curves, etc.
   (b) It is a critical element of the CBTC system. Therefore, the required availability is very high, so it is very common to have some redundant ones in parallel.
   (c) It interfaces with both train subsystems (brakes, traction system, TCMS, etc.), and it also has a user interface (UI). This UI is mostly for drivers and maintenance staff.
5. Interfaces with other onboard systems (brakes, traction, HMI, etc.).

2.5.3 Degraded Modes and Other Scenarios

The normal operational scenario for a CBTC line is to have all trains with a CBTC system, everything working properly and no issues at all. However, there are many situations that can take a turn for the worse. Therefore, it is required to define degraded modes of operation to minimize the impact of failures and allow the safe operation of the railway line [32].

Among the degraded modes we can find the following:

1. Mixed (CBTC and non-CBTC trains) operation. In this scenario, which can be either a degraded mode or a normal situation, it is needed an alternative train detection system to handle non-CBTC trains, plus a complete backup signaling system (wayside signals, etc.) to operate safely.
2. One or more trains are not able to communicate with the ground systems. Until this failure is solved, this train acts as a non-CBTC train.
3. Unavailable wayside equipment in a section of the line. Consequently, trains cannot be controlled by the CBTC system.
4. Any other failure onboard the train that makes its CBTC system unavailable.

On the other hand, there are some other failures that do not impact the system performance, due to the redundancies and the system design. These failures must not be addressed but to account RAMS KPIs. For example, mean time between failures (MTBF) calculations must include all types of failures, no matter their impact on system performance [32].

If for any cause, a CBTC line experiences any of the failures or any other with impact on system performance, there are three possible scenarios:

1. Stop operating the line completely. That is, stop the whole line. This is not desirable at all and shall be avoided for obvious reasons.
2. Protect the train operation by operational rules. Operational rules are those that do not rely on signaling technology. For example, before leaving a station, the driver must call the OCC. Very often these rules imply a remarkable delay because they are inherently slower than the normal CBTC-based operation. They also rely on human performance and, therefore, are more prone to human errors. These operational rules must be designed carefully, and railway staff must be trained accordingly.
3. Protect the train operation by another ATP system rather than CBTC. This is something very expensive because it implies to maintain two ATP systems in parallel: CBTC and the backup one. However, in those lines which were upgraded to CBTC from an older technology, this is a potential solution, that is, keeping the old system to serve as a backup for CBTC or to allow non-CBTC trains.

2.5.4 CBTC Functional Requirements

The following requirements are an example, just for guidance for CBTC systems' design. They are included in the standard as an informative annex and have no implications at all. However, we consider that they are illustrative of the implications of having a CBTC system in terms of both performance and requirements for other subsystems (including communications). They are the following [32]:

1. Maximum number of trains handled by a wayside controller: 10–40 trains.
2. Train location accuracy: that is, the maximum error in normal operation ±5 to 10 m.
3. Train location accuracy for station stops: ±0.05 m.
4. Resolution of movement authority limits: ±0.25 m to ±6 m.
5. Train speed determination: ±0.5 km/h to ±2 km/h.
6. Wayside reaction times: 0.07–1 s.
7. Onboard reaction times: 0.07–0.75 s. Please observe that the maximum reaction time for the onboard equipment is shorter than for the ground.
8. End-to-end communications delays, that is, train-to-ground and vice versa, up to 2 s.

Another requirement off the standard is the MTBF, usually around 250 thousands.

Notice that in most cases, the requirement comes as an interval. For example, it is not the same to have 10 trains on a zone controller than 40 trains. Usually, most CBTC systems are designed with an intermediate figure.

All these requirements are the main input for system engineers. Some of the requirements are exceeded by the real implementation (like end-to-end delay, because current technology can largely outperform standard CBTC requirements). In the next section, we provide a non-exhaustive list of requirements for the communications system based on real-world implementations (and not in the standard, but they do not contradict it).

2.5.5 CBTC Requirements for Communications Systems

CBTC communications' requirements are the following:

1. Bidirectional data transmission: that is, both train-to-ground and ground-to-train.
2. Handover shall be as seamless as possible.
3. Data rate requirements are not very high: 20–30 Kbps per consist. This figure accounts for the aggregated uplink and downlink.
4. Security: authentication, integrity of data, and end-to-end encryption. The trend for authentication, as of 2023, is to use SHA-256.
5. Dual coverage with frequency diversity. Typically, the required availability is 99.999%.[5] This availability is achieved by network architectures with a high level of redundancy. More of this can be found in Chap. 4 and 5, when we explain networks and radio communications systems.
6. Given that the traffic is both safety and non-safety related, there must be a mechanism to prioritize safety data over non-safety. Safety data is usually called "critical data" as well.
7. Connection setup time of around 1 s. This is the maximum time a train needs to get connected in the network to communicate with the ground CBTC systems.
8. End-to-end delay as short as possible, but typically is around 100–200 ms. Perhaps shorter.

[5] 99.999% is usually referred as "five nines", too.

9. Registration latency for a train: up to 1 s.
10. As usual, the communications must be all-IP. That is, non-IP standards are not recommended. However, in some cases, they are still widely used, like train communications networks (TCN), which still rely on legacy buses and are discussed in Chap. 6.

2.6 ERTMS (European Rail Traffic Management System)

ERTMS is the state of the art in traffic management. It comes from a European desire to have interoperable trains across the EU. As we discussed in Chap. 1, signaling systems are not the only issue when dealing with interoperability but one of the most important. For example, in Fig. 2.12, we can see a map of Europe showing the most common ATP systems on each country. Notice that a train on an international route must have installed many of these ATP systems. And define overlapping areas in the wayside, to allow the safe transition from one ATP system to the next one on the train route. Otherwise, the train will stop at the border between these two systems.

ERTMS specification comes from several documents [44], but there are two which are very relevant:

System requirements specification (SRS) [45]: This document is an outcome of the EIRENE project, led by the UIC, and it is mostly about interoperability. EIRENE, as we see in Chap. 5 when we discuss GSM-R, the ERTMS radio bearer provides a specification for a digital mobile radio for both voice and data transmission. Therefore, in this document, a detailed (system-level) specification for the GSM-R system is provided, including requirements for the network, interfaces between elements, handovers, mobile (that is, onboard and handheld) devices, and other specific operational details (i.e., numbering plans, addresses, etc.). In this chapter, we will cover only the requirements for lower (communications)

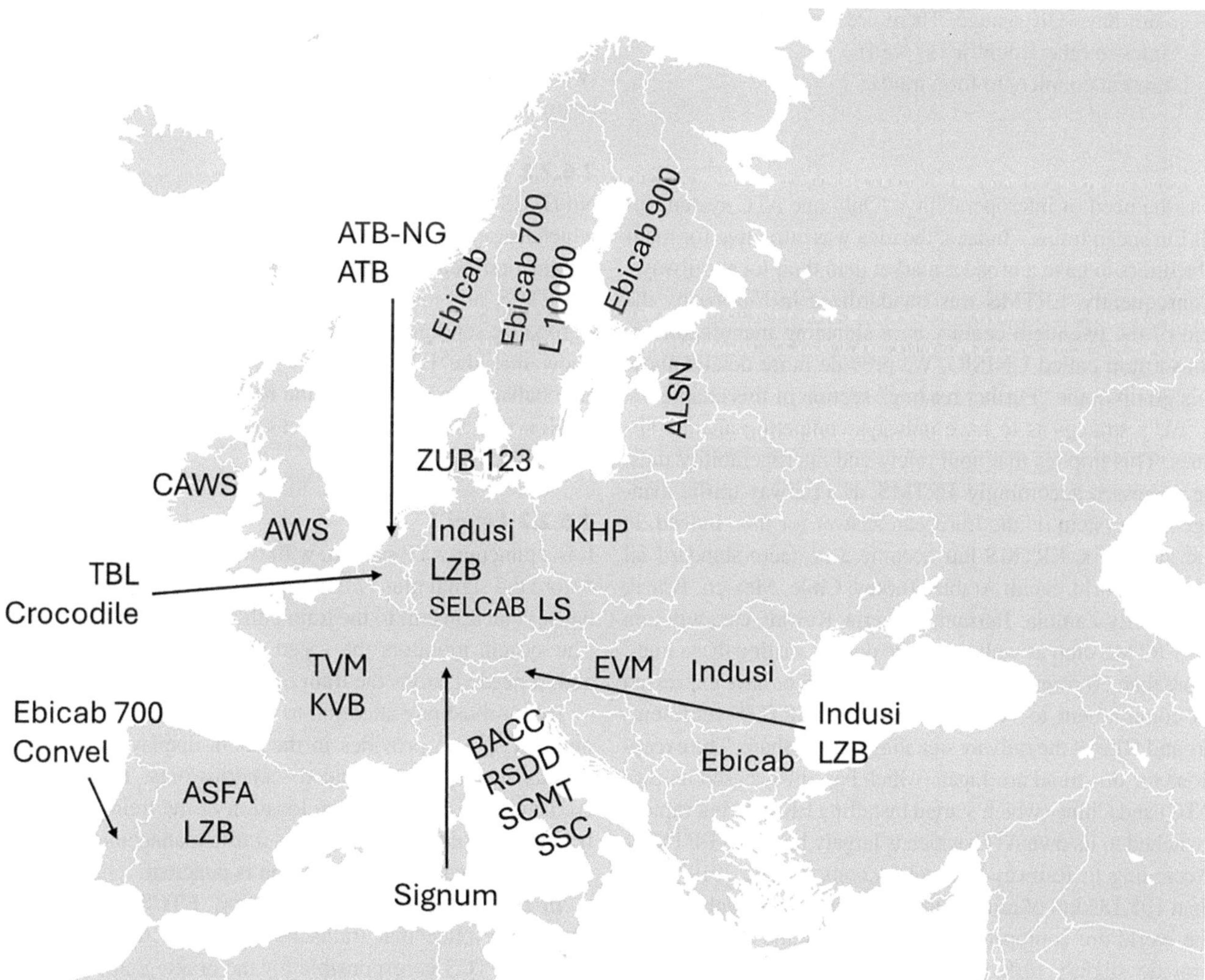

Fig. 2.12 Railway signaling systems in Europe. (Own elaboration based on data from [47, 39]. Blank map by [48])

levels that come from this specification. All the other contents are explained in Chap. 5 altogether with GSM-R.

Functional requirements specification (FRS) [46]: This document specifies high-level operational requirements for ETCS implementations. It has both mandatory and optional requirements (as most standards). This document is the basis for the following sections on ERTMS:

Therefore, the main trigger to start working on ERTMS was the need of interoperability. "Only one ATC system for all European trains." Indeed, the idea was attractive for manufacturers to have a broader market than their local railways. Consequently, ERTMS was standardized in Europe by the end of the twentieth century by a signaling manufacturers' consortium called UNISIG. We provide more details about this group in the "Further reading" section of this chapter.

Barriers for Interoperability

Different ATC system

Different operational rules (including train homologation and driver training)

Different railway electrification systems

Different radio communications systems

Different languages (drivers and onboard staff need to communicate with OCC many times)

Different rail gauges (more in particular Spain, Portugal, Ireland, Russia, and Ukraine, which do not have UIC gauge. There are many other examples in other countries as well)

Different couplers to form trains

EU's strategy is to have railways competing against airlines. This implies that both safety and interoperability must be addressed accordingly. ERTMS, as a railway traffic management system, is the European answer for this. Indeed, in the last years, ERTMS has become a de facto standard all over the world. Saudi Arabia, Turkey, Chile, Mexico, Brazil, and shortly Zambia, Tanzania, Nigeria, Kuwait, etc., will join the ERTMS club as well. As of the date of writing these lines, more than 50 countries either have ERTMS or have expressed its commitment to do so. Furthermore, ERTMS represents around 50% of the railway signaling market share. The exceptions for this trend are Japan (which has always used its own ATC) and China (which started installing ERTMS but rapidly switched to its own ATC, which is largely based on ERTMS). According to Railway-Technology.com, "As of 2019, more than 105,185 km of railway tracks and 13,219 vehicles across the world are equipped with ERTMS." Some other sources provide slightly different figures. In Europe, Spain is the country with the highest ERTMS deployment so far.

2.6.1 ETCS Concept

ERTMS has some things in common with CBTC, so given that we have already explained CBTC, here we focus only on the differences between them.

First, ERTMS is formed by two components:

1. ETCS (European Train Control System): That is, the ATC.
2. GSM-R: That is, the communications system. We explain GSM-R in detail in Chap. 5.

A very common mistake is to address ERTMS and ETCS as if they were synonymous, when, in fact, they are not. From now on, we are discussing only ETCS.

Different from CBTC, ETCS is structured in levels.

2.6.2 ETCS Levels

We are introducing ETCS in two different ways: ETCS operational modes (more often called levels) and its architecture (which includes both wayside and onboard subsystem). Let's begin with ETCS levels:

2.6.2.1 Level 0

No ETCS at all. It is assumed that an ETCS L0 line is a line which has a junction with an ETCS-equipped line. Therefore, a transition between the non-equipped line (ETCS 0), and the ETCS one must be set. This transition is done with a Eurobalise sending a telegram to the train to let the onboard know that the ETCS area begins (or ends). More about Eurobalises can be found in the next section.

Non-ETCS lines have a strict speed limitation of 200 km/h in the EU.

2.6.2.2 Level 1

It is a punctual ATP system, with two balises (around) 500 m before the signal plus two more just before the signal. These balises can transmit to the train either static or dynamic data. The system monitors the speed of the train on each balise and, if needed, stops the train before the signal.

Driver must pay attention to the maximum speed that the onboard system provides in the cabin displays, and lateral signals are kept for degraded scenarios only. Track circuits are needed to determine the location of the train. A centralized traffic control is needed for traffic management.

Notice that the train protection is punctual.

In Fig. 2.13, a summary of how ETCS L1 works is depicted. Notice that trains have a balise reader, and the onboard ETCS is responsible for the braking curve calculations (based on the information from the balise).

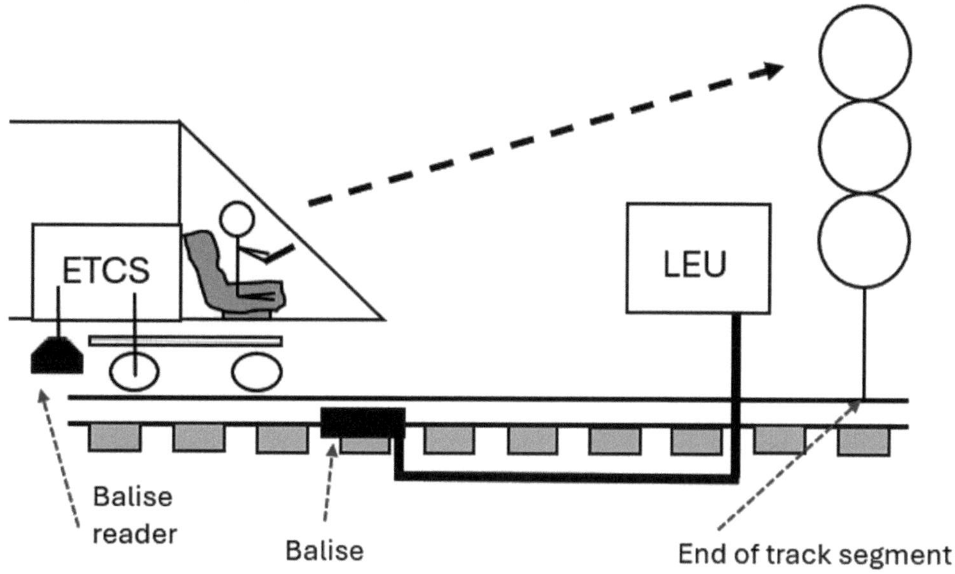

Fig. 2.13 ETCS level. (Adapted from [49])

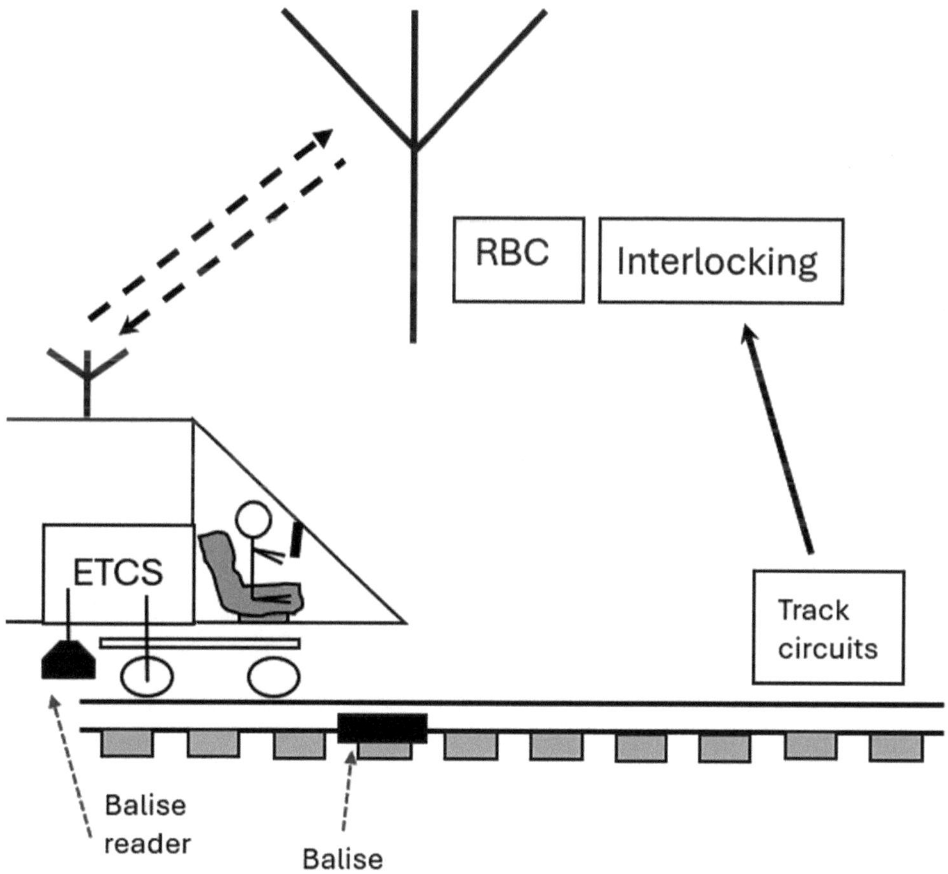

Fig. 2.14 ETCS level 2. (Adapted from [49])

2.6.2.3 Level 2

In this level, the information is provided by the radio communications system (GSM-R) and continuous, that is, available at any location.

Having a continuous communication between the wayside and the onboard subsystems allows the former to transmit the safety information, so no balises are needed in this level. At least, those who transmit variable information. That is, ETCS level 2 is very similar to DTG systems.

Track circuits are still needed but not lateral signals. Therefore, the train driver has all the safety-related information in the cabin. Figure 2.14 summarizes how ETCS level 2 works. Please notice that the driver does not pay attention to trackside signals but to onboard screens. Track circuits are still needed.

Signals are usually not removed when upgrading to ETCS level 2 to have a backup, but in normal operation, they are not needed. In railway slang, the upgrades of an existing line are called "brownfield" projects. New lines (that is, "greenfield" projects) skip a relevant number of lateral signals if they have ETCS L2 from the beginning.

ETCS level 2 is a continuous ATC, and it is also the current state-of-the-art in railway signaling for HSL and long distance. This railway service may benefit from interoperability. Providing an accurate performance figure is challenging because it depends on many factors, but ETCS L2 can have a headway of 2.5 min with trains running at 300 km/h. In ETCS L1, we have 8 min with trains at 200 km/h in the best case.

2.6.2.4 Level 3

In this level, track circuits are removed, and the information is delivered by the radio. That is, train transmits its location to the wayside (in level 2, the location was determined by the TCs) and wayside transmits movement authorities to the train.

It is a pure moving-block technology. However, this ETCS level has not been installed on any railway line. Only some tests have been performed.

The reasons for this are many, but perhaps the most relevant ones are:

1. With some exceptions, it is not worth it to invest much money on a system that would improve headway down to few minutes only.
2. It implies more complexity for the onboard system, which is responsibility of operators. It allows the removal of TCs from the wayside, which are the responsibility of the IM. This means very little incentives for operators to assume more complexity to benefit IM. See Fig. 2.15 for more details.

2.6.2.5 Other Functions

Apart from the typical ATP functions already explained, ETCS is intended to transmit to the train information to avoid stopping the train at viaducts, neutral zones (sections of the track without electric supply), or bridges.

2.6.2.6 Imperfect ETCS Standardization

In theory, ERTMS was conceived to be fully interoperable, but in practice, this has revealed to be much more difficult. Given that the standard marks some requirements as mandatory (those who impacted interoperability) and some others as "designer choices" (no impact on interoperability), it was only a matter of time that many different software versions arise.

This is something that can complicate largely the operation (but not safety) on transitions between national ATC systems and ERTMS, and even for transitions between ERTMS implementations of different companies. This is a very deep world, but a good example of how complexity is managed on an everyday basis. The following reference (in Spanish) provides more details about this [41].

2.6.2.7 ETCS Architecture: Onboard and Wayside

Figure 2.16 shows the complete ETCS architecture, which includes both wayside and ground subsystems.

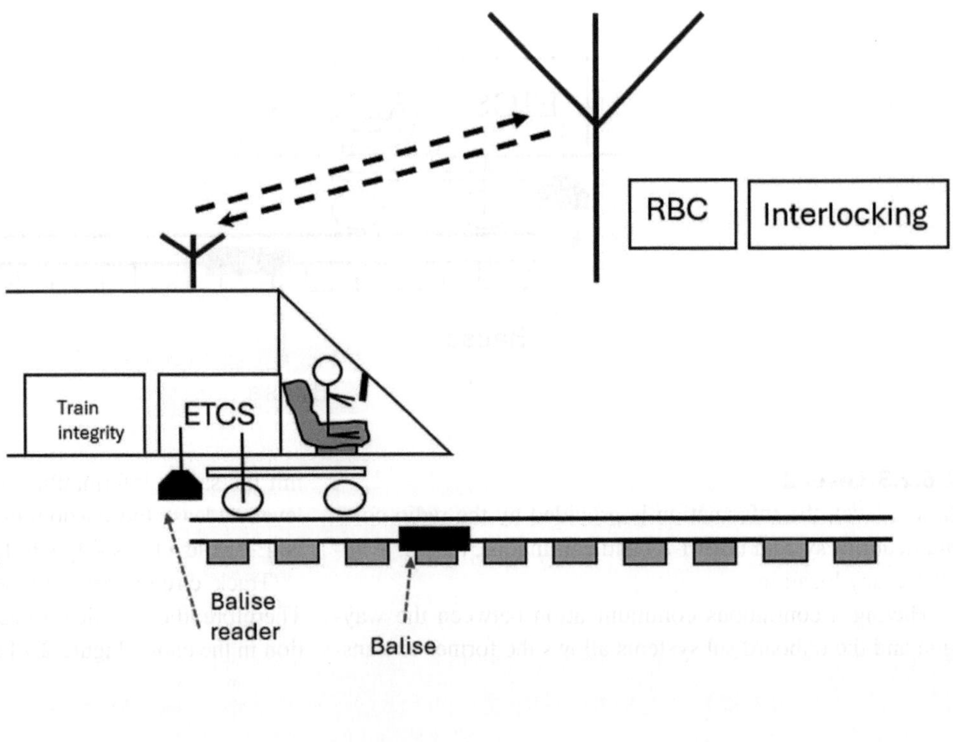

Fig. 2.15 ETCS level 3. (Adapted from [49])

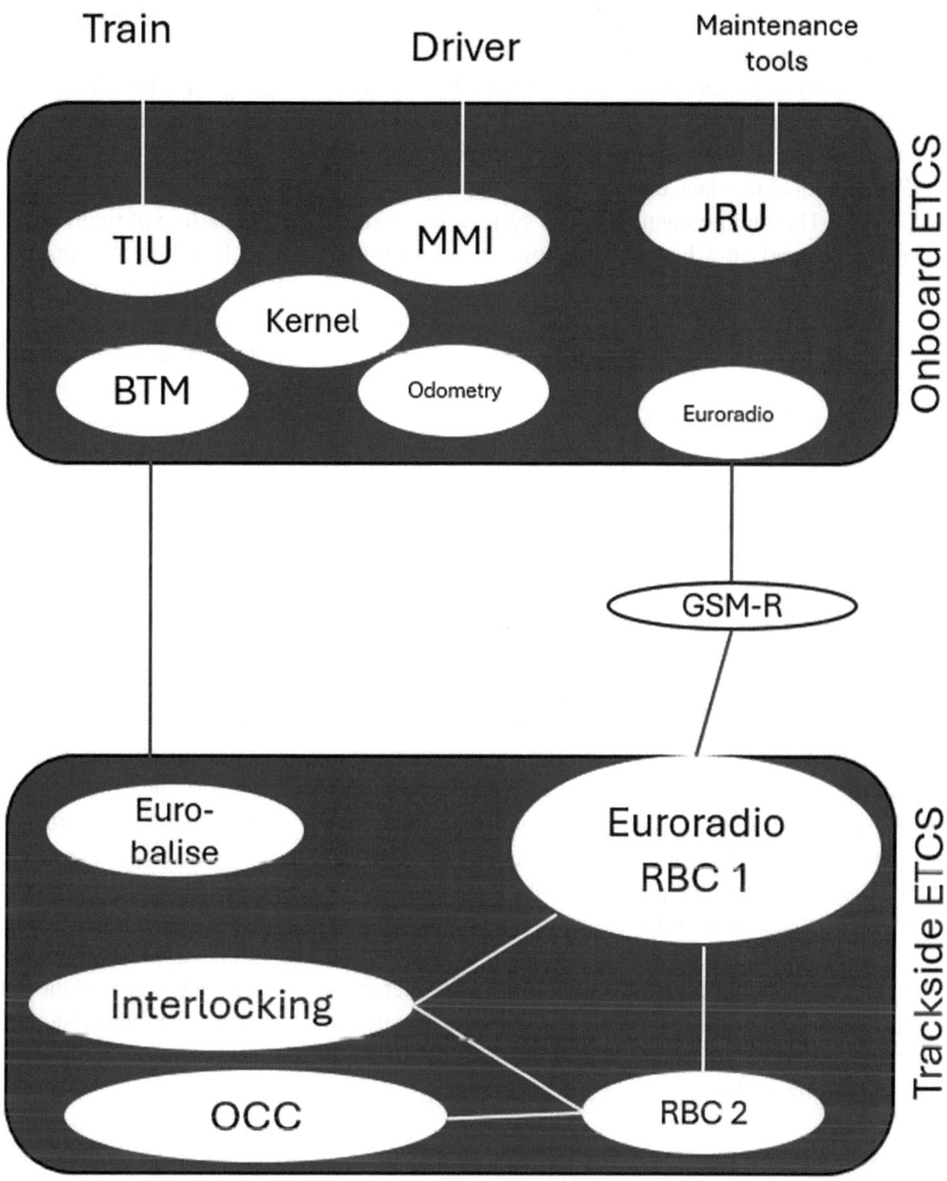

Fig. 2.16 ERTMS block diagram

In the **onboard** subsystem, we find the following elements:

1. Balise reader: receives the information from the balises installed on the track.
2. GSM-R antenna.
3. Train GSM-R radio: for the train-to-wayside communications (both voice and data). It is connected to the EVC and has one antenna.
4. BTM (balise transmission module): receives data from the balises and sends it to the EVC.
5. EVC (European Vital Computer): onboard processor, interface with other subsystems, and peripheral control.
6. DMI (driver machine interface): to show all the safety information (including the movement authority) to the driver.
7. Speed sensors to measure the train speed, and are connected to the EVC. They can be Doppler radars, tachometers, or any other sensors.

In the wayside, there are the following:

1. LEU (lineside electronic unit): receive information from interlockings and send them to the balises. This information is sent in "telegrams."
2. Eurobalises, which send the "telegrams" to the train. There are two types:
   (a) Static balises: transmit a static telegram usually related to location, distances to signals, or maximum speed for a specific section.
   (b) Dynamic balises: for the transmission to the train of non-static information, like movement authorities, speed limits, signal status, etc.

The balises require no power supply because they are powered by the train balise reader.
3. RBC (radio block center): It is the central ETCS system, responsible of the ATP functions (movement authority calculations based on information received from onboard EVC, interlocking data, etc.).
4. GSM-R: The radio communications system. It is extensively covered in Chap. 5. It allows train-to-ground communications.
5. Track circuits (except in level 3) and interlockings: They are needed to provide ATP functions.

2.6.2.8 Benefits of ERTMS

Among ERTMS benefits, we find an increased capacity for the railway line. If we compare ETCS L2 against legacy regional ATC systems, the capacity increases up to 40%. Regarding speed profiles, ERTMS is designed for train speeds up to 500 km/h. Consider that current HST commercial speeds is 350 km/h in the best-case scenario, so this implies that ERTMS will not become obsolete for this reason in the short term.

On the other hand, ERTMS allows lower maintenance costs. In theory, in ETCS L2, the IM can remove lateral signals and in L3 even TCs can be skipped. Theoretically, this means relevant savings in maintenance costs. Actually, no one removes lateral signals in L2 and L3 has not been implemented yet. Moreover, the predictions about cheap ERTMS deployments due to standardization have not been fulfilled at all.

Finally, an essential benefit from ERTMS is the interoperability across regions (only for ATC, and it does not apply for other interoperability barriers, like electrification or gauge limitations, obviously).

2.6.2.9 ETCS Requirements for Communications Systems

These requirements come from SRS [45] and represent the minimum necessary and are many times over requested by the national IM. We address them in Chap. 5 in the GSM-R section.

2.7 Centralized Traffic Control (CTC) and Operational Control Centre (OCC)

In this section, we describe a key piece of modern railways: the operational control centers (OCC). OCCs have several systems within them, but the most relevant for safety is the centralized traffic control (CTC). The CTC is a large part of OCCs, which may include many other subsystems. In Chap. 3, we explain in detail some parts of OCC like energy control centers and security control centers.

2.7.1 CTC Functions

Among CTC functions, we find two types: (quasi) real-time and batch functions. Within the first group, we have [39]:

1. Regulation: that is, to coordinate both the arrival and departure of trains at every station.
2. Scheduling of trains, numbering, etc.
3. Routing of trains' movements, including parking trains in the depot or yard.
4. Supervision of interlockings (and its peripherals: switches, signals, etc.).
5. Remote control of the energy systems (substations, transformers, etc.), and supervision of energy consumption. This function is essential for the availability of the system [39] to deal with failures in some sections, which will be supplied by the neighboring ones.
6. Remote control and supervision of wayside systems, like elevators, escalators, ventilation shafts, etc.
7. Integration with security forces (police, security agents, railway police, etc.), emergency, paramedics, and civil protection units.
8. Passenger information: both automatic (i.e., scheduled or periodic) messages and based on events.
9. Interfacing with track detectors (lateral wind, objects, vertical impacts, earthquakes, etc.).
10. CCTV displays and other security devices.
11. In GoA4 lines, without staff onboard trains, the operational model is significantly different, so OCC staff must provide some of the functions that drivers used to do, like train supervision or dealing with degraded modes, answering passenger calls, etc.

Notice that the first four are pure ATS functions, but the other seven are operational ones. The essential concept here is coordination: if an emergency arises, having all systems integrated in one place makes operation much more robust and efficient.

In a railway system, many emergency situations may happen. Some examples are failures in trains or wayside elements, floods, antisocial behavior, large gatherings, etc.

For all these functions, CTCs are the operational core of railway lines. Among the batch functions, that is, those functions than can be performed without strict time constraints, we find the following [39]:

1. Operational planning: scheduling train services, timetables, train parking for maintenance tasks, etc.
2. Simulators to train OCC staff, to validate new operational rules, etc.
3. Gears to optimize train's energy consumption, allowing more efficient coordination
4. KPI calculation and report generation

The typical way an OCC is constructed is using large screens to display real-time data from the railway network, with computer workstations for OCC staff, which manage certain aspects of the operations (energy, detectors, CCTV, etc.). Also belong to OCCs the control systems that support CTCs are Supervisory Control and Data Acquisition (SCADAs), which is a control system architecture, with graphical UI and several interfaces with subsystems, sensors, etc. Finally, all the underlying infrastructure (like networks, databases, etc.) allow the proper functioning of CTCs. More details on this can be found in Chap. 4.

Summary of the Chapter

Automatic train control systems are formed by three subsystems: ATP (which is mandatory), ATO, and ATS. In this chapter, we have provided general details about ATP, ATO, and ATS functions.

Due to the maturity of both signaling and non-signaling systems, automation of railways is becoming more and more complex. Unattended (GoA4) lines sit atop this ranking, while GoA3 (no driver required but staff stays onboard for degraded scenarios), GoA2 (driver needed but some functions like speed regulation and some others are performed by the ATC), GoA1 (driver needed and the ATC protects the train from overspeed or running a red signal).

The core of the chapter is about the most relevant TBS: CBTC for urban trains and ERTMS for all the others. We have seen what they have in common (radio transmission of commands from the wayside to the train, train speed, and location transmission to the wayside) and some of their differences (the most important one ERTMS has been designed to be interoperable and not CBTC). We have also paid attention to the requirements that these TBS systems have for lower levels of the communications stack.

Then, we have examined how railway traffic is managed from a control center. Finally, the most important high-level concept the reader must retain is "there is no zero-risk technology."

Exercises

1. Which is the maximum hazard rate allowable for a SIL4 system?
2. Which is the maximum hazard rate allowable for a SIL4 system per month?
3. To which GoA level belongs a system where the driver needs to do everything but setting the maximum speed?
4. To which GoA level belongs a metro system with no human staff on board?
5. What are the differences between ETCS L2 and L1?
6. Consider a train able to run up to 300 km/h and to brake at 1 m/s^2. Consider that the train braking curve follows the Uniformly Accelerated Motion (UAM) equations. Calculate the distance and distance needed to fully stop the train in the worst-case scenario. Now, recalculate both time and distance if the error in speed determination is 5%.
7. Read sections for ETCS L1, L2, and L3. Suggest a potential L4 and sketch the architecture. L4 should be an improvement from L2 on at least one indicator.
8. It is required to implement an onboard PIS based on CBTC data. Please describe the information that these two systems must exchange and how often, to have the system working.
9. There are punctual and continuous train protection systems. Explain the implications of having a punctual train protecting many different points instead of a continuous one.
10. Why 100% safety is not achievable? What would be needed to reach that figure?
11. Consider a safety-critical system with very bad maintainability numbers. Does this have an impact on the safety of the system?

Further Reading

The fundamental references for this chapter are the technical information from signaling companies, like Siemens, Alstom, the former Bombardier Transportation (now part of Alstom), Hitachi, and some others. Some of them have a lot of documents worth reading on their websites and interesting videos in their YouTube channels as well.

On the other hand, a very useful reference is the book by F. Javier González, unfortunately only in Spanish, titled "*Señalización Ferroviaria*" [39]. Sections about ATP and movement authority estimations are largely based on this book.

Sections for ATC functions are based on IEEE 1474. This set of standards deserves a careful read in case the reader wants to deepen in CBTC requirements, UI and CBTC testing. The same applies for other standards mentioned in this chapter, like EN 62290 for train automation grades.

Alongside this chapter, we have mentioned "safe-braking," "safe-braking model," and other related terms. We have only introduced these topics to allow the proper understanding of the main TBS concepts. It is out of the scope of this book, but it is interesting to notice the subtle implications of these braking models, how they are implemented in signaling systems, and also their implications in (ERTMS, for example) standardization. The ERA has released a document explaining the math behind these models [50] and also an Excel spreadsheet to simulate these curves.

Given that we have explained CBTC on a generic way, avoiding vendor-specific details, the reader may find useful some details about the CBTC market plus some references to real-world implementations of this technology. Almost the complete CBTC market share comes from the following manufacturers, which are (in alphabetical order):

Alstom: A historic railway company (which also manufactures rolling stock, including HST), France-based, responsible of many CBTC-related milestones, like the first operational CBTC (Singapur, 2003), or the first CBTC over LTE (instead of IEEE 802.11). Its CBTC solution is called Urbalis and has around 100 metro lines across the world.

Bombardier Transportation (now part of Alstom): Another historic railway company, acquired by Alstom in 2020–2021, both signaling and rolling-stock manufacturer, has installed its CBTC CityFlo 650 [51] solution in more than 15 metro lines.

Hitachi STS (formerly Ansaldo STS): Has more than 350 km of CTBC-equipped lines, many of them GoA3/4, like Copenhagen, Ankara, or Xian [52].

Siemens: A historic German company present in many fields (including train manufacturing), among many others not related to railways, has two CBTC brands: Sirius and Trainguard. Its technology has been installed in more than 20 metro lines.

Thales: Is a French company for aerospace, railways, transportation, and some other sectors. Regarding CBTC, its solution is called SelTrac and is available in many subway networks all around the world.

An interesting way to go deeper on ATC technologies is to study some of the national ATC systems available in Europe. ASFA, LZB, Ebicab, Crocodile, and so on.

The most notable countries that decided not to push for ERTMS development in their railways are China and Japan. Here China is different from Japan because in the early days of its HSL construction plan, they installed ERTMS in many lines, helped by European signaling companies. However, more recently, they decided to support its national equivalent, called Chinese Train Control System (CTCS). More precisely, CTCS is an alternative to ETCS, because GSM-R has not been replaced in China. It has subtle differences with ETCS (i.e., four levels instead of three in ETCS), but it is largely based on some ETCS aspects. Here [53], the reader can find a reference that explains differences between these two systems.

In Japan, most HSL operations rely on a moving-block TBS system called Advanced Train Administration and Communication System (ATACS). It is not widely extended in Japan but in some lines operated by JREast [54]. Digital Communication & Control for Shinkansen-Automatic Train Control (DS-ATC) is another ATC technology employed in Japan HSL. The signaling companies that have larger market shares for these systems in Japan are both Mitsubishi and Hitachi.

In the USA, the most prevalent ATC system is called positive train control (PTC). This system is less advanced than ERTMS, CTCS, or ATACS but is prevalent in many railway lines in the USA. Their freight rail has a great importance, and some freight railway companies have adopted this system, like Florida East Coast or Union Pacific. PTC retrieves train location from GPS signals and protects trains from overspeed and provides other basic ATP functions. The US Department of Transportation website has some interesting documents and links about this technology [55].

Regarding ERTMS, there is a lot of websites with useful information about this system. For example, ertms.net updates periodically its "facts and figures" section with statistics on the usage of this system. As of 2020, there were ETCS-equipped vehicles in the five continents, but most of them were in Europe. In this reference, it is possible to check the deployment by country (notice that many countries have a step-by-step rollout plan for ERTMS, instead of a "big bang" approach).

As we mentioned in this chapter, ERTMS specifications are two main documents: SRS and FRS, that is, system-level and functional requirements, respectively. Reading the last version of both documents can be harsh for beginners, but the amount of information is remarkable. Notice that ERTMS documents from signaling companies is usually not available to the public. Latest FRS' version is 5.00 (2007) [56] and SRS' is 16.0.0 from December 2015 [57].

SRS and FRS are the main ERTMS documentation corpus, but there are many other documents that should be mentioned as well, like interface specification documents (FIS, FFFIS, which are more than 60 documents) and other informative-only documents. The UIC released in 2008 an interesting document [49] titled "ETCS Implementation Handbook" which summarizes many technical aspects of ETCS on a single reference and includes a detailed list as well of all ERTMS documentation.

ERTMS suppliers are more or less the same as the CBTC ones, plus three more: Mermec, an Italian company that is mostly focused (apart from signaling) in railway inspection solutions and diagnosis; AZD Praha, a Czech signaling company, and *Construcciones y Auxiliar de Ferrocarriles* (CAF), a Spanish company with a large tradition of rolling-stock manufacturing.

All these companies form the UNISIG group (AZD Praha and CAF joined recently) which is responsible of ERTMS specification together with the ERA and many railway companies. UNISIG is a working group within UNIFE. It may be interesting to take a look on how UNISIG works and other related aspects [59].

3 Non-safety Services for Railways

3.1 Types of Communications Services

In Chap. 2, we explained safety-related services, which are, under our point of view, signaling services only. Therefore, in this chapter, we address all the other services: operational and nonoperational services (both not related to safety). Nonoperational services are usually called customer-oriented services or passenger-oriented services.

Regarding safety, there is an open discussion about some use cases and services. For example, CCTV[1] systems are generally assumed to be not safety related but security related. However, if an operator uses a CCTV system to monitor if somebody falls to the tracks, the reader may argue that this is a safety-related use case. Another example is voice communications are very helpful for operations, to the point that some operators move a train into the depot if the cab radio breaks down. We want the reader to understand that every taxonomy has its borderline scenarios. As we explained in Chap. 2, in this book, we consider only the signaling services as safety related. Everything else is either operator oriented or customer oriented. Let's see some examples of each one:

Operator-oriented services (OOS): They are those where the passenger is not the end customer, but the operator, infrastructure manager (IM), or any other agent. For example, CCTV cameras are for passenger protection, but there is no need of interaction from them to work. Here, the final customer is the security agency, policemen, operator staff, or whoever is the user of the real-time CCTV, video recordings, etc. Some other examples for this are intercoms, remote maintenance, remote diagnostics, and ticketing systems (indeed, this chapter is focused on everything related to the additional examples). We consider them as not safety related, but this discussion is not relevant.

Customer-oriented services (COS): It includes all the others, that is, not related to operator services but to passengers. Some good examples are passenger information systems (PIS), Internet access, gaming, video on demand, etc. Notice that some of them may involve the operator, IM, etc., to some extent, but the end customer is the passenger. There is no controversy in saying that these COS are not safety related. These services may generate incomes to operators, or perhaps are provided for free to increase passenger comfort.

3.2 Operator-Oriented Services

In this section, we explain the most relevant OOS in railways. On each subsection, we provide some functional details about the service, a description of the architecture (in both wayside and trains, if both apply) and finally, the requirements for lower layers of the communications stack. As usual, in the "further reading" section, the reader will find papers, videos, and other useful information to know more about the topics addressed in this chapter.

3.2.1 CCTV (Closed-Circuit Television)

There are two main groups of security services in railways: CCTV related, which include newly video analytics (both AI- and non-AI based), and all the others, like access control, anti-intrusion, security patrol control, etc. In this section and the next one, we cover the first group and the second one, respectively.

[1] CCTV systems are being renamed as video surveillance systems (VSS). Many standards use the new notation but given that it is still not widely accepted among professionals, we have opted in this book for the old-fashioned term.

Supplementary Information The online version contains supplementary material available at https://doi.org/10.1007/978-3-031-82247-6_3.

J. Moreno García-Loygorri, *Telecommunication Systems for Modern Railways*, Textbooks in Telecommunication Engineering,
https://doi.org/10.1007/978-3-031-82247-6_3

Everybody has a more or less clear vision of what CCTV is. In this section, we highlight the differences between typical CCTV setups and other specific details for railways. In general terms, CCTV is about the transmission of video from its source (the camera) to another location (a monitor, storage device, etc.). In this sense, CCTV is equivalent to video surveillance. In railways, it is a key element in several scenarios, like stations, passenger areas onboard trains, and other assets worth monitoring, like technical rooms, viaducts, escalators, etc.

Therefore, CCTV enables many different use cases that allow both operators and IM to provide a more secure and efficient transportation service to passengers. Thus, we can say that CCTV is a pure operator-oriented service (OOS), perhaps the most paradigmatic. The ideal application for CCTV is crime prevention. Having CCTV cameras distributed over a place helps to prevent crime and antisocial behavior. It seems that this is more relevant in certain places like parking lots or stations, rather than residential areas. For railways, these cameras are an essential element to avoid littering, harassment of other passengers, trains from being graffitied, glass scratching, traveling without ticket, or, simply, theft. Graffiti vandalism has become a scourge, with a significant impact on operators' budgets due to the costs of both graffiti removal from trains and the reduced availability caused by train stoppages for cleaning.

Depending on the assumed delay, we can find two main groups of functions for CCTV-related use cases: first, real-time video surveillance, where video can be transmitted to one or more screens or processed to perform other functionalities, like motion detectors, objects in the track, facial recognition, etc. For example, the train driver can see the CCTV streams from the passenger area of the train, upcoming platform or front/rear cameras of the train, or operations control center (OCC) operators can watch what is happening onboard the train or at a station. Second, analysis of video recordings: usually, the video is stored just in case it is needed afterward. Recordings are stored for a period (typically, a few days or weeks) and then overwritten by new recordings. Some countries have strict regulations for the maximum time the recordings can be stored. Typically, operators and IMs are enforced to maintain the chain of custody of the recordings, just in case they are needed in court.

Both are usually provided together and need a communications network to work properly (with some exceptions, as we will see). For example, to watch onboard CCTV videos in the OCC screens, at least a train-to-ground system is needed.

Depending on the solution for storage and other specifics about the service, this has an impact on the network and the price of the solution. We can see two extreme scenarios: first, if all video is stored on the edge, which is much more expensive than the cloud (even more if the storage is onboard a train); second, if the video is stored in the cloud (centralized): cheaper storage, but very demanding requirements for the network (which implies large expenditures as well). In this case, due to regulation on information privacy (like the European General Data Protection Regulation or GDPR) perhaps the storage could not be in the cloud or maybe it is not recommended. Important remark: when we say cloud, we are talking about public cloud (that is, accessed via Internet) and on-premise or private cloud (accessed via the private network of the IM or operator).

All options require specific tools to manage video, either for real-time surveillance or for recordings. If an incident arises unnoticed, there is always the chance to retrieve the related recordings. Every year, security forces and courts require a significant number of recordings from railway companies, to investigate many incidents. Typical use cases for video streaming ("real-time" video) are periodical visualizations of cameras of stations, trains, and other locations, random visualizations, chosen on demand by human (or non-human) operators, etc. There are many ways to do video surveillance of railway networks.

Another relevant aspect of CCTV networks is alarms, which can be either based on information from the video or from other sensors and linked to video data. Finally, there are other security-related functions or maybe not security related but based on the video streams. For example, passenger counting can be done using video cameras (even in some cases running the application in the camera), which send the results to a centralized management platform where they are aggregated and visualized.

Operators and IMs have a "security area" within the OCC, to manage not only CCTV operation but also other security-related functions. While some operators do not have security very much integrated with other parts of the OCC, the trend is to put all of them together. Notice that in many countries, there are specific laws and regulations that may have an impact on the responsibilities of the security area of railways.

The most relevant aspects of a modern CCTV system are that it is digital (that is, cameras provide a digitalized video stream as its output), and it is Internet Protocol (IP) based. IP is a network-layer protocol which allows the interconnection of computers and packet transmission (more details on Chap. 4).

The general architecture of a digital CCTV system is formed by the cameras, switches, management systems, clients, and recorders. Let's know more about them.

Video cameras: It is the essential component of the system in charge of the acquisition of video streams and their transmission to the network. We will explain them in the next subsection.

Switches: Strictly speaking, switches are part of the network and do not belong to the CCTV system, but they are much needed because they provide connectivity for a system, which is strongly network based (that is, it can work without a network but will be severely degraded on its performance). The other optional (but very common) function is to use switches to power up cameras through the network wires. This is called "Power over Ethernet" (PoE) and follows two IEEE standards [60, 61, 62]. Please see the text box below for more details about PoE.

Video management system (VMS): It is a software component where all the operational and management-related functions of a CCTV system are done. That is, all the integrations with other systems, attending client requests, video analytics, etc., are performed here. Recording management is performed in network video recorders (NVRs), which are explained below.

Clients: Everything below "he general architecture of a digital Closed-Circuit Television (CCTV) system is formed by the cameras, switches, management systems, clients, and recorders. Let's know more about them" is an enumeration. These are all the parts of the system.

Network video recorder (NVR): It is where the recordings are stored. There may be one or more, cloud based, on-premise, centralized, distributed, etc. In large networks with many cameras, the NVR can be very complex and costly. NVR can be considered part of the VMS or a different entity. Recordings can be stored centrally, in a distributed manner or a mix between them both.

Centralized: This is storing all recordings in a data center or a cloud-based location, that is, separated from the camera locations. This approach is much more efficient in terms of storage costs but requires higher network capabilities to transport all the video data from the cameras to the network storage. Notice that cloud storage is not necessarily centralized, but it is the opposite of the distributed solution, that is, edge (by definition). Due to the increase in network capacity and the popularity of cloud storage technologies, the network-storage approach is getting more popular in recent years. However, for trains, it is still more popular than the edge storage alternative.

Edge: This is storing very close to the sensor. In railways, this means on-camera storage (typically, a SD or microSD card) and disk drives near the camera (onboard the train or in the station or location where the camera is). In this scenario, recordings are retrieved on demand only. Notice that this setup implies having several NVRs distributed in the network (one per train or station, for example).

Hybrid approach: Edge storage in some cases and download some recordings to the cloud or data centers only in case of alarm, or some specific hours every day (for example, the morning rush hour in subways). This implies having two layers of NVRs: a centralized one and distributed across the network.

Finally, data from other sensors rather than video cameras, like audio devices, fire detectors, anti-intrusion, etc., may be needed in the CCTV system. These devices can be managed together with the CCTV system and other security systems on a specific platform. We address some of them later in this chapter, together with an essential system for security management called physical security information management (PSIM).

3.2.1.1 CCTV Use Cases in Railways

Here we provide a list of CCTV-related use cases in railways. It is impossible to have an exhaustive list and to keep it updated, but the reader will find it useful to understand how CCTV has become a central piece in railways operation. The most relevant CCTV-related use cases are online surveillance of stations, passenger areas in trains, platforms, and other assets; online surveillance by the train driver of the upcoming platform; automatic pinning of a camera when an alarm is triggered (fire, smoke, train door opened, etc.); and automatic pinning of a camera when a call is placed. Examples of pop-up video streams in case of alarms are when a passenger requests assistance on a station; help access control systems, e.g., in degraded mode or complementary to regular access-control systems, information desks, etc. To increase the automation of stations, more cameras should be installed and accessed remotely; detection of abandoned objects (luggage, etc.); presence detection in the tracks; platform screens to allow metro drivers to see the complete platform; in automated stations, when an escalator stops and can be switched on remotely, a CCTV camera is needed to perform this function safely; crowd monitoring in stations, automatically switching on/off some systems to help the crowd; train-driver rollback assistance: when the driver needs to move the train backward, front/rear cameras are helpful; rear-view cameras to monitor passengers; accident analysis, together with data recorders and logs from other systems; condition-based maintenance (CBM). For example, monitoring of the pantograph-catenary or the wheel-track contact can be used to know the condition of the assets and plan maintenance tasks accordingly. More on this is in Chap. 7, driver behavior monitoring, etc.

3.2.1.2 Legacy CCTV Systems

We have explained IP digital CCTV systems as they are the most prevalent solution for CCTV in railways. However, legacy CCTV systems are still present in some railway networks, so we will briefly explain them. These legacy systems are usually analog, that is, the cameras do not provide a digitalized output stream, but an analog one. Some other CCTV systems are digital, but they are not standardized as the modern ones are, for example, not being IP based.

Legacy CCTV systems can be full analog, that is, visualization and storage can be analog as well (using video tapes, for example) or the video stream can be converted into a digital stream by a video coder (which requires an extra device to perform the video codification). In this case, we have a mixed CCTV system: part analog and part digital. Very often this setup was used to store (digital) recordings on a file system instead of tapes, allowing some management functions up to a certain level. The main drawback is the extra complexity plus less flexibility, but the early days of the security era that started just after the 9/11 were full of analog cameras that today are still being replaced for IP-based full-digital ones.

3.2.1.3 CCTV Cameras

As we said, CCTV cameras are the key element of every security system, in railways and everywhere. In this section, we explain some specific aspects of these devices, assuming that we are on a digital CCTV system. The most relevant characteristics of a camera are video standard, type of camera, PoE availability, Open Network Video Interface Forum (ONVIF), and other certifications. Let's learn more about them.

Video compression standard: The de facto standard is H.264, or more precisely, MPEG-4 Part 10. In 2019, a remarkable 91% of the video industry developers were using H.264. It has some limitations, like a maximum display resolution of 8 K (that is, 7680 × 4320). Typically, the de facto standard for cameras is 2 MPx (1920 × 1080), but it is very likely that this will go up in the future (5 MPx or even higher).

Type of camera: Depending on the application, there are several configurations for CCTV cameras in railways (see Figs. 3.1, 3.2, and 3.3 for depictions of each one of them) like pinhole, domes, rear-view, bullet, and even phony cameras.

Pinhole cameras: They are usually hidden from view, largely used in cash machines, but still present in some locations across railway networks, like old trains.

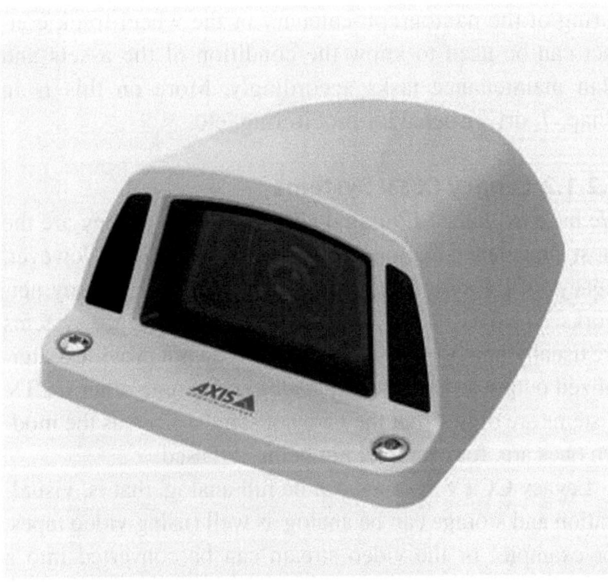

Fig. 3.1 Rear view CCTV camera by Axis

Fig. 3.2 Bullet-type camera by Axis

Fig. 3.3 Example of dome camera by Axis

Dome and mini dome cameras: They are cameras with dome-shaped housing that can be either large or compact and are prevalent in many scenarios around the world. They are very much used onboard trains and in stations (that is, indoor scenarios only) and other scenarios where a broad angle is needed.

Rear view cameras: They are installed on both extremes of the trains, are used to allow drivers see (on a screen inside the cabin) the platform and even the surroundings of the train. Sometimes they have a retractable housing and only pop up when the train stops.

Bullet cameras: They come with a long housing, typical for outdoor scenarios, like railway crossings, viaducts, and others.

Phony cameras: It is still possible to see some phony cameras, that is, empty housings that pretend to be a real working camera. The objective of these phony cameras is to provide a higher feeling of security for passengers, resembling a wider coverage than the provided by the real cameras. Obviously, these cameras do not need any kind of connectivity or power supply to work properly.

Power over Ethernet (PoE): If the camera can be powered by a PoE-compatible switch or not. See PoE text box below.

ONVIF-compatible: Open Network Video Interface Forum (ONVIF) is an open industry forum, which enforces a standard for the IP connection of CCTV products (not only cameras, but mostly), de facto standardizing many functions, like IP configuration, end device discovery, device management, etc. This is the fulfillment of the ONVIF requirements allow interoperability on a CCTV system, for example, having cameras from different vendors on a network or a VMS from another one. It is strongly recommended to use ONVIF-compliant cameras. The ONVIF forum was founded by three major companies in the CCTV business: Axis, Bosch, and Sony. Now it has more than 400 members on different categories, and there are more than 20,000 ONVIF conformant products.

Certifications: Depending on the application, it is a good idea to use cameras with specific features and certifications. A typical example is the EN 50155 certification [63] for cameras onboard trains. In Chap. 6, we provide more details about this standard. Another key characteristic is the IK factor, which is the protection against mechanical impact, defined in the IEC 62262 standard [64], and the ingress protection (IP) factor. IP defines the level of protection against solid and liquid particles (typically, water). The reference for this is the IEC 60529 [65]. Depending on the location of the camera, exposition to rainfall, etc., a different IP factor may be needed.

Other relevant features are motion detection, infrared vision, pan-tilt-zoom (PTZ) to steer or zoom the camera remotely, etc.

Bitrate for Video Cameras

When dealing with CCTV systems, it is very important to have a good estimation of the resources needed to work properly:

1. Data rate: The output data rate from each camera. It depends on:
   - The resolution: The size of the pixel matrix to be displayed on screen. Some examples are 1920x1080 (also known as FullHD), 1280x720 (HD or 720p), etc. The larger the size, the more resources are needed.
   - The compression format: The video format. As we stated before, the most common is H.264, but there are many more, such as H.265, MPEG-4, MPEG-2, and M-JPEG.
   - The compression quality: Many compression formats like H.264 allow different video qualities (typically: high, medium, and low). The higher, the more resource consuming. For example, a H.264 video, 24 fps in HD resolution, requires 0.8 Mbps, 1.5 Mbps, or 2.8 Mbps if the compression quality is low, medium, or high, respectively. However, this also depends on the image being captured by the camera. As the coding algorithm is differential, it is not the same as the camera on a crowded station hall and a camera surveying an alley with no activity at all.
   - Frame rate: Number of frames per second (fps), that is, how frequently the image is updated. It is widely accepted that above 12 fps [66], the human eye perceives it as motion and below that figure, we do not. However, the usual standard in security systems is 24 fps. Sports video transmissions generally use at least 60 fps, having 120 fps for slow motion playbacks but depends heavily on the specific case. As usual, the trend for this figure is to go up.
2. Storage: The capacity (measured in gigabytes or terabytes) for the storage system to allocate all the required recordings.

$$C = N_c \cdot T \cdot 24 \cdot 60 \cdot 60 \cdot \frac{B}{8 \cdot 1024 \cdot 1024} \approx 100 \cdot N_c \cdot T \cdot B$$

   where C is the capacity measured in TB (terabytes), B is the output data rate from a single camera (in Mbps), T is the number of days required, and N_c is the number of cameras.

For example, if we have 50 cameras in our network and we want to keep the recordings for 1 month (30 days), assuming we have H.264, 24 fps, HD resolution, medium quality, we need:

$$C \approx 100 \cdot N_c \cdot T \cdot B = 100 \cdot 50 \cdot 30 \cdot 1.5 = 225000 TB$$

which is near 225 PB (petabytes) of storage. It is an impressive figure with a remarkable cost and not easy to handle.

Therefore, the higher the quality of the video, the more resources are needed, and the more cameras, the higher the data rate. These two aspects are essential to properly dimension the whole system and its connection to the network.

Finally, a good reference for CCTV systems is the EN 62676 "Video surveillance systems for use in security applications" family of standards [67]. EN 62676 standards are worth reading for those interested on specific aspects, such as the minimal requirements for a security system, defining several degrees of protection depending on many factors, interfaces, protocols, and many other useful information to understand the state-of-the-art of modern security systems.

Finally, as we have seen, there are so many railway-specific aspects on a CCTV system for railways that we can conclude that it is much more than a sole CCTV system placed in railway locations.

Power over Ethernet (PoE)

PoE is a standard to transfer both data and electric power on an Ethernet cable. This means that only the Ethernet cable must be installed to connect the powered device and the power sourcing equipment (PSE, typically a switch, but not only, because there are also PoE injectors). Therefore, the power supply and the associated wires can be removed.

There are three standards approved so far:

- IEEE 802.3af: Called PoE, able to provide up to 15.4 W on each port (which is no less than 12.95 W available at the powered device). These are called type 1 devices.
- IEEE 802.3at: PoE+, up to 30 W on each port (25.5 W at the device end). These are called type 2 devices.
- IEEE 802.3bt: PoE++, up to 60 W or 99.9 W, depending on the mode (51 W and 71.3 W at the powered device, respectively). These are called type 3 and type 4 devices, respectively.

CCTV cameras do not have high power needs, so the first one is enough for them. PoE+ is for wireless access points (more details in Chap. 5) and PoE++ is for high-performance wireless access points and other devices.

There are some requirements for the wires (maximum impedance, etc.), but the most important thing to notice is the number of required pairs on each mode: two pairs for PoE and PoE+ and four pairs for PoE++, that is, four wires for PoE and PoE+ and all eight wires for PoE++.

Examples of PoE devices are CCTV cameras, intercoms, WiFi access points, speakers, VoIP phones, routers, clocks, etc., that is, many devices that need network connectivity and do not demand much power.

Be careful when choosing a switch for your network, because many switches do not have PoE in all its ports (which can be a problem if there are several devices), and do not forget to consider the total available power in the switch. That is, the sum of all the power needed on every port must not exceed the maximum power the switch can transfer. Another important aspect is the distance, which shall be lower than 90 m between the switch and the end device. In the end-of-chapter questions, we have an exercise for this. Finally, there are some nonstandard PoE implementations, so make sure your equipment does follow the IEEE 802.3af/at/bt standards to avoid further trouble in your designs.

3.2.2 Access Control and Others

In the previous section, we explained CCTV but we have mentioned that it is not the only security-related service (or group of services) in railways. There are many more, like anti-intrusion systems. The technology behind these systems may differ, but the working principle is always the same: when somebody enters a certain location, an alarm must be triggered. This may apply for train cabins, technical rooms in stations, and many other locations. When an alarm is triggered, usually in the OCC, the nearest camera is pinned to provide more information to the security staff.

Another relevant use case is security patrol management: security guards patrol stations, depots, and other facilities. To make sure that their routes are walked adequately, without skipping any milestone, some railway companies request their security staff to register these steps (for example, with an radio frequency identification (RFID) card on a reader conveniently placed).

Access control: It includes technical rooms, OCC, data centers, trains, etc. Only authorized personnel must get into them. As usual, in large enterprises, there may be subcontractors in charge of some of these duties, and they may change over time. Usually, there is a centralized tool that deals with access lists (periodically sent to the access machines) and event logs (events are sent by the distributed access machines to the centralized platform).

Management of visitors: OCCs and other railway spaces are heavily guarded. Visitor management tools and processes are therefore needed.

Interfacing with the police, other security forces and civil protection: A large percentage of the security tasks in railways is performed by the companies (and subcontractors), but some of them require the intervention of policemen and other security forces. Having an adequate integration that may include having a representative in the OCC (acting as a "liaison officer") can help a more efficient actuation when a security problem arises. Or perhaps, policemen having compatible handheld devices (i.e., "walkie talkies") with the railway radio telephony can be useful as well.

Smoke detectors and other sensors: Integrating these and other sensors with the centralized CCTV can allow a quicker response. This is very common in some trains and in some stations. For trains, an integration between the onboard CCTV and the onboard fire-detection system is required, which is usually implemented via Train Control Management System (TCMS), as we explain in Chap. 6. This CCTV-TCMS integration may have an impact on the required CCTV hardware, capping the market and making the solution more complex and costly. On the other side, the benefits of such an integration are obvious in terms of safety.

There are some auxiliary functions that must be considered when designing security systems for railways, like:

1. System administration functions that include the permissions to access recordings, video streams, delete files, etc.
2. Audits of the system, to see who has accessed a camera, configured a device, downloaded a recording, etc.
3. Maintenance of the system: configurations, updates, fixing issues, etc. Usually, recording systems are redundant, with periodic backups. If information privacy regulation applies, another important feature is to make sure that the recordings are not accessible for more time than specified by the regulation.
4. NAC (Network Access Control-802.1x) systems for access control and encryption Secure Sockets Layer/Transport Layer Security (SSL/TLS) communications is very recommended as well in railway systems.

What we said about CCTV remains valid here: There are specific laws and other regulations that may apply here and have an impact on both the design and the operation of the system. For example, in many countries, only certified companies can install or maintain security systems. This is a guarantee for end customers but has its impact on the price and the market as well.

Due to the increasing complexity of security systems, not only in railways, the trend is to have a physical security information management (PSIM), which is a software platform that integrates security-related applications into one interface. It is very useful for security operators because it helps them to manage the situation more efficiently and intuitively. It allows a unified response to situations that many times require accurate information and steady pulse to make the right decisions. For example, step-by-step procedures can be provided by the PSIM UI to fix some well-known situations that may require an advanced degree of coordination or training (i.e., the evacuation of a station or a train in a tunnel, etc.).

Finally, to sum up, having a secure railway is essential for every country as this transportation mode is critical for both the economy and the day-by-day routine.

3.2.3 Digital Train

There is another point of view of the "oriented-services" paradigm that we have presented across this book: safety, operator, and customer-oriented services. We can call it the connected asset-staff-customer vision, and very similar to the other paradigm except for the safety aspect, which is missing in this new notation. This notation is more common on digitization slang, while the other one is more common in railway forums. Let's see the railways through the lens of digitization. We find three things connected here:

- Connected customer (passenger): In this book, we have seen COS services. We see more of these later in this chapter, when we explain access to the Internet, passenger information systems, gaming, and some others.
- Connected staff: Radio telephony and other services oriented to the operation itself, to make it more efficient, fluent, and secure, that is, OOS in our notation.
- Connected asset. Where the railway assets (trains, switches, ticket machines, and many other elements) send information to a centralized platform or to another systems. In this section, we explain the most notable connected-asset example, which is the "connected train," but this concept can be extended to many more with only few changes. That is, the digitization is not limited to trains only but also to any other asset in the railway world, like track switches, stations, level crossings, freight containers, warehouses, etc.

The workflow for this "connected train" service requires at least four steps: sensing, processing, transmission, and cloud persistence of the data.

Sensing and acquisition of the signals: This is the sensing of a physical phenomenon. Typical examples in railways are:

- Accelerometers in train bogies to determine the vehicular condition
- Temperature sensors for train brakes or bearings
- Humidity sensors in stations
- Smoke detectors
- Meteorological stations
- Wayside detectors (see Chap. 1 for more details)
- Train communications networks, like multifunction vehicle bus (MVB) or wire train bus (WTB) (see Chap. 6 for more details)

Notice that each one of them requires a careful study to determine the most suitable electronic device, sampling frequency, power supply, communications technology, location, etc. It is out of the scope of this book, but we provide some insight in Chap. 5 (radio communications systems), Chap. 6 (train networks), and Chap. 7 (maintenance). In the further reading section of this chapter, the readers will find some references about this interesting (and trending) topic.

Edge processing (if any): Data is processed before being consumed by the end customer. This processing can be done on the edge (that is, near the sensor), in the cloud (far from the sensor), or in both.

Transmission to the centralized data platform (or cloud): Collected data from one or more sensors, processed on the edge or not, must be transmitted to be consumed. This action involves a train-to-ground system always. In Chap. 5, we present several of these systems which are an essential part of the topic we address on this book.

Data capture, validation, persistence, and visualization. Once the data has arrived at the data platform, the reader may be wondering if the job is already done. The answer is no. Once in the data platform, there are many tasks to be done, like data collection, validation, and usage.

Data collection: Data can come from several different sources and can have different destinations on the platform. For example, some data will serve to feed quasi-real-time use cases and must be quickly processed and routed to its destination; meanwhile, some chunks of data are to be stored on a database and perhaps batch processed at the end of the day. Therefore, data platform resources must be allocated accordingly to have everything on its side and all tasks performed efficiently.

Validation of the data: Once the data has arrived, before being consumed, it must be a valid piece of data. This is to check that it does not contain errors, is correctly formatted, it is not outdated, etc. In case it does not pass the exam, it must be discarded; otherwise, it is ready to be consumed.

Data usage: This can be done by several means:

- Data persistence: Data is stored on a database or by any other mean. Depending on the future uses for the data, the engineering of the database must be carefully done. It is out of the scope of this book, but there are many different types of databases, because some are for real-time data processes, others are oriented to time series (very typical when data comes from sensors), others are non-relational (which are more flexible than its relational counterparts), etc.
- Batch processing: Opposite to real-time processing, batch analysis implies a delayed processing of larger chunks of data, to extract certain features. Batch processing can be very efficient due to the less tight requirements in terms of delay. There are many examples for this, but a very simple one is to compute a day average of a variable and compare it with the day before.
- Real-time consumption: Data is quickly delivered to its final destination.

Types of Data
There are many ways to classify data, but perhaps this one could be both simple and useful to navigate this topic:

- Nonstructured data: Typically, files of any kind which do not share a common structure, binary data, videos, images, audio files, etc.
- Semi-structured data: Logs from processes or applications, JSON files, CSV, XML, etc.
- Structured data: Typically, relational databases, that is, heavily structured data, with its headers, presented on a tabular form (i.e., rows and columns)

We have explained the generic picture of digital train services. Now we go through some specific details of this service to illustrate how useful it is for railway operators and infrastructure managers. Knowing the condition of an asset is always better than being in the dark. This is the key idea here. It is obvious, but commissioning everything needed to have the data ready to be consumed is much harder.

As we have seen in Chap. 1, trains have many systems and subsystems, like doors, brakes, communication networks, bogies, etc. that are continuously working, exchanging information between them (for example, the train control makes sure the train is stopped to open the doors), reporting their condition, keep alive functions, etc. But data does not need to leave the train because before digitization, trains in many senses were like islands. These isolations make some authors claim that trains are indeed "data silos," which is very bad news because it implies that data is no longer useful. In our humble opinion, this statement is wrong, because even if data does not leave the train, it is indeed very useful and necessary for the normal operation. Of course, if a digital train infrastructure is available and trains can transmit its internal data, it is much better.

3.2.3.1 Digital Train Use Cases

Now let's see which are the end customers of digital train use cases. We have three relevant ones: railway operators and IMs, maintenance companies, and passengers.

Railway operators and IMs: There are many examples of interesting use cases for operators, but there is one that clearly illustrates the advantages. When a failure happens on a train, many times the driver can solve it and keep going, but very few times this is not possible, and it may cause delays or, perhaps, service disruptions. Some operators dictate that the OCC staff help the driver to solve the issue, but the reality is that this is very complex to do because the OCC is almost blind. They do not know what's happening inside the train unless the driver tells them or the train sends this information to the ground systems. There are many ways to implement this kind of solutions, but the general idea is that with this information, the OCC can detect the problem and help the driver. In Fig. 3.4, the reader can see a dashboard that shows the OCC staff the train controls and other useful information of a particular train, just the same way as if the OCC operator were in the train cabin, with the driver. This service is called "remote desk" and can be implemented with data from the train buses, without extra sensors.

Maintenance companies: Remotely knowing the condition of a train enables them to do a more efficient maintenance, rather than being only reactive to failures. Furthermore, predictive-maintenance strategies are close to the optimal, decreasing costs and raising train-fleet availability. The key idea here is to uncover noncritical issues that can have impact later or help to predict future breakdowns. For example, a train door that takes more time to open or close than the others can have a problem on its engine. Therefore, monitoring times to open/close doors can unfold more severe failures (and allow maintainers to fix them before there is some kind of impact on the service).

Passengers: For example, knowing in advance the less crowded car of a metro train allows passengers to travel on a more comfortable way.

Regarding the requirements for communications systems, digital train use cases are very diverse. In some cases, the data volume will not exceed some bytes per second, but in some others, like acquiring the raw data from the train buses may imply a large volume. Consider that modern trains (since 2000) can have several thousands of variables on their buses, each one of them being updated several times per second. In Chap. 6, we describe these buses and train communications networks (TCN) in more detail.

Therefore, the needs in terms of data rate can range from some bytes per second to several Mbps. However, a proper planning of the data to be sent should prioritize the quasi-real-time information to be sent quickly, and then transmit later the rest of the variables (for example, when the train is stopped at the end of the day).

3.2.4 Passenger Counting and Others

In this section, we cover a very specific subtype of digital train use cases: those who have specific hardware to acquire the data. Strictly speaking, all data needs a sensor, but here we are talking about data that does not come from the TCN, but from other sensors, like passenger counting systems, accelerometers, CO_2 sensors, and some others.

Passenger counting: Knowing the number of people onboard a train is very useful for the operator. In high-speed train (HST), mainline or regional, passengers usually have a seat assigned on the ticket, but in urban rail, this does not happen. Moreover, depending on the location of the platform

Fig. 3.4 Dashboard for the remote driver's desk of a metro train. (Courtesy of Smart Motors (TM))

exits, passengers tend to concentrate in some cars of the train, leaving some others almost empty or almost. Indeed, large crowds can be very dangerous, so the security staff at stations make great efforts to avoid or remove them. As usual, the most effective approach is to prevent them. The occupancy of the train can be obtained by several methods, like video analytics, load sensors (needed for brake and traction effort estimations), automatic people counters (APC), or counting Bluetooth and/or WiFi cards from passenger cell phones.

In stations, all methods but load sensors are widely available. Access turnstiles are much used as well, but only provide information about entrances and exits from stations, not particularized to a train or a platform.

Accelerometers are sensors to measure vibrations and are a key element in trains to address both comfort and safety issues. Large vibrations can be not only uncomfortable or unhealthy, but also dangerous for train stability. Therefore, is very common to install these sensors in many train locations, like bogies, axles, axle boxes, passenger area, etc., to acquire this data. There are two standards to measure vibrations and to obtain suitable validation KPIs for both comfort [68] and stability [69]. With accelerometer-based systems and suitable processing, an estimation of the condition of the train can be retrieved, triggering alarms in case something goes beyond an acceptable level. A common problem is the large data volume, because sampling frequencies are between 200–10 kHz (depending on the specific application) and the number of channels can be high.

CO_2 meters: Due to the COVID-19 pandemic, many operators had CO_2

sensors installed onboard trains to have an estimation of the air quality inside trains. Many times, these sensors were deployed separately from both train networks and managed apart, using dedicated apps or even displays in the sensor case.A special category is test trains or laboratory trains, which are trains only for infrastructure-health monitoring, that is, not for commercial service but for two main purposes: first, sensing the condition of several wayside elements, like catenary, rails, etc. To measure all these parameters, test trains must have some specific equipment, plus storage drives, UIs, and other auxiliary elements. This sensing is carried out periodically, for example, once or twice per year. A report of the infrastructure condition is generated and delivered to the maintenance department which will fix whatever is needed. Second, test new lines or systems. When the construction of a new line is completed or it is upgraded to have another signaling system, it must be tested before opening it to the public.

Both tasks can be done with "normal" trains, but doing so has two main problems: the first one is that IMs should rent a train from the operator,[2] and all specific portable sensing equipment must be installed and uninstalled for the tests. This equipment is expensive, hard to install, and calibrate. Depending on the use case, it is almost impossible to do it on a portable manner (i.e., catenary displacement or measurement of rail defects). The major drawback is that test trains are very expensive to manufacture and to operate. They are costly to manufacture because many times they are train prototypes, formed by short series of trains (this is, without economies of scale) with large engineering costs, costly equipment, complex system integrations, etc. On the other hand, it is expensive to operate because sensor calibration and analysis of the retrieved data require skilled profiles and measurements are retrieved usually in night hours.

▶ This special category of trains is not owned by operators but by IMs. In Fig. 3.5, the Séneca train owned by ADIF (the Spanish IM) is depicted. It is a high-speed train, based on a prototype of 102 series by Talgo (this is a convenient approach: transform a train prototype into a test train, rather than manufacturing only one). This train has the speed record for the Madrid-Valladolid HSL (336 km/h) and has the following measurement systems:

1. Track condition inspection
2. Catenary condition inspection
3. GSM-R coverage measurement
4. ASFA and ERTMS
5. Aerodynamics
6. Energy consumption measurement to validate speed profiles in the railway line

3.2.5 Voice Communications and Radio Telephony

Voice calls are an essential tool in railway operations. Even if it is not a safety-related service, if there is a breakdown on the train radio used for voice calls in some railway networks, the train is evacuated and moved to the maintenance facilities. This is an extreme scenario which can probably be attenuated with operational rules but illustrates properly the importance of such a system. Apart from safety, efficiency improved largely.

Voice calls are bidirectional, and they are not limited to a station or train but to the railway line (including stations, technical rooms, and buildings along the line) and the complete railway domain. That is, the coverage has the ambition to be as much complete as possible. Bidirectional voice communications are usually addressed with the generic term "telephony," which includes both wired and wireless telephony.

We usually call this service radio telephony, because it is like a telephony system but involves a radio communications system to communicate with trains (and other terminals, like

[2]Only in those countries where IM and operator are not the same company.

Fig. 3.5 High-speed SENECA test train, owned by ADIF and manufactured by Talgo [70]

handheld phones). Being able to have bidirectional voice communications between the OCC and train drivers; plus, some other railwaymen, implied a remarkable step forward in the 1960s when it was first put into service. Later came almost every single railway line in the world.

However, the systems that we can find today are not the same as the ones that pioneered this service. Nowadays, most of them are digital systems, integrated in complex infrastructures and also onboard modern train control systems. Nevertheless, some of the early day systems still survive as a backup just in case the modern one fails. In this section, we focus on the state-of-the-art technology, but we mention briefly an example of legacy system (but still in use) to illustrate how complex railway are sometimes with its multiple layers of redundancy and interconnection between systems.

In this book, we do not cover legacy telephony systems as they were replaced by digital VoIP (voice over IP) systems. However, due to the importance of telephone in railways, we include a text box with some historical anecdotes on this interesting topic. Consequently, our focus here is twofold: operational radio telephony and VoIP telephony.

This section is for voice communications systems, but the next one is for another voice-related systems, with subtle differences with voice calls. Intercommunication (intercoms system) is for bidirectional voice communications but within a specific area, like a station or a train. Typically, intercoms are largely used in railways for either emergencies or requesting assistance (for example, ticket machines or access to stations or platforms). Intercoms are not restricted to railways as they are present in many other scenarios like airports or private houses and offices, for example (Fig. 3.6).

Public announcement systems (PAS) are for unidirectional announcements on a restricted area, like stations or trains. Several countries or regions have tight regulations about PAS systems for safety reasons, like evacuations. As the reader may see, the border between safety and non-safety services gets blurrier as we get closer. In railways, announcements can be broadcasted from a central location (typically, the OCC) or from a local entity (station chief, train driver, or any other authorized staff). Another frequent scenario is to have prerecorded audio messages that can be selected and broadcasted by an operator.

Both intercoms and PAS are addressed in the next section, but we wanted to explain their differences with voice calls. In the remaining of this section, we focus on voice communications only. Unlike some other services already explained in this chapter, voice calls have many railway-specific features which make them a true niche technology. With an associated toll in terms of price and

Fig. 3.6 TETRA walkie (by Motorola). In railways, voice communications may involve intercoms, mobile devices, onboard handsets (typically in train cabins), and some others. People in stations or on the wayside use walkies very often, for safety reasons

Telephony

Telephony is a communications system still present in railways. Do not think that it has disappeared in favor of radio telephony or other mobile communications systems like Global System for Mobile Communications-Railway (GSM-R).

Indeed, railways developed at the same time as telegraph and telephone since the 1840s and the last decades of the nineteenth century,[3] respectively. Earlier, it was the telegraph which was installed together with new railway lines across many countries [71]. A remarkable example of this joint development was the United Kingdom and an exception was Spain. In Spain, railway lines were constructed slower than the telegraph but, in the United Kingdom, both infrastructures were built together.

In 1851, for the first time, telegraph was used for railway dispatching, in the Erie Railroad (USA) by Charles Minot, its superintendent (who had learned telegraph and wanted to use it in railways).

Telegraph was soon used around the world to transmit train movements, and telephones replaced them shortly after its invention in 1875. In 1877, Pennsylvania Railroad was the first successful implementation of railway telephone but was not for dispatching trains. This was to come later (in 1879 on a Massachusetts line).

Indeed, telephone dispatching remained the prevalent technology for train dispatching until the 1950s when modern centralized traffic control (CTC) replaced them. Nowadays, some remain in service as a backup system but are rarely used. It is out of the scope of this book, but there are subtle differences between American and European dispatching systems [71].

complexity. Niche technologies are very expensive due to the absence of economies of scale and are harder to maintain and upgrade. In Chap. 8, we address some of these aspects.

3.2.5.1 Radio Telephony

We find three radio telephony technologies in railways: GSM-R, Terrestrial Trunked Radio (TETRA), and many other legacy systems.

GSM-R (Global System for Mobile Communications-Railway) for mainline, high-speed, and many others. It is a digital system, based on GSM and is explained in detail later in this book.

TETRA (Terrestrial Trunked Radio) for urban rail (both metro systems and tramways). GSM-R is a digital system but comes from the public safety world (that is, security forces, police, etc.) and was adopted by those railways that were not allowed to use GSM-R.

Legacy systems are typically analog systems and most of the times still in use as a backup system for the two others. There is no standardization at all in these systems, so we can find many of them around the world.

[3] Telegraph was invented by Samuel Morse in 1845 and telephone in 1875 by A. G. Bell. The famous "Come here Mr. Watson, I want you" was not pronounced until March 1876 [217].

In Chap. 5, we explain both TETRA, GSM-R, and we present an example of an analog train-to-ground system for voice communications. Digital T2G are the evolution of the early analog ones and as of 2023; they are starting the final stage of their lifecycle. In particular, GSM-R will be replaced by a future system which is known as Future Railway Mobile Communications System (FRMCS) and is explained in Chap. 8.

Talking about the functionalities (which is the main focus for this chapter), the most common one is voice communication between the train driver and the OCC. But it is very useful to be able to talk with other persons involved in the operation and maintenance, so they are given a handheld device to communicate as well.

So far, there are no railway-specific requirements, only bidirectional calls between the OCC and railway agents, some of them onboard trains, some others not. But there are many requirements which do not apply for regular cellular customers. When a regular customer wants to make a call, they dial the number and press a button. That's all. In railways, this procedure can be a little different. Before moving on, let's introduce two essential concepts here: functional and regional addressing.

Functional addressing: Calls are not routed to a specific endpoint (terminal) but to a terminal or group of terminals that are dynamically assigned. This is better understood with an example: the OCC does not call a train by its fixed ID (103–001), but by the train number (which can change every day or within a day, or represent a train formed by some coupled vehicles). These train numbers are more useful for the OCC as they represent railway services and are related to an itinerary, etc., rather than the fixed IDs which are only useful for maintenance. Therefore, the OCC will not call the 103–001 train, but train 214 of the Circle Line. To make the call using functional addressing instead of static addressing, a middleware is needed to map functional addresses to static ones.

Regional addressing: This is more or less the same as functional addressing but based on the location of the trains. Imagine that the OCC wants to tell something to all trains that are approaching Munich Hauptbahnhof coming from Nuremberg. OCC does not want to talk to all trains in the line, only to some of the trains in one of the tracks. If the system has already defined regions, the OCC will establish a call between the OCC and some trains that fulfill a certain condition (i.e., being on a specific track or location).

Moreover, there are some other specific requirements that radio telephony systems must fulfill:

- Group calls: One-to-many calls instead of one-to-one calls.
- Pre-emption: Low-priority calls can be removed from the system (emptied) to allocate their resources to high-priority calls. For example, OCC has usually the authority to be considered the highest priority rank for this.
- Direct communications between terminals: Usually mobile communications are routed through the network (core), but it is possible to directly set up calls between end devices, not using the network. This is the same as walkie-talkies and can be extremely useful in case of emergency if the network goes down. However, it has a severe limitation coming from the distance between terminals. Another important limitation is that voice calls cannot be recorded, which is a very common IM's policy.
- High security standards: As we explain in Chap. 4 when we talk about cybersecurity, low security standards can be a burden, so encryption, authentication, etc., must be carefully considered for these systems.
- Set-up time: Voice calls must be initiated very fast (much faster than normal cellular calls which can take more than 5 s): this is called PTT (push-to-talk) functionality, and 500 ms is a suitable figure for this parameter. Indeed, this is a very challenging requirement.
- Integration with other systems: Typically, with both audio systems and Train Control and Monitoring System (TCMS) onboard trains, and in the wayside, voice systems plus CTC.

Finally, although radio telephony is mostly for voice communications, digital systems are also able to send short data packets. For example, some operators send commands from the OCC to a train using the TETRA system, to switch on or off a train remotely, raise the pantograph, etc. This is common in some Unattended Train Operation (UTO) lines, because TETRA is an ultra-reliable system, with a high level of redundancy.

3.2.5.2 IP Telephony

IP telephony is more known as VoIP (voice over IP). This is a technology for voice calls over IP protocols. It is the evolution of the old telephony system, based on analog signals over copper wires and called Plain Old Telephone System (POTS). It is also the evolution of Public Switched Telephone Network (PSTN). However, POTS and PSTN are not the same at all. POTS is about the full-legacy analog system, whereas PSTN works over different physical channels (copper, fiber optic, satellite, etc.), and its core network can be digital or even IP based. VoIP works over an IP network, independently from all underlying layers, including the physical one. This is in theory only. In some cases, there are some scenarios that can make VoIP calls almost impossible due to large delays and jitter. For example, geostationary satellites that orbit Earth at almost 36,000 km make round-trip delay very high.

VoIP is not only about voice calls because it can provide other services like short message service (SMS), but its main strength comes from being IP based. Indeed, as an IP technology, it is packet switched, whereas POTS and PSTN are circuit-switched systems. The difference is important because circuit-switched networks assign resources on a fixed and dedicated one and are, therefore, less flexible and efficient, but perhaps more reliable. Given that GSM-R is circuit switched, in Chap. 5, we pay some attention to this dichotomy and its impact on railway systems.

VoIP calls require a digitalization of the voice signal using audio coders, and there are many possible alternatives. VoIP also requires signaling procedures (as many other telephony systems, including PSTN or POTS) to register, initialize, and manage the call. As with audio coders, there are many alternative protocols, but perhaps the most accepted one is Session Initiation Protocol (SIP). See text box below for more details on SIP.

As in old telephone systems, an endpoint is needed, that is, a phone. There are three alternatives for this: first, a dedicated VoIP phone, connected to the network; second, a non-IP phone with an adapter to connect it to the IP network, and, finally, another device running a VoIP application. Or, more conventionally stated, a softphone app running on a cell phone, tablet, or PC.

There are even dual phones, able to connect to two different networks, like cellular and WiFi.

VoIP technology is cost-efficient and highly scalable, so it has become widely extended. The main reason for this is both voice and data can be transmitted on a single network. Actually, this is a major trend in communications (not only in railways): unification, which implies a reduction of technologies, networks, and end devices. In Chaps. 4 and 5, we explain some technologies that are still in use but being replaced by others that can concentrate more services (thereby simplifying architectures and networks).

Regarding QoS, VoIP systems can suffer from three different problems called delay, jitter, and packet loss.

Delay: Voice and video calls are highly sensitive to end-to-end delays that can come from many different sources, like routers, coders, or the physical channel itself. To have good-quality calls, end-to-end delay must be as much as 100–150 ms.

Jitter: It is variation in delay. It is usually caused by network congestion. There are some ways to fix this (as we explain in Chap. 4, where we discuss network technologies), but sometimes it is very hard to minimize. A reference for the maximum jitter in VoIP calls is 30 ms or below.

Packet loss must be below 1%. Notice that packet retransmissions can work for some other applications but not for VoIP, due of its delay-sensitive nature.

In recent years, the technology trend has been to make video calls instead of voice calls only. This is the same as in instant messaging apps like Telegram or WhatsApp. However, for railway use, video calls are not that usual, and voice calls remain more extended. This does not mean that it will continue to be this way in the future.

SIP

SIP can be used either in unicast or multicast calls and it is for:

- Call initialization
- Call termination
- Adding more participants to the call
- Modifying the session (changing ports, addresses, etc.)
- Notifications
- Messaging

Every SIP entity,[4] like phones, voice mailboxes, or routers, is addressed by its Uniform Resource Identifier (URI), which follows the same syntax as webmail. This is:

```
prot_name:user_name@domain_name
```

where `prot_name` is the identifier of the protocol (SIP in this case), `user_name` is the ID for the user, and `domain_name` identifies the domain and can be either a Domain Name Service (DNS) name or an IP address.

Regarding the communications protocol itself, SIP follows the same approach as Hypertext Transfer Protocol (HTTP) does. That is, clients request something from a server and the server sends a response. Conveniently, this is called request-and-response model.

Finally, SIP is a complex technology, and for the sake of simplicity here, we do not have more room for this. We suggest interested readers to check some useful references on this topic.

Finally, VoIP technology has been implemented in railways without specific requirements, that is, on a standard way. However, this does not mean that VoIP solutions are fully standardized and interoperable. There are many different technologies coming from several vendors that are not completely compatible. Therefore, engineers and tech strategists must be careful when choosing a particular solution for VoIP in their railway network.

[4] A more suitable term is user agent (UA), which includes both servers and clients.

3.2.5.3 Intercommunication and Passenger Announcement Systems

In this section, we discuss two systems that are different in nature but share a common audio aspect: intercommunication and passenger announcement systems (PAS).

Intercommunication is a bidirectional voice communications system to be used on a clearly defined area, like stations, trains, and tunnels.

PAS is a unidirectional audio system, to be used on a clearly defined area.

Both have some things in common and some differences.

Starting from the beginning, intercoms are audio systems with a microphone (that acts as the audio input) and a loudspeaker (audio output). They need a power supply and need to be connected to a network (otherwise the intercom will be isolated which does not make sense). See Fig. 3.7 for some examples of railway intercoms.

Intercommunication is typically an emergency channel in those trains who do not have a driver, or the onboard staff is limited in number. If something happens, passengers can push the intercom button (or trigger the handle) and talk to the driver or the OCC. On stations, it usually follows the same approach but can be used for nonemergency situations, like helping passengers to buy tickets, request wheelchairs, etc. Regarding tunnels, it is needed to have intercoms if the tunnel length exceeds a certain value (in Europe it is 1000 m). All these intercoms allow communication with the control center if an emergency arises.

Modern intercommunication systems are IP based, with PoE-powered intercoms and, like VoIP, usually SIP based. Legacy systems are analog, with an audio line instead of an IP network. For rolling stock, the reference here is the UIC 558 standard for cable and connectors. Anyway, the most important part of railway intercoms is the requirements and the consequent integrations between other systems. However, these requirements are very specific to the railway company, so here we address only the most common one, noticing that even those are not implemented everywhere.

Fig. 3.7 Train intercom by Commend

The usual functional requirements for intercom systems are the following:

Intercoms can work in two different ways: as a normal intercom and for emergency calls. In some trains, the intercom is activated by a trigger (or "handle"), which can be configured to stop the train in some circumstances. That is, it acts as an emergency brake for passengers. It can save lives if, for example, somebody's coat gets caught in between the train doors, and the train resumes its journey.

Therefore, intercoms are part of the passenger alarm system (do not confuse "passenger alarm system" with "passenger announcement system" even if they both share the same acronym). This alarm system must be designed and installed following some strict considerations in terms of user interface, like size, buttons, colors, etc. For mainline trains, there is a standard [72] and another one for urban rail [73].

Intercom calls are routed to a specific endpoint, where the call is to be received. In case it is not answered, an alternative jump can be defined. This approach allows answering calls near to their origin (typically, a station, train, depot, or a tunnel) and if the first jump is not available, forward it to the OCC or a more centralized location.

Redundant architecture: To avoid losing the intercom system due to a single failure (like a switch), intercoms on a specific location (like a train car, platform, or a tunnel) are usually connected to the network (and powered) by different switches. This way, if a single switch fails, there will be others able to keep working on some intercoms. The same can be said to power amplifiers: a centralized approach (having an amplifier serving several intercom speakers) may lead to a larger service disruption.

Intercoms can have a hearing loop integrated for those passengers with hearing aids. This is an extra feature, but there are many intercoms that provide this functionality, which provide a service of undeniable usefulness. It has an extra cost, and the zone to be covered must be designed carefully to avoid interference from other zones and from other non-audio sources (like lamps, for example).

Intercoms should have an automatic power control that senses the surrounding noise level and adapts its output power to it. Intelligibility can be affected by background noise, bandwidth of the voice signal, volume, and some others. Intelligibility is measured by the Speech Transmission Index (STI) parameter, according to standard IEC 60268–16 [74]. The ideal value for this parameter is 1 (perfect intelligibility), but a more realistic value (higher than 0.5) is the objective in real-world scenarios. Regarding limitations in bandwidth, for example, voice in the legacy phone system (POTS) has 3.1 kHz, which clearly introduces distortion, but saves bandwidth. Some intercoms can work up to 20 kHz to decrease this distortion, but probably at a higher cost.

It is very useful that intercoms have self-provisioning capabilities, that is, being able to configure by themselves just after being powered up. This is very useful for any kind of equipment, on railways or elsewhere. It saves time and avoids errors.

Intercoms can be integrated with Private Automatic Branch Exchange (PABX) systems , that is, the telephone network.

Each intercom can be individually addressed and monitored.

Given that in subways, intercoms perform some safety functions (like emergency calls), they are subject to the EN 62267 standard [75]. This standard only applies to urban rail and addresses both the design and functionality of intercoms, and many more things related to safety, like door design, derailment detectors, etc.

In some cases, when an intercom is activated (either in an emergency or not), the conversation is recorded. In some cases, it will be on cloud storage, others near the intercom (in the station or train where the intercom is located), and in some other scenarios, these audio files are stored together with video files from the CCTV system. Moreover, sometimes when an intercom initiates a call, an alarm is triggered, and the nearest CCTV camera pops up to allow the receiver of the call to see what is happening. This functionality requires an integration between intercommunication system and CCTV system, perhaps in the PSIM or by other means (i.e., train TCMS).

Cybersecurity must be considered from the design phase, because intercoms as IP devices are more exposed to attacks than legacy devices.

There are many different potential architectures for this service. We can see several intercoms connected to the network, plus the switches, and routers that form the network. On the other side, a server is needed and some control desks (which may include GUIs) to answer the calls (or any other endpoints where calls could be routed). In case integrations with other systems are needed (as happens in railways), they must be considered as well.

Passenger announcement systems are for broadcasting audio messages from a specific location. They are just like intercoms but not bidirectional (the endpoint in PAS are speakers instead of intercoms). PAS and intercommunications have many things in common and some differences. Let's see them:

Redundancy needs are very similar, including the amplifier architecture. Indeed, in PAS, amplifiers are more critical and must be designed even more carefully, to allow intelligible audio messages everywhere.

PAS can be used for safety functions, like evacuating stations or trains. The reference standard for this is EN 50849 [76]. Everything we said before for hearing loops, self-provisioning capabilities, PoE, and automatic power control remains the same for PAS.

In trains, there are speakers inside the train for passengers' area and speakers outside the train for passengers in the platform. This approach is very common in urban railways, and the train does not broadcast an audio message to the speakers outside if it is not stopped at a station.

Speakers and intercoms are on one side of the system; dispatchers are on the other side.

3.2.6 Station Remote Control

Railway stations are a key part of railways because they are the place where passengers get on and off the train. Due to its specific architecture, they have several systems, like heating ventilation and air conditioning (HVAC), escalators, elevators, CCTV cameras, other security devices, ticket machines, access turnstiles, platform screen doors, (passenger information system (PIS) devices, like screens, LED displays, intercoms, fire detection and suppression system, ventilation shafts in underground stations (and tunnels), and so on.

An efficient way to operate stations is to connect all these devices or a large part of them to the network and control them remotely. The usual approach is to use SCADA (supervisory control and data acquisition) systems in the OCC or at the very station (which supervises only that specific station). SCADA systems imply an UI for high-level interaction (that is, both supervision and control) and a network to connect all the required elements to the platform.

Typically, industrial systems (elevators, escalators, etc.) are monitored by Supervisory control and data acquisition (SCADA), but nonindustrial ones (Closed-Circuit Television (CCTV), intercoms, etc.) are typically integrated into more complex monitoring tools. The difference is that SCADAs are bidirectional, and are oriented to make decisions and actuate on field elements, whereas monitoring tools are oriented to gather logs and let users know about the condition of the elements in the field (but not actuate on them).

Some fully automated metro lines use SCADA systems for systems onboard the trains.

PLC

Programmable logic controller (PLC) are computers used for industrial control, usually in harsh environments, like factories or, closer to us, trains and railway stations, with strict real-time requirements. PLCs are connected to other PLCs or SCADAs, and they have several inputs and outputs that can be digital or analog. This is large systems interface with both sensors and actuators using PLCs.

The architecture is as follows:

CPU for all the processing.
Memory to allocate both program and data from sensors. The programming language can be graphical or text based.
I/O module for sensors and actuators interfacing.
Communications ports (Ethernet, RS-232, etc.).

Oversimplifying, PLC software works in the following way: device is initialized, runs a continuous loop which reads inputs, does some logic operations, and writes outputs, over and over again. The logic operations include the communications with other PLCs or with the SCADA.

Many trains since mid-1990s have redundant PLCs at the heart of its Train Control Monitoring System (TCMS). Modern stations have many PLCs too.

IEC 61131 [77] is a standard for PLCs, with several parts. For example, part 3 [78] is about programming languages and part 5 [79] about communications.

PLCs are convenient when the application is specific, or the number of required devices is not very large. For example, automotive industry (which deals with a huge number of vehicles) does not use PLCs but microcontrollers, which have larger development costs. They manage to keep unit costs low due to the large number of manufactured units.

SCADA is a very mature technology which has several generations. The UI in modern SCADAs is Web based (following a trend common to many other systems) and the connectivity is based on TCP/IP.

3.2.7 Energy Control Center

The control center is formed by several parts, including the centralized traffic control (CTC), security, detectors, centralized station control, and energy. As its station counterpart, the energy CC is a SCADA, which oversees two main functions: first, management of primary energy sources, typically AC redundant drops: The ideal scenario is to have fully redundant drops to avoid outages in case a company fails. This is not always possible because in some locations, there is only one alternative available. Second, control and monitoring of substations for traction: AC substations in HSL, most long distance, and DC for subways, tramways, commuter trains, and some long distance. The objective is to allocate resources to fulfill all the needs from the trains. In case a substation fails, collateral ones will take over. Moreover, there can be several elements within substations that can be remotely controlled and monitored, like circuit breakers, disconnects, etc.

From a corporate point of view, there are many different degrees of integration between the parts that form the control center. In some cases, all parts are together in the same place, but the respective staff belongs to different departments, with different profiles. Other IMs, perhaps less mature on its operational organization, have a lower degree of integration between the CC parts.

Finally, notice that energy management is a critical task because if it is not available, the service will be severely disrupted. Of course, there is uninterruptible power supply (UPS) that can provide energy supply for an extra time period for critical systems, but never for traction (that is, trains). This solution is very common in wayside equipment.

3.2.8 Auxiliary Systems for Operational and Maintenance Staff

As we said in Chap. 1, when we introduced railways, this business demands many people for both operation and maintenance. However, the trend is to decrease these needs with technology. Instead of spending several human hours on visual inspections, installing sensors to predict the future condition of an asset, more technology is used to leverage the need of workforce. This is not always possible because of some limitations (please see Chap. 7 for more details), but the idea is to use some tools to help both operational and maintenance staff. The common acronym for them is operation and maintenance (O&M). In railways, it is very common to have a clear distinction between them, but in this chapter, we have put them together.

Operational staff are those in charge of the day-to-day duties of stations, trains, and OCCs. For example, selling tickets, assisting passengers, GoA3 (non-driving staff onboard the train) tasks, degraded modes in stations due to failures in some systems, etc.

Maintenance is usually separated by systems (catenary, track, rolling stock, elevators, etc.) and structured in levels (level 1 for onsite maintenance; level 2 for workshop repair, etc.). Therefore, level 1 requires access to databases or technical specifications to solve an issue, do some tests, or validate other worker's job. Some of these tasks can be done more efficiently with a computer or a tablet that may require connectivity.

In GoA3, in some railway networks, their onboard staff has corporate tablets (or even smartphones) to assist passengers more efficiently (for example, answering questions about destination, connections with other train services, etc.). Another interesting use case is for train diagnosis in case something is wrong on a GoA3 train. Staff can gather extra information from the tablet and perhaps fix the problem, or interface with the OCC. All these techniques are focused on decreasing service-disruption times, which is a major KPI everywhere in railways. This approach is very common onboard trains for GoA3 modes, but in highly automated stations, it is even more common no matter which GoA applies for the train operation.

3.3 Customer-Oriented Services

In this section, we cover the final type of services, which includes passenger information systems, entertainment, Internet access for passengers, and some others. Opposite to safety-oriented services, this category has stronger requirements for data rate, but not for reliability or availability. A service disruption for any of these systems may cause small annoyances, but not to the rail service itself. However, COS are very marketable, attracting more passengers (especially young people) and improving the corporate image of the railway company.

Another important type of COS is related to selling tickets to passengers. Ticket machines, which can be either onboard the train (like in tramways) or at the stations, need connectivity for the payment methods or to react to changes in the service (station closures, etc.) and to download logs. See Fig. 3.8 for a ticket machine onboard a train.

Closely related to ticket machines are the ticket validators. Usually, RFID readers are connected to a central unit that relays the aggregated data to a ground database. See Fig. 3.9 for an example.

Fig. 3.8 Ticket machine onboard a tramway in Spain

3.3.1 Passenger Information Systems (PIS)

Operators and IM work very hard to make railways an easy-to-use transportation mode. Many times, this is absolutely true, and everybody is happy. However, there are some situations that can complicate the user experience and make it difficult to deal with, like different tariffs can make very complex the purchase of train tickets. This is more frequent in urban rail but can happen elsewhere. Another important information to be provided to passengers is the normative, like baggage size, pet regulations, etc., onboarding/offboarding the train, navigate through the stations, different destinations for the same railway line, etc. Finally, giving information about special events, like renovation works, closings, etc., is very relevant in terms of customer satisfaction.

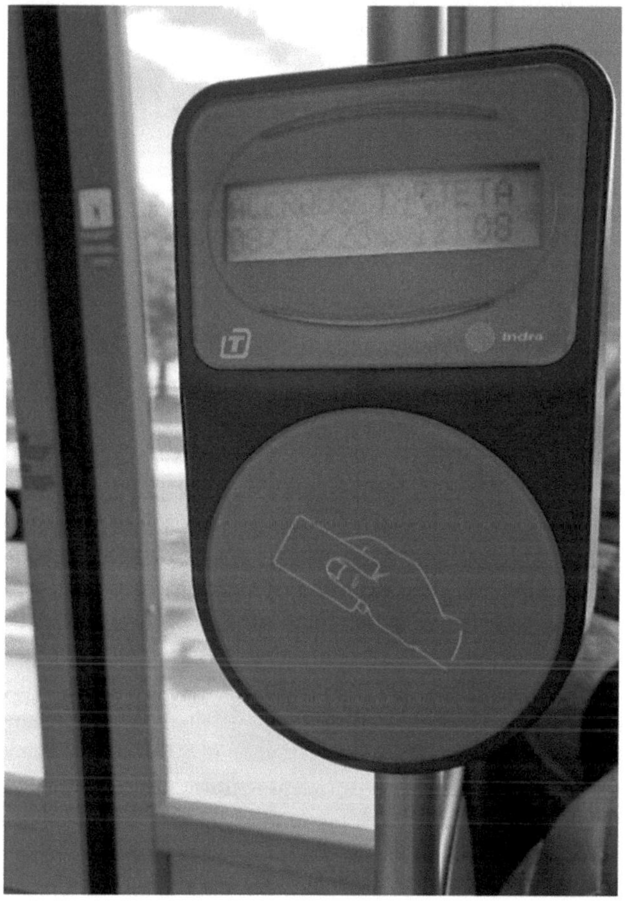

Fig. 3.9 Onboard ticket validator

The solution is to make sure passengers receive all the needed information. Depending on the scenario, many different media could be useful (ads on newspapers, signage, posters, broadcast audio messages, etc.), but in this section, we focus on the PIS.

This system provides information through the following elements:

1. Station and platform screens.
2. Onboard screens and displays, both in the passenger area and in the train car bodyshell (See Figs. 3.10 and 3.11).
3. Mobile apps and other websites.
4. Audio messages are either automatic or originated by the train driver or the OCC.

Therefore, the objective is to have an informed passenger correctly routed to the right platform, timely onboarding its train, and noticing on advance when one must get off the train to reach its destination. Moreover, passengers must know any special circumstance, like delays.

As we mentioned in Chap. 1, fast passenger onboarding/offboarding is critical for train operators. Therefore, passengers must be aware which side of the train can be used to get off the train, know in advance that the train is approaching its destination, etc.

To explain PIS, we identify two main scenarios: wayside and onboard. For the wayside PIS, it displays information on screens placed at stations, platforms, etc. The information is usually departures/arrivals of trains, indicating platform, expected time, etc. In subways, occupancy of each car of the upcoming train (typically with a green/yellow/red color code) is a useful information to for passengers waiting in the

Fig. 3.10 Platform display showing passenger information (typically, destinations, stops, waiting time, etc.)

Fig. 3.11 Examples of PIS displays onboard trains. Typically, these displays provide information about current date and time, next stations, arrival times, and sometimes, but not in this case, train speed

Fig. 3.12 External display for onboard PIS displays

platform. In mainline and high speed, displays can show the number of cars in the platform, to ease passengers to get in the train. For lines that go to the airport it is useful to have information about flight departures and arrivals. This use case requires an integration with the air traffic manager entity. Weather forecast and many other useful information like this is sometimes shown on the screens as well. Finally, any other useful information for passengers, like failures, delays, if the next train does not accept passengers, service disruptions, etc., could be useful to show on screens as well.

The onboard PIS is slightly different, as the train first needs to locate itself to provide accurate information. Once it is located, the onboard PIS usually lets passengers know the next station, which needs to know where the train is. Not very accurately, but the reliability is very important here, especially for disabled people. Accuracy is not very relevant if the next station is announced 2 min or 90 s before the stop. But making no mistakes is very relevant. Therefore, PIS systems go silent if they do not know accurately where the train is.

Another information provided by the onboard PIs is the side of the doors to get off the train, destination of the train, speed of the train (this is very common in high-speed trains), location of the train; in subways, information about platform exits, to allow passengers get off the train through a convenient door; same as in stations, information about flights if the train goes to the airport, any other useful information for passengers, like failures, delays, if the train must be vacated, etc.

There is even a third scenario, which is for mobile apps. This is known as the bring your own device (BYOD) paradigm and implies that the wayside-PIS information is displayed in the passengers' mobile app. The operator or the IM must have an app, and some engineering must be done to provide the related information (see "Digital Train" section in this chapter for more details about these use cases).

For the onboard, information is a little different (it is not straightforward to relate the train and the mobile phone locations, especially in dense lines, like subways or commuter trains). Some operators use QR-code stickers placed onboard the trains to help to relate them and allow passengers to have PIS information on their devices. In Fig. 3.12, the external display of a tramway is shown.

3.3.2 Internet Access

Apart from PIS, the other notable COS is providing Internet access for passengers, in both stations and trains. In PIS, there are no safety issues here, but perhaps many cybersecurity concerns. In Chap. 4, we address the topic of cybersecurity in railways, and we explain the challenges related to providing Internet access to passengers (and many other things).

There are several ways Internet access could be provided to train passengers, and they involve several technologies that are covered in more detail in Chaps. 4 and 5, like train-to-ground communications, like satellite, LTE, 5G, WiFi, etc., as well as some network protocols and tools to enforce security and quality of service (typically, MPLS, as we will see).

Apart from the cybersecurity challenges, providing Internet access for passengers does not look like a big deal, does it? In this section, we explain why it is a challenge or, more precisely, something that makes more expensive the electronic equipment. Let's go into the reasons for this.

We are in an extreme environment because of vibrations, high temperatures, low temperatures, presence of strong EM fields, dust, etc. All of them can damage electronic equipment which is responsible for the communications link. This is especially true for onboard train equipment but not only, because wayside elements are also exposed to some of them. In Chap. 6, we go deeper on this topic.

From the physical-layer point of view, trains are large metallic masses. These metallic walls attenuate largely the incoming radio signals. In Chap. 5, we provide more details on other physical-layer issues and their potential solutions.

Tunnels are very common in railways (not only subways) and represent a special scenario for train-to-ground communications for two reasons: first, remove the possibility of using satellites to provide connectivity and second, for terrestrial links, tunnels are a complex scenario which needs to be considered separately, with a different approach.

Frequent handovers: This problem is more prevalent in IEEE 802.11 or LTE than in 5G, but trains running at high speeds jump from one cell to the next one very fast, so the handover procedure must be triggered very frequently. This issue does not only happen in terrestrial networks but also on low-orbit satellites. More details about handovers and non-terrestrial networks in Chap. 5.

Old antennas were large and represented a challenge to be installed onboard a train. Fortunately, recent developments have removed this problem, which was twofold: train gauges were already very tight, so antennas were drastically limited in volume, and its shape had strong aerodynamical impact. Indeed, the radome of some of the very first prototypes flied away from the train when running at high speeds.

In some stations and inside some trains, there may be large crowds. Providing connectivity in ultra-dense environments is a remarkable challenge, radically different from normal scenarios, as we explain in Chap. 5. For the moment, consider that in some cases there are up to six people per m^2 onboard a metro train or waiting for the next one on a crowded platform.

High costs: Fulfillment of railway standards implies higher costs, not only for CAPEX but also OPEX. Furthermore, railway operations usually demand night shifts or traveling long distances to get to the train for maintenance. Moreover, networks for passenger connectivity need many more resources than other networks for operator or IM use. The reason for this is the number of clients and the traffic volume.

Business model and regulation: This is a WiFi-specific problem because railway companies can face some trouble if they intend to provide free Internet access or financially support Internet operators. Moreover, the Internet is widely seen as a basic service and more people are reluctant to pay for it.

As we mentioned before, cybersecurity remains a major issue when dealing with Internet access for passengers. Not only the technical threats must be considered but also the reputational cost for the railway company must be accounted for when providing such a service.

However, all these problems have been recently solved to a large extent, and now there are many trains that deliver a high-quality broadband Internet access to their passengers, and for sure, many more will imitate them in the future.

3.3.3 Mobile Operators' Connectivity

Mobile operators are key players for customer comfort as well. Indeed, they do not provide any railway-specific service, but it is very relevant for passengers to have an adequate coverage along the tracks.

The role for IMs is to give space for mobile network operators (MNOs) to have their equipment (including antennas, power supply, masts, towers, etc.) and, in some cases, connectivity as well. That is, the backhaul of the mobile network (Public Land Mobile Network, PLMN) is provided by the IM network.

The usual way to do this is by unused optical fibers. As we explain in Chap. 4, IMs deploy a fiber network, which is usually oversized, and some of these fibers can be leased to third parties. It is very common to install more wires than needed, because the physical installation takes a large share of the costs, so having a 32-strand optical-fiber costs almost the same as a 64-strand one. Therefore, the best strategy is to go for the 64-fiber cable and resell the spare resources. This concept is the key idea behind the "dark fiber" business, to which many railway IMs are very devote.

In our case, these optical fiber network will be deployed in the railway domain, very close to the tracks; so nearby

MNO facilities are the perfect candidate to have dark fibers to connect their equipment to their core network.

Regarding the radio access network (RAN) of the MNOs, it is usually conditioned by some requirements from IMs, like tower heights and other limitations. IMs allow (and encourage) MNOs to install their equipment, renting them spaces in stations or near the tracks, but have some rules for this. In Chap. 5 of this book, we address mobile technologies, and we see the impact of the height of the tower where antenna equipment is placed.

3.3.4 Infotainment Platforms, Gaming, and Advertising

Like airliners, some long-distance and high-speed operators have an infotainment platform, with movies, series, and also games, to entertain their passengers when they are onboard trains. These platforms are well stocked, and usually the contents are stored on a server onboard the train. Therefore, the challenge here does not involve train-to-ground systems but logistics (to load new contents from the train to keep them updated). These multimedia files can have a large size and may take much time to be downloaded, so it is critical to do this task efficiently, saving costs and time. Notice that in these platforms not only new contents are added, some others are removed and cannot be played after a specific date. Therefore, the operator (or its subcontractor) must ensure that these old contents are no longer accessible, which imply having up-to-date software and multimedia contents.

Another strategy to update these multimedia servers is to do it automatically but only on some specific moments where trains are stopped and can have a stable coverage to download new contents easily. This is usually done by WiFi networks at certain points (terminal stations, for example), but the progressive 5G deployment will make this approach obsolete (as we see in Chap. 5).

Infotainment servers also host game applications which are played locally onboard the train. Many games have very strict requirements in terms of communications, like low latency, which is very important for user experience. They depend heavily on the type of game, but perhaps 50–100 ms are needed in some cases. Another one is the data rate, which may depend on the details of the game, but may demand some Mbps, and finally, stability, which means low packet-loss rates.

However, if games are restricted to the train local area network (LAN), these requirements are usually not a challenge. If this restriction does not apply and the gaming platform is fully online, the current train-to-ground communications technology cannot fulfill the requirements.

Finally, advertisement in both onboard and station screens is very common in railways around the world. It is a convenient source of incomes for IMs and operators, and the screens are usually shared between PIS and commercials. On the other side, some operators have dedicated screens for PIS and others for advertisement only. This is not a technical issue but a strategic one, heavily dependent on the Service Level Agreements (SLA) that IMs and operators have with advertisement agencies and content providers: the more time on screen for commercials, the more incomes for the railway company.

In subways, for example, it is interesting to provide geo-localized commercials, that is, when a train is approaching a particular station, some specific multimedia content pops up. This functionality needs an integration between the infotainment system of the train and any source of train location (typically, the signaling system or maybe a balise-reader within the PIS).

In stations, this service is easier to handle because no train-to-ground system is needed. However, there are still some challenges, like cybersecurity aspects (cybersecurity issues in infotainment systems can be very costly for IMs and operators in terms of reputation). Content management is usually subcontracted to a multimedia company, which needs some degree of integration within the multimedia platform to manage contents, certificate the number of times a video has been played, etc. Contents may include commercials, news, and other institutional videos.

The requirements for communications are two: data rate and latency/jitter. Data rate is typically 1–10 Mbps but depends on the number of videos to be downloaded. If the system includes news, the list of daily new videos is longer. Usually, videos are downloaded once and played many times. That is, no live streaming is needed, so there are no requirements for latency.

Summary of the Chapter

In this chapter, we have described several noncritical systems, like CCTV, access control, PIS, and many more. The focus was not only on security but on the needs of the operator/IM and the passengers. We have seen that the communications requirements for all these systems are very different from one system to another, and in some cases, they can be very demanding. For example, CCTV systems may require large data rates (or storage), and real-time services, like voice or even video, need short latencies and bounded jitters.

This chapter is very railway-specific because it is essential to understand how do we integrate in railways a system which was not originally conceived for railway use. Moreover, another challenge is to determine how it is operated and integrated with other systems.

Exercises

1. If a CCTV camera is used to assist better passengers in ticket machines, to which type of service does it belong?
2. If a CCTV camera is used to determine the condition of the pantograph-catenary contact, to which type of service does it belong?.
3. Describe the difference between latency and jitter. Why is this important in voice communications?
4. Explain the difference between public cloud and "private" (on premise) cloud.
5. What are the differences between PIS and an intercom system?
6. What does functional addressing mean?
7. What does regional addressing mean?
8. What is the most challenging parameter on a gaming system?
9. We have a switch with 12 ports, 8 of them PoE. The maximum power consumption is 103 W, but the maximum PoE power is 60 W. How many CCTV cameras can be powered over the Ethernet with this switch? Consider that the camera is IEEE 802.3af/802.3at type 1 class 3.
10. Consider a large station near your home. Try to estimate the number of potential IP end devices (escalators, elevators, CCTV cameras) in the station, and group them into types of services. How many IP addresses are needed?
11. Consider a train the reader usually rides. Try to estimate the number of potential IP end devices (CCTV cameras, speakers, intercoms) onboard, and group them into types of services. How many IP addresses are needed?
12. We have a sensing system with accelerometers on each train bogie. The sampling frequency is 2 kHz, the number of bits per sample is 12, and the number of accelerometers is 3 per bogie. If the train has 16 bogies, provide an estimation for the storage system needed to save 10 days of measurements. Consider that the raw data can be compressed up to a 25% of its initial volume, but a redundancy system 1:1 is needed as well. The file formatting overhead and the metadata inclusion imply an extra 30% volume from the raw data.
13. For all the data in Exercise 12, estimate the size of storage system in case we want 30 days of measurements.
14. We have a digital train app onboard that retrieves all MVB data from the train. We estimate that there are around 2000 variables, meaning a (more or less) constant data rate of 25 kbps. If we want to send all the data to the cloud, please estimate the monthly 4G/5G data plan needed to avoid service disruption.

Further Reading

Typically, the best way to go deep on non-safety services is to understand better how all these systems work, that is, getting to know how a digital CCTV sensor works, or a coder/encoder system, RFID devices, etc. that are mentioned in this chapter. This way, the reader will be more able to allocate the needs of all these systems in the network and integrate them better. For example, Bosch has some useful documents [80] and tools (like a lens calculator) to help users understand better how this technology works.

On the same line, data sheets and manufacturer websites are useful. Many of them have a railway line of products, with recommendations to integrate better the devices on a railway system. Finally, the set of standards mentioned through this chapter is worth reading as well.

4 Network Technologies and Systems in Railways

In this chapter, we explain the main technologies and concepts related to communications networks in railways. More precisely, "wired" communications systems, in opposition to "mobile" or "wireless," is addressed in Chap. 5. Moreover, in this chapter, we do not cover onboard train networks, which you'll find in Chap. 6. Consequently, in this chapter, we address the main technologies and systems for the wayside networks, such as synchronous digital hierarchy (SDH) (which is today seen as legacy, but still very much used) and especially, Multiprotocol Label Switching (MPLS).

Unlike Chaps. 5 and 6, that is, mobile communications systems (MCS) and train communications networks (TCN), respectively, "wired" networks and systems in railways follow commercial standards and are not specific developments for railways (as TCN and MCS usually are). This is the main reason why this chapter is much shorter than Chap. 5.

Indeed, apart from preliminary concepts, the main contents of this chapter are two technologies: the legacy one and the state of the art. The former is SDH, which is circuit switched, deterministic, reliable, but not very flexible or scalable. The latter is MPLS, which is not only IP based but much more complex, flexible, and efficient. Railways are moving from SDH and SDH-like technologies to IP/MPLS networks. In this chapter, we explain both and its application in the railway's world.

The structure of this chapter is the following: as usual, we start with some general concepts about networks; then, we explain optic fibers, which are the dominant type of wire across almost every modern railway network; specific network technologies with extensive railway use, like SDH, MPLS, and some others are explained; in the next section, we address a very transversal topic: cybersecurity; finally, we provide a summary of the chapter, some exercises, and a further reading section with interesting links, papers, and videos.

4.1 Basic Networking Concepts

A communications network is a set of interconnected devices or nodes to allow the transmission of information. There are several ways to classify networks, like local area network (LAN) vs wide area network (WAN), or if the network is a general one or a specialized one, like Internet of Things (IoT) networks. In railways, we have a large diversity of scenarios, deployments, services, and networks. A very important differentiation we make is between an access network and a transport network.

Access network is the part of the network for connecting end user's devices to the network itself. Typically, it can be either wired or wireless and has many different aspects, like access points or radio base stations (more about this in Chap. 5), ports to plug a wire or a fiber, and many more.

Transport networks are for connecting access-network elements between them and to the core network. These transport networks can be either wired or wireless and usually receive the name of "core network," which is, by the way, a term related to 3GPP technologies such as LTE and 5G (as we'll see in Chap. 5). In this book, we avoid this name for the transport network because it is also used in mobile communications, and it does not mean the same as in this context.

Transport network is the backbone of the network itself, and it connects several access networks. The data that carries

Supplementary Information The online version contains supplementary material available at https://doi.org/10.1007/978-3-031-82247-6_4.

J. Moreno García-Loygorri, *Telecommunication Systems for Modern Railways*, Textbooks in Telecommunication Engineering,
https://doi.org/10.1007/978-3-031-82247-6_4

is the aggregate of all the network and can be rolled out on a wide area (in railways, hundreds of km).

Typically, transport networks are designed for high data rates and use optical fibers, satellite links, and some technologies that we explain in this chapter: SDH and MPLS.

As usual, network design depends on the requirements from the services (see Chaps. 2 and 3). Some of these requirements are expressed in terms of throughput or latency, but a very relevant one in wired networks is location. In the railway domain, there are multiple elements that need connectivity, such as:

- Technical rooms and racks, typically to host wired network equipment and other electronic devices.
- Stations, with many elements within them, like escalators, CCTV cameras, intercoms, elevators, etc.
- Signaling cabinets near the tracks (that is, not always in the stations), interlockings, etc.
- Signals.
- Programmable balises in the track.
- Detector and other devices in many locations in the tracks (wind detectors, axle box detectors, etc.).
- CCTV cameras.
- Substations and electric transformers.
- Base stations for railway mobile communications (i.e., Global System for Mobile Communications – Railway (GSM-R)).
- Base stations for PLMN, that is, mobile operators.
- Tunnel cabinets.
- Maintenance arcs to inspect the condition of the trains (wheels, pantograph, bogies, etc.).
- Control centers: operational control center, centralized traffic control, energy control, and detectors, plus the replicas if needed. It is common practice in railways to have replicas for the control center just in case something catastrophic happens, and all the operation of the railway line must be done from there.
- Depots and shunting yards.

As the reader may see, anywhere in the railway domain, there could be a device that needs to be connected to the network. Furthermore, there are three main trends for the access network: first, to provide IP connectivity rather than by legacy technologies; second, to connect more devices every day to the network and, finally, to use wireless connectivity if possible. Typically, in safety critical use cases, wireless connectivity is avoided (except for train-to-ground systems, for obvious reasons).

For all these reasons, the "connectivity problem" is solved by engineering the transport network, then the access network, and finally, the optic fiber rollout.

Let's begin with some basic concepts about optic fibers.

4.2 Physical Layer: Optical Fibers

In this section, we introduce very briefly a type of wired cables that we may find in railways: optical fibers.

4.2.1 Optical Fibers

Optical fibers are thin wires made of either glass fiber or plastic that allows a very efficient transmission of optical signals between the two ends of the cable.

Fibers are formed by a "sandwich" of three concentric layers:

1. Core: Where the light is transmitted.
2. Cladding: Helps to have all the energy of the ray confined within the core, that is, enforcing "total reflection" conditions.
3. Coating: Just to protect the other two layers. It has no optical properties.

The light into the fiber propagates with almost no losses because of the "perfect reflection" principle, which implies that no energy is transmitted; therefore all is reflected. This is the energy arrived at the core-cladding interface and gets almost totally reflected (see Fig. 4.1).

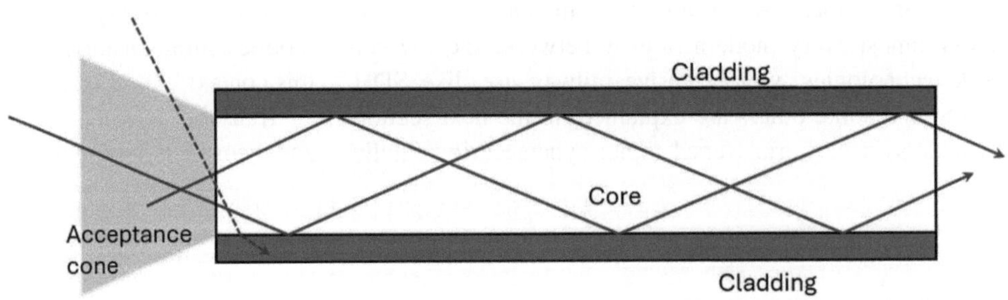

Fig. 4.1 Light propagation through a multimode optical fiber

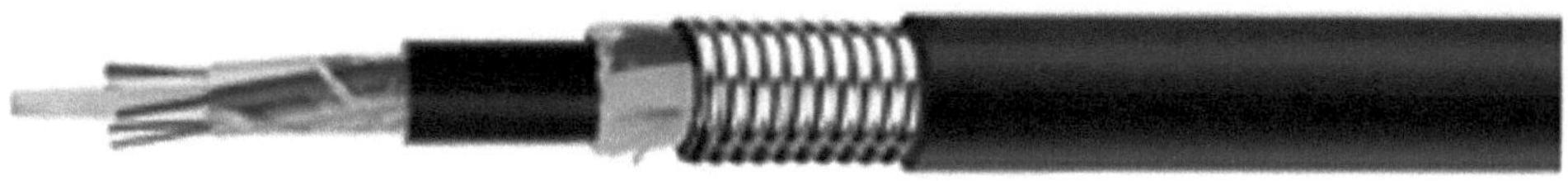

Fig. 4.2 Example of a fiber optic cable. It is open to show the layers with which it is composed. Notice the internal and external jackets (in black), the metallic armor, other protections (including those against water), and, finally, the fibers (at the left, in white). This fiber is manufactured by Prysmian Group for railway use

This means that fibers are segregated from the cable (which has many fibers on it) and connected to the device, which requires to be connected (typically, a switch or a router). For a duplex link, two fibers are needed.

4.2.2 Fiber Cables

Actually, several fibers are aggregated into a single cable, to protect more efficiently the fragile fibers and also to allow an easier installation (Fig. 4.2).

Commercial fibers are usually presented into tubes, each one of them with many fibers inside. That is, the cable can have 96, 32, and 16 fibers, and on each tube between two and eight fibres. There are many different types of cables, not only for the number of fibers, but also for temperature range, indoor/outdoor, rodent protection, and so on.

Optical fiber cables are manufactured in many different numbers of fibers per cable, such as 2, 4, 8, 16, 32, 48, 64, 128, etc. But 96, 32, and 16 are the standard numbers in railways (in some countries).

For the installation of fiber cables, perhaps the best approach is to put a fiber cable on both sides of the tracks, but this means that two cable trays (or "trough") are available, one on each side of the tracks. This way, availability is maximized just in case one of the fibers is broken for some reason. The costs of optical fiber rollouts are highly dependent on the needed workforce, rather than on the number of fibers to be installed. This does not mean that a two-fiber cable has the same price as a 128-fiber cable, but the difference between them is negligible compared to the workforce's costs.

Optical fibers are very popular in communications networks for many reasons like its high capacity; a very low attenuation figure per km; its high-security performance, because intercepting signals from fibers without physical access is almost impossible (unlike copper wires), as fibers do not produce unwanted electromagnetic (EM) emissions. Moreover, fibers are light and thin, which helps drive installation costs low, and have a high degree of immunity against electromagnetic interference, and fibers do not cause short circuits or emit sparks, both very suitable characteristics for railway environments.

Finally, fiber cables adapt very good to the linear nature of the railway tracks. In railways, it is straightforward to install a cable of several fibers and segregate pairs of them when needed (in the associated technical rooms and other locations along the tracks).

But there are disadvantages, like the installation costs, its fragility as optical fibers are more fragile than copper wires, and, finally, its installation, which requires skilled staff and specialized tools.

Sometimes it is required to have a supervision system for the optical fiber network. The objective is to guarantee that the optical fiber network works correctly and the prevention of failures in fiber cables.

Regarding the installation, there are some aspects to consider [81]:

1. It is a good practice to monitor the transmission parameters of the fiber. This is called "fiber supervision."
2. The fiber must be segregated from the cable to provide connectivity to the nodes where needed. This is done in segregation panels.
3. Fibers can't be spliced the same way as copper wires. Fiber fusion is the process where two fibers are aligned, heated, and merged. Doing this the wrong way can imply excessive losses, so once it is done, reflectometry tests must be performed to assure the correct fusion and rollout of the fiber.

There is another interesting aspect related to fiber rollouts in railways (and other environments), which is about the unused fibers allocated for future needs. These spare fibers can be rented to fulfill third-party connectivity needs. For example, when we account for all the fiber needs in our railway line, we may end up with a total count of 30 fibers. The next cable is 32, but instead of installing that one, it is a good idea to jump to a 48-fiber cable or even a 64 one. The costs will be very similar, but the remaining 34 fibers could be rented to get extra revenues. These space fibers are called "dark fibers" and are very much in demand in railways, because the dark fibers connect large cities, as railways do. The same can be said about subways but connecting city districts instead. In this case, it is also convenient to have a network fiber available for third-party services.

Finally, infrastructure managers often have a technical specification about optical fibers, which is the reference for this type of cabling. The same applies for copper wires (multiconductor). In this book, we do not address copper wires as they do not have any railway-specific feature (Fig. 4.3).

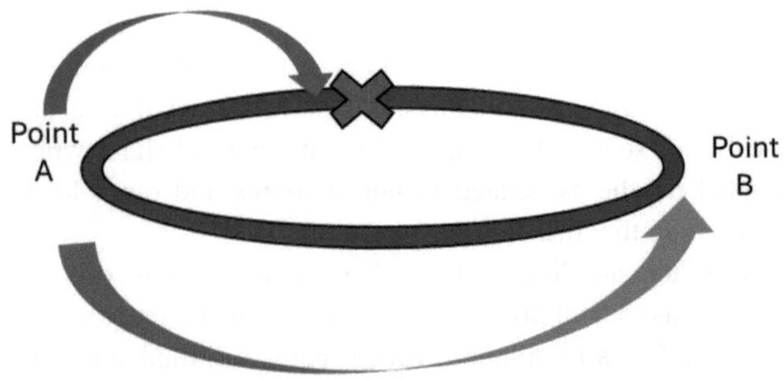

Fig. 4.3 Ring topology. There are two possible paths between A and B, just in case one of them fails. Adapted from

Ring Topology

This is an essential way to arrange network devices in critical scenarios: deploying the wires (optical fibers, copper, etc.) in such a way that allows two separated paths from any source and destination. See Fig. 4.4. In case of failure of a link, the network can redefine an alternative route, and the connectivity is not disrupted. This is the main advantage of the ring topology.

In railways, depending on the criticality of the services, we can find rings of fiber in the transport network and in the access as well.

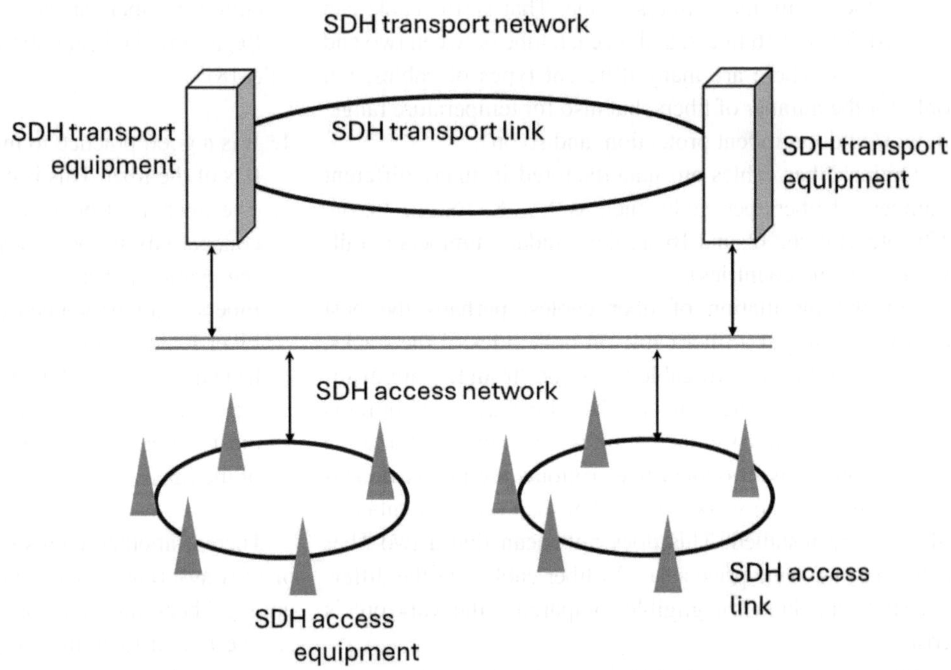

Fig. 4.4 Access and transport SDH networks. Adapted from

4.3 Network Technologies: SDH

A common way to develop communications networks in railways is to have a circuit switching network and a packet switching (typically, IP based) network as well. The first one is typically SDH. The second one is based on IP and MPLS, which are explained in the next section.

Circuit-Switched Network vs Packet-Switched Network

In circuit switched networks, nodes establish a dedicated resource (the "circuit" to exchange data). This circuit is maintained until the communication has finished, no matter the amount of data to be exchanged. This technology was implemented in analog telephones (POTS) and in early mobile communications systems (like Global System for Mobile Communications (GSM) and GSM-R, among others).

On the other hand, packet-switched networks send the data divided in packets that are routed to the destination. A header must be added to the message to have the information delivered. No permanent allocation of resources is made. This technology is the basis of TCP/IP networks, as well as modern mobile communications systems.

Packet-switched networks are more versatile and cost-efficient, but circuit-switched networks are more deterministic. As of 2023, packet switched networks are dominant in terms of market share over circuit-switched ones. However, these technologies have not been completely removed, as they survive in railways (GSM-R), SDH, and some others.

4.3.1 Concept of SDH

Synchronous digital hierarchy (SDH) is a multiplexing technology to transport high-speed data over optical fiber links, on a standardized way. The key concept for SDH is the "hierarchy" of standardized bit rates, for an easy multiplexing/demultiplexing of streams of digital signals. Therefore, the physical layer in SDH is the optical fiber, and the data is structured in synchronous transport module (STM).

The basic data unit in SDH is the STM-1 (STM level 1), which corresponds to a data stream of 155.2 Mbps. STM levels are structured on a 4x basis, that is:

- STM-1: 155.2 Mbps
- STM-4: 622.08 Mbps
- STM-16: 2.488 Gbps
- STM-64: 9.953 Gbps
- STM-256: 39.813 Gbps

However, the minimal data unit that can be introduced on an SDH network is the E-1 frame which is 2048 kbps. Typically, resource allocation is done using only from E-1 to STM-16.

Given that SDH is a circuit-switched network technology, it allows the establishment of dedicated circuits for point-to-point communication [81]. Furthermore, the synchronization of this network is achieved by a Global Positioning System (GPS) signal and a high-performance synchronization device, to generate a synchronization signal.

4.3.2 Transport and Access Networks

Typically, the network in railways is structured into two levels (depicted in Fig. 4.4):

Access level: Usually, an STM-1 ring of fibers, to provide connectivity within a defined location, like a station, technical rooms in the tracks, etc. On each of these locations, there are nodes and devices to insert/extract data into the STM-1 ring. This is the way to integrate end devices in the SDH network. For example, on each GSM-R BTS location, an STM-1 node is installed.

Transport (backbone) level: Typically, double STM-16 or even STM-64 rings, to connect the access level. The configuration depends heavily on the network, as there is no worldwide design policy for this, but a possible scenario is to have STM-64 rings, with STM-16 interfaces.

4.4 Packet-Switched Networks: MPLS

Some years ago, railway networks were populated by SDH networks, and IP technologies were very short in number. However, accordingly with other sectors, IP networks became more and more popular every day, but safety-related services remained in the circuit-switched domain. Why did this happen? The answer is safety concerns. SDH is a circuit-switched technology, reliable, deployed in an environment with few or zero changes, and IP was seen as a modern development perhaps suitable for other sectors but not for railway signaling and train control. Luckily, this old-fashioned approach changed, and now there are many IP networks in charge of many safety and non-safety services in railways. The trick is that these networks are not only IP or IP/Ethernet, as they also include Multiprotocol Label Switching (MPLS) on top of them.

On the packet-switching (which means IP) world, we have more flexible and adaptable networks, suitable for QoS management and other features. Remember all the diversity in the requirements from all services we learned about in Chaps. 2 and 3.

On an IP network is very hard to fulfill all the requirements coming from different services, reroute packets very fast, etc., that is, to achieve the needed QoS for all required. Moreover, this must be done on an efficient, scalable, and manageable way.

4.4.1 Concept of MPLS

The trend in railways is to go more and more every day to packet-switching networks only, but circuit-switched remain alive in many infrastructure managers.

The architecture is very similar to the previous section, with a network in two layers:

Access: A ring of medium-capacity nodes, connected to the transport network. In this layer, networks are IP based.
Transport: A ring of high-capacity nodes that offload the traffic from access networks. The technology here is usually IP as well but complemented with MPLS, which is explained in the next section.

Typically, the access network is implemented with switches (layer 2), to provide connectivity to all devices on every location of the rail network, and the connection to the transport network is arranged with fibers and layer 3 switches. This is not always true, and there are ways to overcome this, for example, using MPLS-based functions, like virtual private networks (VPN) and virtual private LAN service (VPLS). VPNs allow different, isolated paths for each client, and VPLS allow the interconnection of LANs over an MPLS network, creating a virtual LAN formed by all of them. Anyway, we can assume that the general scenario is to have a L2 network in the access domain and a L3 network in the transport one.

4.4.2 Labels

The transport layer is the backbone network, able to connect several locations across hundreds of km. The key technology is MPLS, which has several aspects, which we will explain soon. MPLS can merge the best from circuit-switched networks and the best from packet-switched networks. The idea is to establish labeled static paths (LSP) and to route packets based on these labels instead of on IP addresses. The routing is for only one link (that is, from one node to the next one). Examining only the labels is faster than routing based on addresses. Therefore, MPLS implements an approach like circuits, but over packet-switched networks (Fig. 4.5).

MPLS Header (see Fig. 4.6)
It is a 32-bit header, with the following fields:

Label (20 bits): The label itself.
Exp (3 bits): For the IP precedence field, for QoS purposes.
S (1 bit): "Bottom-of-stack" field. Used to indicate if this label is the last one.
TTL (time to live, 8 bits): Same use as in IP header. It is decremented on each hop and used to prevent loops. If the TTL is 0, the packet is discarded.

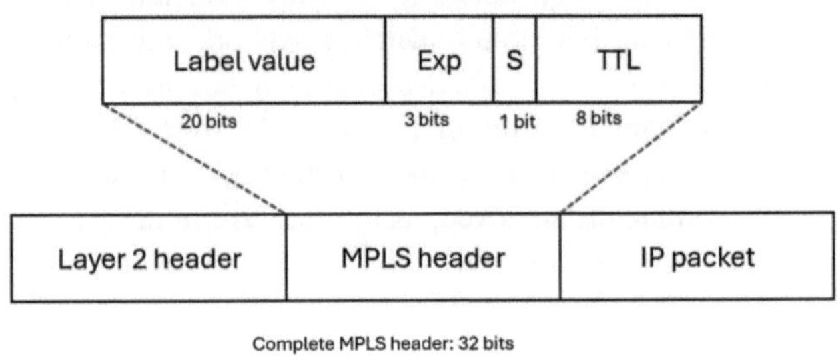

Fig. 4.5 MPLS header

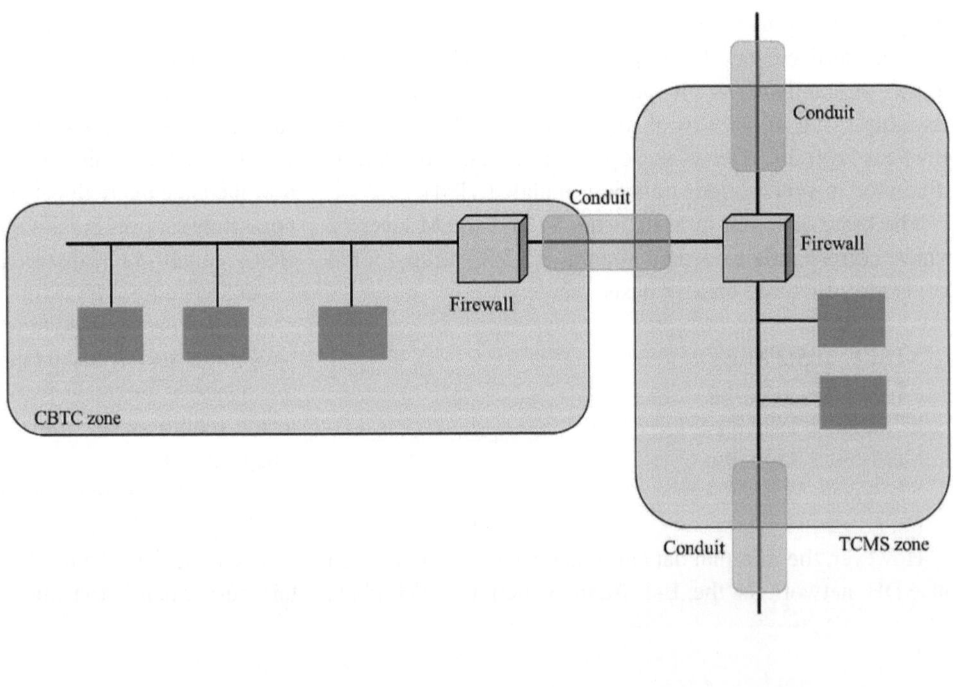

Fig. 4.6 Example of high-level zone model for two onboard networks as defined in TS 50701: CBTC and TCMS. Conduits and firewalls are between every pair of zones, and end devices are within a zone (and depicted in dark gray)

Labels are generated and exchanged based on a certain protocol, like Label Distribution Protocol (LDP), Resource Reservation Protocol (RSVP), or Border Gateway Protocol (BGP). Furthermore, MPLS labels can be stacked one on another, and de-stacked using a last in first out (LIFO) algorithm. The "M" in MPLS if for multiprotocol, as MPLS can encapsulate packets of several different link-layer protocols, like Ethernet and some others.

Routers in charge of label-based routing are called label switch router (LSR). These routers are responsible for removing the label from a packet that arrives and assigning a new one based on the contents of a table. These routers are in the middle of the MPLS network. The contents of this table, that is, the determination of the label, are based on routing algorithms.

On the other hand, label edge routers (LER) are entry (and exit) nodes of the network, assigning ("pushing") the first label to an incoming packet and removing the label ("popping") when the packet leaves the network.

4.4.3 Advantages of MPLS

Multiprotocol Label Switching (MPLS) has several advantages such as fast packet forwarding capabilities; network scalability, because when adding a new node, tables must be updated adding the LSPs from every router in the network to the new node. Another one is traffic engineering, that is, setting paths depending on the type of traffic, which means more efficient resource allocation. Furthermore, the creation of virtual private networks (VPN) by traffic isolation using labels. VPN mean higher security as well, and, finally, the emulation of layer 2 (switching) circuits. From another point of view, enabling the coexistence of different services on a single network core (for example, safety sand non-safety services; CBTC/ERTMS, and CCTV cameras) is very relevant as well. More generally, allowing end-to-end QoS management efficiently, with sophisticated but flexible and user-friendly tools for both operation and maintenance of the network is what makes MPLS very valuable for operators and infrastructure managers.

4.5 Cybersecurity in Railways

Ignored for a long time, cybersecurity is today one of the hottest topics in railways, and probably will in the long term. The reasons for this importance in railways are many, but all of them related to protection of information, the infrastructure or the service, the reputation of the entities, and the prevention of attacks and frauds.

Protecting information from unauthorized access, including data from customers (i.e., credit card numbers), staff (personal data, etc.), or corporate (intellectual property, restricted or confidential information, etc.).

Protecting the infrastructure and railway service from service disruptions of railway operations, avoiding safety issues and other similar incidents. Therefore, cybersecurity can have impact on availability.

Protecting corporate reputation. Sometimes the harm is not related to data breaches, safety issues, or fraud. In some scenarios, the objective is the reputation of the railway company, which can be severely damaged if it is seen as not trustworthy.

Preventing fraud or any kind of attacks, like ransomware, fraudulent transactions such as invalid credit cards to purchase tickets, etc.

In Chap. 2, we explained that major challenges in general terms can't be addressed as a technology issue only but have three essential pillars: technology, processes, and people. This is still valid here, as having a state-of-the-art cybersecurity technology is useless if it is not rooted in the company's culture, and your staff is both well trained and aware of the relevance of the job.

Indeed, railways is a classical operational technology (OT) example, as OT deals with control and automation of machinery (in this case, trains, track switches, level crossings, etc.) instead of an information technology (IT), where the focus is on processing, transmission, and storage of information. Both IT and OT domains exist in railways, but OT is in the core business and IT is not.

In recent years, due to the increase of connected devices, plus the adoption of mobile and wireless communications systems, railways have become more aware of the necessity of paying more attention to cybersecurity. In parallel, attacks toward operators and infrastructure have become more frequent, and the public opinion has become more sensitive to them as well.

4.5.1 Cybersecurity Challenges in Railways

Consequently, there are many challenges for railway companies when working on cybersecurity, coming from the diversity of systems, locations, users, types of services, maturity of technology, etc. In this section, we introduce some of them to highlight its importance only.

Perhaps the most relevant on is the diversity: In railways, there is a large diversity of technologies, locations, wayside and onboard, professional profiles, etc. All this makes more difficult to address the huge challenge that cybersecurity implies to large organizations that provide public (and essential) services.

On the other hand, sometimes railway networks cover a very large extension, meaning that distances are very large, being the challenge to protect (not only physically) all the assets. Moreover, many times data is stored in different and isolated places, instead of a centralized one, making them harder to be protected.

Railway assets last for a long time (for example, trains are designed to live at least 40 years) much larger than the typical IT obsolescence cycles. The problem is the need for support from vendors and integrators for a long time.

Furthermore, there are many OT systems, with few or zero upgrades, based on legacy technologies, which are indeed hard to replace. Cybersecurity processes must include them too. A good example of this is GSM-R and European Train Control System (ETCS), which are both more than 20 years old, and at the end of the 1990s, it was decided to not cypher ETCS emergency stop messages. The reason was the encryption algorithm introduced an additional delay, which has not a critical impact unless the message is an emergency stop. Nowadays, this is now a threat to ETCS operations.

Connected-passengers services are an opportunity to be more competitive and portray a modern image of railways, but also a threat as railway assets is more exposed to attacks.

The complete supply chain must be involved in cybersecurity, which includes integrators, maintenance companies, engineering companies, manufacturers, etc.

Cybersecurity requires periodic evaluations, follow-up meetings, frequent updates, constant training of the staff, etc. It is an essential process that must be intertwined in the day-to-day operations of the whole railway sector, the same way as safety is.

Proprietary solutions are almost everywhere, implementing outdated protocols, sometimes not well documented and more important, not patched with the latest security-related updates. Obsolescence, a major challenge on its own, multiplies the risks for cyber threats.

Attackers can have an easy access to the railway domain and, in some cases, to restricted-access areas as well, both of them implying a threat.

It has improved largely in recent years, but another major challenge is the low cybersecurity awareness in the sector [82]. This has started to change recently, which is very good news.

Finally, another important aspect, common to other industries and transportation modes is the need to balance security, competitiveness, and operational efficiency [82].

Tools for Cybersecurity

Firewall: It is a system that monitors the network based on predefined rules. These rules mostly depend on source/destination addresses, but there could be many more. Typically separate two networks that belong to different domains, organizations, or entities (i.e., a private network and the Internet). Firewalls can be implemented as a software application into a general-purpose hardware or on a dedicated hardware instead. There are several types, based on the communication layer where they work (application, network, etc.). Another firewall functions are authorization management, monitoring, and logging.

Deep packet inspection (DPI): This is a complementary function to firewalls which inspects the contents of data packets to detect malicious contents, attacks, etc. Unlike firewalls, DPI is more related to large organizations, network operators, or, more recently, governments. DPIs are usually present together with a firewall.

Intrusion detection system (IDS) and intrusion prevention system (IPS): IDS are software applications that detect violations and unwanted activities of any kind inside networks. Based on different approaches, IDS are used to detect malware, many types of attacks, etc., and generate an alert in case a malicious activity is present. In case the application could react to the situation, it is called an intrusion prevention system (IPS). These applications, like firewalls and DPIs, usually work under a license service and need frequent updates to work properly.

4.5.2 TS 50701

The reference for cybersecurity management in railways is the CENELEC technical specification TS 50701 [83] "Railway Applications—Cybersecurity, European Committee for Electrotechnical Standardization Technical Specification, Standard CLC/TS 50701, CENELEC, Jul. 2020." From a certain point of view, this specification addresses cybersecurity and safety together [84], due to the importance of safety in railways. TS 50701 provides a seven-step procedure to design safely a system in railways. Here we provide a summary of the procedure, but also we encourage interested readers to read the complete document.

The top-level objective of TS 50701 is to introduce requirements and recommendations for cybersecurity in railways, considering that there are many different types of systems, with different life cycles, safety, and non-safety. The reference is the EN 50126 Reliability, availability, maintainability and safety (RAMS) standard [85], and as this one, TS50701 addresses either wayside and onboard systems, communications and signaling systems. However, this TS does not provide functional safety requirements but intends that RAMS does not get compromised due to intentional attacks.

TS 50701 is largely based on the cybersecurity standard for industrial environments, that is, ISA/IEC 62443 [86].

This standard uses several terms (many of them from [86]) such as:

- Zone: "Group of logical or physical assets" [83] which share common security requirements, like criticality or functionality. For example, onboard a train, several zones are typically defined: signaling network, TCN, multimedia network, train-to-ground network, and passengers' network.
- Conduit: A logical group of communications channels that share common security requirements and serve to connect two zones (never more than two). Conduits are used to control data flow and to separate zones.

Conduits and zones are the most relevant ones, but there are some others that play an important role, such as:

- Countermeasure: Something that mitigates a threat.
- Data diode: Boundary device that makes sure communications are in one direction only.

It is expected that the railway company defines an asset model, with physical areas, systems, and a high-level zone model, which is the main input for the initial risk assessment, zoning, and identification of the system under consideration (SuC). See Fig. 52 for an example provided by the TS 50701 for a zone model.

Then, TS 50701 focuses on the cybersecurity life cycle of railway applications, providing some guidance to railway companies about the activities to be performed on every stage. This is not a detailed list of activities but a mere guidance on setting up processes to work in cybersecurity management.

From a system-level perspective, TS 50701 specifies how to do the impact assessment of the SuC, addressing both the qualitative impact and its likelihood. After this, once the zoning and conduits are defined, next is the detailed risk assessment, which is specified on a very similar way to EN 50126. That is, for each zone and conduit, threats and vulnerabilities must be identified, a code of practice must be defined for them and the risk assessment should be reevaluated. If the code of practice is not available, a reference system to compare with it can be used as well. Then, for all threats not covered in the code of practice, an evaluation of the risk must be done, assigning a security levels (SL). SL go from SL0 (no security protection is needed) to SL4 ("protection against intentional violation using sophisticated means" [83]).

Once the SL is determined, the necessary countermeasures must be determined. There are many other aspects covered by the TS 50701, like cybersecurity assurance, validation, etc.

Although not railway specific, there are at least three other documents that are worth considering when dealing with cybersecurity in railways: "IEC 62443: Industrial communication networks - Network and system security" [86] which is composed of 11 parts (1-1, 1-2, etc.). TS 50701 is based on IEC 62443, adapted to railways, and certainly written on a much simpler way. IEC 62443 is far more ambitious and complete than TS 50701, so some authors [87] see this standard as the true reference for security in railways, and TS 50701 as a source for specific adaptations to railways, not more. Perhaps this view may change in the future but, anyway, IEC 62443 will remain useful for the railway sector. Second, ISO/IEC 27001:2022 [88] information security, cybersecurity, and privacy protection; and finally, NIS (Network Information Systems (NIS) directives, from the European Union. Directive (EU) 2022/2555, known as NIS2 Directive, replaced the original NIS directive in December 2022. The first one was Directive (EU) 2016/1148. These directives are for the implementation of common policy and law regarding cybersecurity.

Regarding security standardization in (European) railways, we have the following bodies and entities working together:

- IEC Technical Committee 9, Working Group 26.
- ENISA: European Network and Information Security Agency, which not only for railways and transportation.
- The European Railway Agency (ERA).
- Other standardization bodies, like ISO, ISA, ETSI, and IEC.
- Informally, some research projects like X2Rail-1, Connecta-1, -2, and -3 include work packages related to cybersecurity and have also contributed to standardization.

Finally, collecting information about cybersecurity incidents in railways is not always easy, but we provide here a list (mostly retrieved from [89]):

1. January 2008: A teenager in Lodz (Poland) derails four trams, injuring several people [90].
2. 2012: the CBTC system of a subway line in Shenzhen (China) was reportedly jammed [91].

3. July 2016: Network Rail, the UK infrastructure manager, acknowledges four significant attacks in the last year [92].
4. November 2016: San Francisco (USA) tram system, not able to sell tickets for a day due to a cyberattack [93]. Several other operators attacked using ransomware: DB (Germany) in 2017, STM (Montreal, Canada) in 2020, and many more [89].
5. August 2023: 20 Polish trains were stopped by an intentional attack [94].

In the future, TS 50701 perhaps will evolve into a strict sense standard, rather than a technical specification, as cybersecurity becomes a more relevant discipline within railways.

However, as you may have noticed, cybersecurity is not only about networks but also about processes and continuous learning and improvement. This is a great challenge in large railway companies, with fragmented technology, several human roles, and geographic dispersion.

Summary of the Chapter

In this chapter, we have introduced the basics about wired networks in railways, explaining how networks are structured in certain domains or levels. In railways, typically, there are access networks and transportation networks, plus a physical layer formed by optical fibers and copper wires.

Regarding other technologies used in railways, we see that there are two domains: circuit switched and packet switched. The most relevant technology for the former is SDH, and we've seen how it is used in access networks (to connect end devices to the network) and in transport networks (backbone network). For packet-switched networks, the key technology is MPLS, which allows fast routing.

Finally, we have seen how cybersecurity is managed in railways, paying special attention to some recent standards like TS 50701 [83].

Further Reading

In this chapter, there are clearly three main topics to read about: SDH, MPLS, and cybersecurity. For SDH, which certainly is an old technology, there are many good books, like this one by Siller [95]. Regarding MPLS, a very popular reference is De Ghein's book [96]. However, Internet is full of tutorials about this technology.

For cybersecurity, the first recommendation is to read TS 50701. It is the essential tool, and engineers interested on cybersecurity must know this technical specification. There are many other resources available, like [97], which is a very good summary of the situation. The reader may find the presentations and videos from an ERA conference held in 2021 interesting [98] and perhaps some of the outcomes of the EU-funded 4SECURAIL Project [99].

Exercises

1. If we want to provide bidirectional connectivity to an end device using fibers, how many of them we need to segregate from the main cable? Why?
2. Which is the minimum circuit that can be established on an SDH network?
3. What is an MPLS label?
4. What is label stacking in MPLS?
5. How are GSM-R BTS connected to the network? Explain which technology is used, to which domain are connected, and which link is used.
6. Explain the differences between OT and IT systems and why this has an impact on cybersecurity.
7. Which type of MPLS router is in charge of adding the label to the packet?

5 Radio Communications Systems in Railways

Radio communications or wireless systems are an essential technology in modern railways. In Chap. 4, we addressed wired communications systems, so in this chapter we focus on wireless systems.

This chapter focuses on train-to-ground (T2G) systems, which are an essential part of this book. Among these T2G systems, we find some very popular ones, like Global System for Mobile Communications-Railway (GSM-R), WiFi, and long-term evolution (LTE), while some others still not in service or not very much extended, like 5G (especially in millimeter-wave bands) or Future Railways Mobile Communications Systems (FRMCS) are explained in Chap. 8.

Before going into technical details about the systems, we introduce radio communications (without the "systems"), or more precisely, to mobile communications in railways. This is not a general introduction to radio communications (which can be found in some excellent monographies like [100] and [101]), but to some specific aspects of railways that have an impact on the radio channel (that is, the physical layer).

This chapter is one of the core parts of this book, together with Chaps. 4 and 6. However, Sects. 5.1, 5.2, and 5.3 are focused on the foundations of mobile communications, a topic that we consider essential to understand properly the mobile communications systems in railways (which are described just after, in Sect. 5.5). Section 5.4 describes both general aspects about antennas and antennas for railway scenarios (wayside and onboard). The reader may skip Sects. 5.1, 5.2, 5.3, and 5.4 if they only have interest on the systems and not on the underlying technology.

As usual, in this radio communications chapter, we do not pay much attention to historic details, but we absolutely encourage readers to go find them in other books and papers, like [102]. Just let us mention the very inspiring pioneering work of some true giants of science and engineering like Michael Faraday, James Clerk Maxwell, Heinrich Rudolf Hertz, Nikola Tesla, and Guglielmo Marconi. Many more deserve almost equal credit, like Christian Doppler, Claude Shannon, Martin Cooper, etc. and some others who did not work on strict-sense radio technologies, but their contributions are essential to radio engineering, such as Joseph Fourier, George Boole, John Von Neumann, Harry Nyquist, or Alan Turing.

5.1 Basic Concepts of Radio Communications

Radio communications are those done through electromagnetic (EM) waves which do not propagate along guided media. This definition can be tricky,[1] but it works, so we leave it as it is. To separate pure radio waves from optical waves, a frontier around 2–3 THz is usually accepted, with radio waves below and optical waves above that threshold.

Radio systems must have a transmitter (which radiates EM waves into the air) and a receiver (which gathers those waves from the air as well). The transmitter is usually composed of some subsystems, like modulators, filters, amplifiers, etc. The receiver is usually composed of the counterparts of the transmitter subsystems (demodulator, etc.) plus some others.

Between the transmitter and the receiver lies the channel, which is usually beyond human control and is responsible of the introduction of many perturbations, like noise and interference, signal power attenuation, and many more effects which are described later in this chapter.

Typically, we find several uses for the radio resources (like spectrum), such as [100]:

[1]Some may argue that radio waves inside a tunnel are indeed guided, and they truly are. Therefore, the concept here could be the ratio between the wavelength of the radio wave (usually denoted as λ) and the size of the wave guide. Please see more on this on the section about tunnels.

Supplementary Information The online version contains supplementary material available at https://doi.org/10.1007/978-3-031-82247-6_5.

J. Moreno García-Loygorri, *Telecommunication Systems for Modern Railways*, Textbooks in Telecommunication Engineering,
https://doi.org/10.1007/978-3-031-82247-6_5

1. Broadcasting: A unidirectional service where a station sends information to multiple receivers (typically, TV or radio receivers).
2. Mobile communications: Bidirectional service where at least one of the ends can be in motion.
3. Public safety and emergencies: This service is a variation from the mobile one [100], with some extra functions like group calls and direct communication between devices. Notice that some of these functions are present in railway-specific communications systems, as we highlighted in Chap. 3.
4. Amateur radio: For noncommercial and recreational use. This service can be either satellite- or terrestrial based (as almost all the other services in this list).
5. Military: Radio communications for defense.
6. Radioastronomy: This is a subfield of physics that studies celestial bodies in radio bands. Radio astronomy devices are mostly receivers, while stars, quasars, pulsars, etc., act as radio transmitters. A large array of antennas for radio astronomy use is shown in Fig. 5.1.
7. Satellite communications: Non-terrestrial communications that involve at least one end orbiting the Earth (or other planets or stars). A more suitable name for this service is non-terrestrial communications, as this name includes communications with high-altitude platform stations (HAPS) and other devices. A good example of satellite-based communications system is the SpaceX Starlink.
8. Satellite location services: Where the satellites are used for location purposes, or more precisely, GNSS (Global Navigation Satellite Systems). Some examples of this technology are the American GPS (Global Positioning System), the European Galileo, Russian GLONASS (GLObal'naya NAvigatsionnaya Sputnikovaya Sistema), or the Chinese Baidou, among others. In railways, this technology is not used for train location determination but for auxiliary purposes (telemetry data, remote maintenance, etc.).
9. Aviation services: For communication between civilian aircraft and ground stations.
10. Unlicensed use: Strict sense not a service, but a way to assign spectrum on a non-exclusive use basis, for several potential services. Some very good examples are wireless local area network (WLAN) and Bluetooth. In this book, we cover non-licensed radio because WLAN (more in particular, Wi-Fi networks) are very extended,

Fig. 5.1 Very-large array (VLA) in New Mexico, USA

for safety and non-safety services. For the moment, it is enough to understand that this use has some spectrum allocated in some bands. Notice that both non-exclusive and unlicensed do not mean a pure libertarian access to the media, since some restrictions apply, like maximum radiated power and so-called spectrum masks (which specify the admissible levels of power out of the band).
11. Fixed wireless access (FWA): A wireless link where both of its ends are static. Typically, large towers on the top of a hill with many antennas are used for this purpose.

In this chapter, we mostly focus on "mobile communications," plus "public safety and emergencies." Non-licensed use has also some relevance due to the lack of spectrum for urban railways. The reason for this is that railway communications services are not pure mobile services but also demand some public safety functions as we saw in Chap. 3.

5.1.1 Radio Communication Service Requirements

Depending on the application and the circumstances, there are different requirements like data rate, range, number of users, mobility, energy consumption, and spectrum use [100]. The key idea here is to design a suitable system that fulfills the requirements of the applications, minimizing the needed resources (typically, spectrum, and energy).

5.1.1.1 Data Rate

Number of bits per second (bps) delivered: This is perhaps the best-known feature of a communications system, but it is not always the most important one. In the three types of services in railways (safety oriented, operator oriented, and passenger oriented), we have three different data rate requirements, as safety services usually require very low figures (typically kbps), operator-oriented much larger figures (tens of Mbps, due to CCTV and digital train use cases), and passenger-oriented are usually based on the "the bigger, the better" anthem.

5.1.1.2 Range

Length of the communications link: This length can be meters or even less than 1 meter (body area networks, BAN), 1–100 m (wireless local area networks, WLAN), macrocells (hundreds of meters up to a few km) like the mobile base stations around us, 5–100 km for FWA links and hundreds or even thousands of km for satellite links. As a fun fact, the distance record for a communications link is between the NASA deep space network (DSN) and the Voyager 1 probe, with an outstanding 24 billion km as of October 2023 (since contact has yet to be lost, this record will be very likely surpassed when the reader encounters these lines). This range is determined by many factors, one of them being the sensitivity of the receiver and the available power on the transmitter side. But the desired level of interference to and from adjacent cells is a limitation as well.

Notice that coverage (of the mobile communications system) is not the same as (cell) range. The coverage is the aggregated surface within the range of all the cells of the system. In railways, the coverage is usually limited to the railway domain, and the cells are distributed along the tracks.

5.1.1.3 Number of Users

Number of users served by the mobile communications system: This is a figure that has been growing since the very early days of mobile. Second-generation GSM systems had very few users per cell (typically, less than 100) and 5G mMTC (massive machine-type communications) intend to connect up to 1 million of users per square km. In railways, the number of connected users is, for the moment, not very large, except for crowded stations and platforms, but rarely exceed 1500 people. For safety-related services, the number of connected users is usually very low (one per train, plus some railwaymen on the ground, which in average could mean less than one per cell). Please notice that there is an important distinction between active users (that is, active users, sending and receiving data) and idle users (users registered in the network but not sending data at this precise moment).

5.1.1.4 Mobility

The expected speed of users: For several reasons that we explain in the mobile radio channel subsection, it is not the same to communicate with a user walking in the street, on a car (typically, up to 120–140 km/h) or onboard a high-speed train (for the moment, up to 350 km/h). Therefore, the speed of the users must be considered when designing the network. A suitable example for this is IEEE 802.11 systems, which were initially designed for nomadic users (that is, very low speeds, typically for offices and homes) and not for strict-sense mobility [100]. Later, some amendments came to the rescue and allowed its use for vehicular communications.

In high-mobility scenarios (like HST), the Doppler effect can have a noticeable impact, so it must be considered [100]. Typically, mobile communications systems are designed and optimized for a regular mobility profile (between 10 and 120 km/h) but not for higher speeds. Modern mobile technologies are better at high mobility scenarios, but still not optimized for them. Obviously, the slower the MS is, the better the quality of the transmission will be, even at very low speeds.

5.1.1.5 Energy Consumption

Radio equipment consume energy, especially transmitters. This metric has an important impact on both sides of the link, because mobile devices are usually battery powered (like cell phones) or rely on vehicular power supplies (train radios are a good example for this); on the other hand, base stations have a relevant power consumption with a strong cost impact (consider that a typical three-sector LTE BS can use 2.5–10 kW of power [60]). Finally, there is another type of device which is formed by IoT devices and other sensors powered by non-rechargeable batteries, whose life cycle ends when the battery is exhausted. In this scenario, the energy-efficiency requirements are tighter than the others. In railways it is common to have battery-powered sensors like these to monitor temperatures or even RFID balise readers onboard trains, to serve as tags to identify assets.

5.1.1.6 Spectrum Use

This is the spectrum bandwidth needed for the mobile communications system. Due to the high cost of spectrum, especially in the bands suitable for macrocells, systems are designed to optimize this resource.

The spectrum can be dedicated for a certain service or unlicensed. In general terms, the most usual approach is the former, but not for railways. We will discuss more on this later in this chapter.

5.1.1.7 Direction of Transmission

The intuitive idea for this is to assume that both directions transport data. However, this is not true. There are some services that are unidirectional (i.e., broadcasting) and there are also some other considerations. Therefore, we have three main types of services depending on the direction of transmission [100]: Simplex, where information is sent in one direction only and half-duplex, where information is sent in both directions but only one at a time. A good example for this is PTT devices, often used in railways. Finally, full duplex, where information is sent in both directions at the same time.

5.1.1.8 Links and Multiplexing

Typically, for wireless communications services, two different links are defined:

Uplink (UL): In terrestrial services, transmission is originated from the mobile station (MS) and delivered to the base station (BS); however, in non-terrestrial services, the UL is originated in the ground station and ends in the satellite.

Downlink (DL): The opposite to UL. In terrestrial services, transmission is from the BS to the MS; in non-terrestrial services, the DL is from the satellite to the ground station.

Recently, with the rise of vehicle-to-vehicle communications, a third type has been added: the side link (SL).

Regarding the duplexing, there are two different types:

Frequency division duplexing (FDD): there are two different frequency bands for the UL and DL. This type of duplexing is more efficient when the traffic UL/DL pattern is symmetrical. GSM-R, as GSM, is FDDbased.

Time division duplexing (TDD): The opposite of FDD, as the same frequency band is used for UL and DL use. In cases where the traffic pattern is not symmetrical, TDD may be more efficient than FDD. This is because more time slots can be allocated for either the UL or the DL, whereas in FDD, the same number of resources is allocated to each one of them. Actually, due to the nature and limitations of MS in comparison with BS, the UL is more sensitive to interference and power available is more limited. For this reason, in FDD, the UL is usually assigned to a lower band than the DL. GSM-R in Europe (876–880 MHz for the UL and 921–925 MHz for the DL) is a good example for this.

LTE is defined both for FDD and TDD, but as of 2023, LTE-FDD is much more extended than LTE-TDD.

5.1.2 Concept of Spectrum

We have identified spectrum as a resource for wireless services. We have understood that it is expensive (and strategic), so we need to avoid wasting it. But what is spectrum?

Spectrum is the range of frequencies occupied by an electromagnetic signal. In Physics, this definition is usually addressed in terms of wavelength instead of frequency, but the concept is the same because one magnitude is the inverse of the other ($f = c/\lambda$), where f is the frequency (measured in Hertz or Hz), c is the speed of light in vacuum, and λ is the wavelength (meters).

We are usually accustomed to think about signals in terms of time functions, but in communications, dealing with signals in the frequency domain (the spectrum) is far more convenient many times. The way to switch domains, that is, to convert a time-domain signal into a frequency-domain signal is through the Fourier[2] transform, which is:

$$X(f) = \int_{-\infty}^{+\infty} x(t) e^{j2\pi ft} dt$$

[2]For Joseph Fourier (1768–1830), who was thinking more on a tool for thermodynamics and vibration analysis, rather than on the (by that time) futuristic EM waves. Anyway, as usual, our recommendation is to learn more about his life and career, which includes working for Napoleon Bonaparte and many more things.

where $x(t)$ is the time-domain signal and $X(f)$ is the frequency-domain expression of $x(t)$. There is a whole subfield of Math that is based on this equation, but for the purpose of this book, it is enough to understand that every signal that we want to transmit occupies a certain bandwidth, which is defined in the frequency domain.

The main concept to understand about spectrum is that all bands are not equal at all. Not only is there a trade-off between bandwidth and range, as high frequencies can have higher bandwidths, but attenuation is larger at high frequencies. So, which bands are better? The right answer is "it depends," but there are some bands better than others. For example, the 600–1000 MHz band allows relatively large bandwidths and does not suffer very large attenuations; moreover, it penetrates buildings without large power losses. That is why these bands are used for one of the most profitable communications markets: public land mobile communications (PLMN).

To allow the most efficient usage of the EM spectrum, authorities define two types of spectrum allocation licensed (or administrative concession) and unlicensed.

Administrative management, where the authority decides which services and, maybe, which technologies are allowed on a certain band, to an exclusive user (for example, a mobile operator or a TV station). This procedure is usually related to a license or authorization for years. In some cases, these licenses are auctioned by the government. The main advantages here are the total control over the spectrum and the guarantee that the license-holder is protected against interferences. The main drawback is the need to manage the spectrum. Once again, this is a battle between laissez-faire and statism. But it looks like here the best option is to strongly regulate most of the spectrum and leave unlicensed some specific bands. This is the current approach worldwide.

Unlicensed spectrum: To allow certain services without administrative allocation, some bands are left unlicensed, so everybody can use them under certain restrictions. That is why we can set up a WiFi network at home without a lot of paperwork (and is the reason behind the popularity of many other technologies like Bluetooth and some other short-range wireless communication systems). Please see the box below for more details about unlicensed bands across the world. The main problem here is the presence of interference which cannot be easily removed and the absolute lack of control. Wireless communications systems for unlicensed bands are designed in a radically different way than systems for licensed bands.

There are two different notations for spectrum bands: the ITU and the IEEE. Both are depicted in Tables 5.1 and 5.2.

Indeed, this is the way spectrum is allocated by regulators, like the Federal Communications Commission (FCC) in the USA [104]. That is, for a certain spectrum band (i.e., 2.400 GHz and 2.483 GHz), a particular service is allocated (for this case, unlicensed use). It is very insightful to look on how the spectrum is assigned to understand the strategy behind it. Remember that, if semiconductors are strategic for a country, spectrum is even more than that. Once this chapter is finished, the reader should be able to understand which bands are better for a particular purpose and why. Therefore, the reader should be able to choose a suitable band for the mobile communications systems in railways.

Table 5.1 ITU frequency bands

Band	Meaning	Frequency range
VLF	Very low frequency	3–30 kHz
LF	Low frequency	30–300 kHz
MF	Medium frequency	300–3000 kHz
HF	High frequency	3–30 MHz
VHF	Very high frequency	30–300 MHz
UHF	Ultrahigh frequency	300–3000 MHz
SHF	Super high frequency	3–30 GHz
EHF	Extremely high frequency	30–300 GHz

Table 5.2 Frequency bands as defined by the IEEE

Band	Frequency range
L	1–2 GHz
S	2–4 GHz
C	4–8 GHz
X	8–12 GHz
Ku	12–18 GHz
K	18–27 GHz
Ka	27–40 GHz
Q	33–50 GHz
V	50–75 GHz
W	75–110 GHz

Unlicensed Spectrum Bands

In this box, we present some of the unlicensed spectrum bands available for railway use, among others. There are many others, but the most important ones for mobile communications services are:

433 MHz in Europe.

868 MHz band: Only in Europe 863 to 870 MHz, for short-range communications and sensor networks, which means low power (typically, 10–25 mW, 500 mW in the best case) and low data rates.

915 MHz: in the USA, the equivalent band for the 868 MHz is the 915 MHz, which ranges from 915 to 928 MHz.

2.4 GHz: 2400.0–2483.5 MHz, mostly for WiFi, Bluetooth, and some other short-range communications systems like LoRa (long range).

5 GHz band, divided into the 5.150–5.350 GHz, 5.480–5.730 GHz, and 5.735–5.895 GHz, with slight differences between countries and regions.

And there are some other in millimeter wave, like 24.00–24.25 GHz and 57.24–74.52 GHz.

More recently, the 6 GHz band has been included as well. The use of this band differs across the world as some countries have adopted the 5925–6425 MHz band as some others extend this use up to 7125 MHz.

Some of these bands are addressed as "industrial, scientific, and medical" (ISM) bands. But not all of them belong to this taxonomy, like most of the 5 GHz bands listed above.

Unlicensed spectrum has cybersecurity implications because if everyone is allowed to access, it is much easier to attack the wireless communications system.

Spectrum allocated for railways is very limited. There are only few exceptions like the GSM-R band (4 + 4 MHz): 876–880 MHz for the uplink and 921–925 MHz for the downlink. In some countries, this band has been enlarged up to 873–876 MHz and 918–921 MHz. It is called the E-GSM-R band, but it is not harmonized at an international level, which means no interoperability at all. This extension increases the capacity of the GSM-R system in crowded areas like large stations or shunting yards.

In China and other countries, the 4 + 4 band for GSM-R is different: 885–889 MHz for the uplink and 930–934 MHz for downlink [105]. Furthermore, there are other bands allocated for railway use, like 1800 MHz band in Australia (1770–1785 MHz for the uplink and 1865–1880 MHz for the downlink).

DSRC Band in the USA: 5.850 GHz to 5.925 GHz and its European counterpart, the ITS Band: 5.875-5.935 GHz. However, this does not mean that there are 60 MHz reserved for railway use. This allocation gives only 20 MHz to railways, but ten of them are shared with road traffic, while urban rail only has priority in 10 MHz (5.915–5.925 GHz) [106, 107].

There are some nonstandard narrowband allocations for railways in many countries, like in Spain the *"Tren-Tierra"* system (447.550–448.650 MHz for the uplink and 457.600–458.600 MHz for the downlink, in 25 kHz channels), which are assigned on a national (or even regional) basis. The problem here is the lack of standardization, and the very narrow bandwidth. This system is explained later in this chapter as an example of legacy train-to-ground system still in use.

There are some specific bands for public safety and emergencies allocated for railway use on a regional basis. For example, for TETRA systems (typically at 380 MHz) or, more recently, private LTE networks at several bands.

5.2 Mobile and Wireless Communications Systems in Railways

In this subsection, we introduce some helpful definitions about mobile communications systems.

5.2.1 Concept of Cell Coverage

The coverage is a region where the received power is above a certain threshold. That is, where the communication service is possible with a certain quality.

In mobile communications systems, coverage area is divided into cells, being each one of them served by at least one transceiver or base station (BS). This division into cells means that a network design (or planning) must be done, in terms of BS placement (to choose suitable spots where BS can transmit enough power to the MS within the whole cell), frequency planning (to avoid interference between neighboring cells), etc.

The cell concept has several advantages over the no-cells approach (that is, no division of the coverage area into small parts), like higher capacity, higher spectral efficiency (higher data rates for the same bandwidth), less MS power consumption (cells are smaller than the whole coverage area), scalability (BS can be added, more or less, easily, without rearranging all the other cells).

5.2.2 Types of Cells

There are many types of cells depending on its radius:

1. Macrocells: 500 m or higher. Outdoor use.
2. Microcells: 100–500 m. Mostly outdoor but can be indoor as well.
3. Picocells: 30–100 m. For indoor use.
4. Femtocells: Less than 30 m, for indoor use. Typically deployed by home or office owners instead of mobile operators.

Cells are connected to the core by the backhaul, which is typically wired.

Typically, cells are designed to provide the most efficient coverage, and there are some ways to do this: cell splitting and cell sectoring.

Cell sectoring: Instead of having an omnidirectional antenna (that is, able to transmit/receive in all directions of the horizontal plane), the 360 degrees are divided into more than one directional sector (typically three, each one of them for 120°), each one of them with a more directive antenna. Notice that 120° is not in strict sense a directive antenna, but it is more directive than an omnidirectional antenna (which is not directive at all).

Cell splitting: The macrocell is divided into one or more than one small cells, each one of them with all the BS equipment. This approach is very helpful in crowded scenarios, allowing higher capacities.

5.2.3 Power and Capacity

Capacity is the data rate that the BS can provide to all the users within the cell. It is limited by several parameters, like power, interference, number of users, and access to media schemes.

Link budget is a method to estimate the power at a specific location and frequency. Therefore, it is used to for coverage planning. The typical approach for this calculation is to consider the following parameters: transmitted power, antenna gain (in both ends), losses due to other subsystems (combinators, wires, etc.), and propagation losses: using a suitable channel model (more on this later in this chapter).

Link budget calculations are often carried out in logarithmic units (dB and dBm, among others).

Usually, it is not only needed to determine the received (for the signal) power but also the noise power, to calculate the signal-to-noise ratio (SNR).

$$snr = \frac{P_s}{P_n} \rightarrow SNR[dB] = P_s[dBm] - P_n[dBm]$$

where P_s is the power of the signal and P_n is the noise power. Detailed calculation of noise powers of communications systems is out of the scope of this book, so we only provide some useful intuitions. For more details, please check any of the reference books on this topic [100].

Noise is random by nature. So, we don't try to predict it, but to characterize it, based on statistical tools, like average, variance, probability distribution functions, etc.

The most relevant noise source is the thermal one. This is due to random movement of charged particles (like electrons) in conductors, and it is present in all electronic devices. It is ideally white (which means that its spectral density is flat) up to 2 THz.

The higher the bandwidth, the higher the noise power. This is because the noise power is computed as the product of the noise spectral density and the bandwidth. That is why filters are so relevant in reducing noise levels.

There are other noise sources rather than thermal, like man-made noise, atmospheric, cosmic, phase, shot, and flicker noise. Always consider the noise sources that are present in your environment.

The noise figure of an element provides a measurement of the additional noise when the element is inserted on a chain with other elements. The global noise figure can be computed using the Friis equation for noise:

$$F = F_1 + \frac{F_2 - 1}{G_1} + \frac{F_3 - 1}{G_1 G_2} + \ldots$$

where F is the global noise figure of the complete cascade of elements, F_i and G_i are the noise figure and the gain, respectively, of the ith element of the chain. This expression computes F in linear magnitude, but the reader will find F in data sheets usually expressed in dB instead.

This expression has some implications, such as on a receiver, the global noise figure is affected by the order of the subsystems. Notice that if G_1 is very high and F_1 is low, $F \approx F_1$, which will be also very low. These devices with high gain and low noise figures are called low-noise amplifiers (LNA) and are perfect candidates to be the very first in the chain.

Another one is that every element in the chain always adds noise, as F_i, $G_i > 0$.

What about wires? They attenuate the signal, so the longer they are, the more attenuation they have and the higher their noise figure will be. That's why always the shortest possible cables must be used.

Do not confuse noise with distortion. We have seen where noise comes from, but what about interference and distortion? Distortion is something internal to the communications systems that alters any parameter of the signal (like amplitude, phase, or frequency). A typical example is an amplifier working on a nonlinear way (due to saturation), which may "square" the signal, distorting it. It is important to notice that, unlike noise, distortion is not random in nature. It is absolutely deterministic.

Do not confuse noise and interference. Interference comes from external sources (typically, other BS working in the same channel or close to the same channel). In mobile communications, interference is the main limitation for performance of the system and the main driver for planning.

5.3 The Mobile Radio Channel in Railway Environments

The radio channel in railways is especially challenging due to diverse factors, like the diversity of environments (tunnels, viaducts, cuttings, etc.) and the high speed of the trains, among others. To design radio communications systems, it is very important to have a proper understanding of the physical layer. Therefore, in this section, we briefly explain some of the physical effects that impact the signal propagation and some mathematical tools to address them. Perhaps this section can be summarized as follows: "never, never, never underestimate the radio channel, even less so if it is mobile."

The radio channel comprises all the effects on the signal between the transmitter and the receiver, such as attenuation, reflection, scattering, diffraction, and waveguiding [100].

5.3.1 Attenuation

Between the transmitter and the receiver, the signal loses power. Attenuation is that loss of power, and it is proportional to the square of the distance and to the square of the frequency too. It follows the Friis' Law [100]:

$$P_{RX} = P_{TX} G_{TX} G_{RX} \left(\frac{\lambda}{4\pi d} \right)^2$$

where P_{RX} and P_{TX} are the received and transmitted power, respectively, G_{TX} and G_{RX} are the gain of the transmitting and receiving antenna, respectively, λ is the wavelength of the signal, and d is the distance. Perhaps this equation is more convenient in logarithmic form:

$$P_{RX}[dB] = P_{TX}[dBm] + G_{TX}[dB] + G_{RX}[dB] + 20\log\left(\frac{\lambda}{4\pi d} \right)$$

This equation is valid only in the far-field, that is, for distances larger than the Fraunhofer distance: $d_{FR} = \frac{2L^2}{\lambda}$, where L is the largest dimension of the antenna. It is also required that $d \gg \lambda$ and $d \gg L$.

Moreover, it is only valid in free-space conditions, as some other scenarios, like indoor (i.e., tunnels) the path loss exponent which here is 2, may have a different value.

From this expression, there are two important parameters: the equivalent isotropic radiated power (EIRP): $EIRP = P_{TX} G_{TX}$.

And the free-space loss factor: $\left(\frac{\lambda}{4\pi d} \right)^2$.

5.3.2 Multipath

If the transmitted signal follows more than one path to arrive to the receiver, we say that we have multipath. This implies different times of arrival to the receiver (some paths can be longer than others), which means that we transmit a signal with a particular duration, but the received one has a longer duration. Therefore, the energy has dispersed in time. This may cause inter-symbol interference (ISI) which degrades the signal quality. Moreover, having more than one copies of the signal (i.e., "multipath components" or MPC) implies that the signal is distorted. Each one of the resulting MPC from all these phenomena has a different amplitude and phase (or more precisely: different phase, delay, angle of arrival, angle of departure). The problem comes from the receiver, because it is not able to identify individual MPCs: It only adds signals. Therefore, if the sum leads to an increase of power, we say that we have constructive interference. But if the sum leads to a decrease of power, there is destructive interference (or fading). Only a phase shift of $\lambda/2$ can lead to from constructive interference to destructive interference (see Fig. 5.2).

As we know from high school physics, the Snell Law states that the incident angle is equal to the reflected angle. However, this is only true if the surface is smooth [100] and large enough (i.e., $>10\lambda$). Reflections are very important in mobile communications, because the signal can be reflected into many surfaces until it arrives to the receiver. On each one of these reflections, some power will be reflected, and the remaining will be transmitted (i.e., lost). The ratio between these two amplitudes is the reflection coefficient, which depends on the material and the frequency, and on the polarization of the wave and the incidence angle. Going back to railways world, this explains why the propagation inside tunnels with concrete walls is so different from similar tunnels but metal coated (like many in the USA).

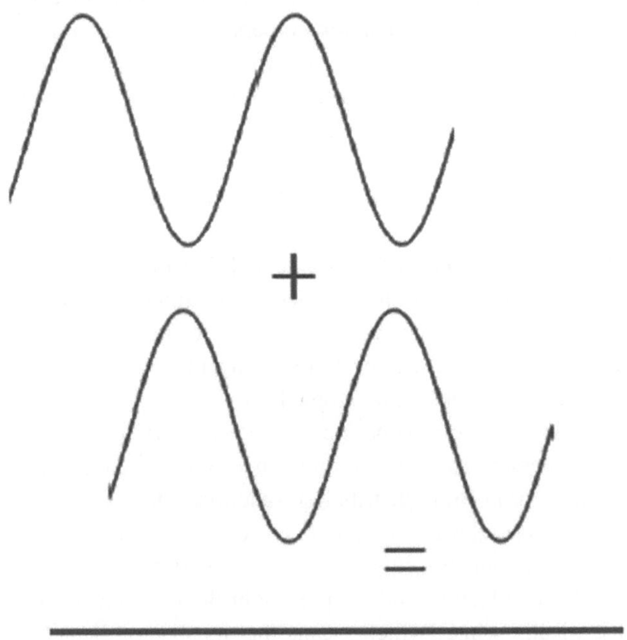

Fig. 5.2 The addition of two waves delayed $\lambda/2$ leads to a total cancellation of the signal. In practice, rather than a total cancellation, we will have a severe attenuation (called deep fading)

5.3.3 Reflecting

Another relevant propagation mechanism in wireless communications is reflection in surfaces. The physics is the following: EM waves bounce off surfaces (either smooth or rough, which depends on the wavelength of the incoming wave) such as buildings, street furniture, tunnel wall, trains, or many others. Therefore, reflected waves could follow several different paths before getting to the receiver, being another source for multipath.

This has also some advantages, like extending the range of the communications link and therefore allow communications in some non-line-of–sight (NLOS) areas.

Typically, reflections can be specular or diffuse. In specular reflections, Snell Law applies (incident angle equals reflection angle, see Fig. 5.3), as the reflecting surface is large enough (typically, its largest dimension must be at least ten times larger than λ). This specular reflection occur when the surface is smooth enough relative to the wavelength. On the contrary, diffuse reflections come from surfaces not smooth enough to the wavelength of the signal, leading to a more complex angular spectrum for the reflecting angles. Notice that even in rough surfaces, a dominant component (associated to the specular reflections) may appear together with some low-energy diffuse reflection components.

5.3.4 Scattering

This happens when the surface is rough, that is, not sufficiently smooth, so there will be no specular reflection. This implies that some power is lost in the scattering, so the "effective" reflection coefficient is lower than in pure specular reflection. This is very relevant in radars, but not in frequencies below 6 GHz, which are the most common for mobile communications (up to 5G). Indeed, the basic principle of stealth aircrafts is to avoid scattering at all costs, so

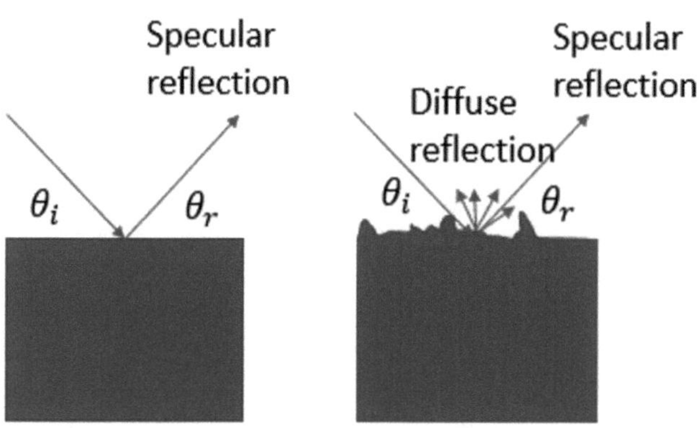

Fig. 5.3 Specular reflection and diffuse reflection

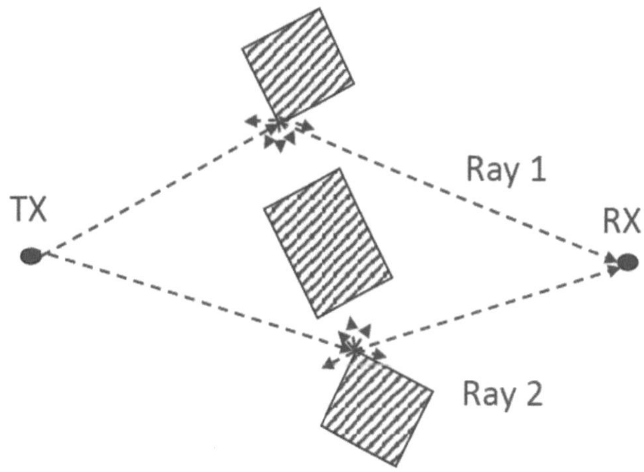

Fig. 5.4 Geometrical representation of the two-ray model without LOS between the transmitter and the receiver

surfaces in the fuselage of the airplane are very flat and smooth. Both specular reflection and scattering are depicted in Fig. 5.4.

In general terms, obstacles within the transmitter and the receiver are called "scatterers," even if they do not actually cause scatter [100]. We use this term in this book.

5.3.5 Diffraction

Diffraction happens when signals interact with sharp objects. In the strict sense, what we have said about the Snell Law and scattering only applies for ideal, infinite size objects. In practice, we can assume that they are infinite if they are large enough (typically, larger than 10λ). Finite objects do not create perfect sharp shadows, resulting in diffraction. The math that explains this physical phenomenon is complex enough to skip it right here. We offer some insights to let the reader understand this qualitatively. Depending on the size, shape, and distance of the obstacle between transmitter and receiver, the outcome is different. As in multipath (more details later), there exists constructive and destructive interference between the diffracted components.

These diffracted components come from the obstacle that acts as a secondary wave source (i.e., sort of a secondary transmitter).

Given that Huygens principle (the main mathematical formulation for diffraction) is not exact in its results, there are several methods to compute the impact of diffraction on wedges, screens, and many other kinds of obstacles [100].

For railways, in general terms, diffraction is not relevant and can be somehow ignored. However, this may no longer apply in millimeter-wave bands or in some other particular scenarios.

5.3.6 Waveguiding

Finally, with waveguiding, we are in the opposite scenario: It is a propagation mechanism that happens in tunnels, which makes it relevant for our interests (waveguiding happens in tunnels. Street canyons, large corridors, cuttings, etc., largely resemble tunnels, and radio waves may experience the waveguide effect). The basic principle is that a tunnel resembles a dielectric waveguide [100] with some differences because tunnel walls are lossy (usually made of concrete or even bricks), they have many discontinuities, etc. The impact of this effect on the signal is twofold:

1. On a tunnel, the losses of the signal are lower than in free space.
2. There are more reflections than in free space, which means higher variability of the signal.

Not all tunnel shapes can be solved analytically using Maxwell Laws, so the usual engineering approach to model this scenario is to apply geometric-optic techniques instead. Both the tunnel cross section and the size of scatterers must be larger than λ to get accurate results [100]. Instead of the typical transverse electric (TE) or transverse magnetic (TM) models, only hybrid EH models are propagated, making all more complex.

5.3.7 Statistical Channel Modeling: Narrowband Channels

When the number of MPCs is large (which is usually the case in real scenarios), it is difficult to describe the channel in an analytical way, and a statistical description is more suitable. Often, it is enough to describe the attenuation, and the delay spread (that is, how the most relevant MPCs are delayed in time). Sometimes, we need a different approach to describe the signal, for example, addressing the variations of the received power, which means both small- and large-scale fading [100].

Small-scale fading are the variations in the received power around the mean due to MPC interference. These fluctuations can be described statistically (typically, using Rayleigh or Rician distributions).

On the other hand, large-scale fading stands for the fluctuations that can impact on a larger basis (typically, 10λ), usually due to shadowing. This kind of fading can be described statistically as well (typically, using log-normal distributions). Path loss influences the large-scale mean (decreases as distance increases) but is usually described on a deterministic basis.

To describe variations in the signal amplitude, the two-ray model is used. This model is the simplest one to describe small-scale fading, and it is based on Fig. 5.4, where many scatterers are between the transmitter and the receiver, but only two MPCs (i.e., rays) arrive at the receiver. Notice that no MPC gets to the receiver directly, so we say that every MPC is non-line-of–sight (NLOS). If there were one direct unobstructed MPC, we would use the term "line of sight" (LOS).

Let us start with an invariant channel (where nothing moves). If we transmit a sinusoidal wave:

$$E_0 = E\cos(2\pi f_c t)$$

which travels a distance "d," the resulting signal we receive will be phase shifted:

$$E_{RX} = E\cos(2\pi f_c t - k_0 d)$$

In base band notation, we have:

$$E(r) = E_1 e^{-jk_0 r}$$

In our model, we have two different paths, with two different distances (d_1 and d_2), which implies two different phases for the resulting signal in the receiver $E(r)$:

$$\tau_1 = \frac{d_1}{c}; \tau_2 = \frac{d_2}{c}$$

$$E(r) = E_1 e^{-jk_1 r} + E_2 e^{-jk_2 r}$$

τ_1 and τ_2 are the transmission delay for the first and second MPC, respectively.

However, vehicular channels like those we have in T2G are time varying, which means that the channel impulse response $h(\tau)$ changes with time t. That is, $h(\tau) = h(t,\tau)$. We call this function the time-variant impulse response [100]. This means that the path length difference changes with time because of movements of either the transmitter, receiver, or even the environment. A usual assumption in this type of model may be that only the receiver is in motion [100]. This implies that the constructive-destructive interference pattern changes with time. Indeed, fadings are spaced a distance equal to $\lambda/2$, so the "fading rate" depends on speed of the mobile station.

On the other hand, this movement also implies a frequency shift, which is called Doppler shift. Let us describe it using the two-ray model again:

We transmit a sinusoidal wave that travels a distance "*d*" and moves at a speed v, so we receive:

$$E_{RX} = E\cos\left(2\pi f_c t - k_0\left[d + vt\right]\right) = E\cos\left(2\pi\left(f_c - v/\lambda\right)t - k_0 d\right)$$

The term $-v/\lambda$ is the Doppler shift which can be expressed in terms of speed and carrier frequency of the signal (instead of the wavelength λ).

$$\upsilon_{max} = \frac{-v}{\lambda} = -f_c \frac{v}{c}$$

If the transmitter and the receiver move away from each other, this term is negative, which means a shift to lower frequencies and vice versa. If there is no alignment between the trajectory of the MS and the transmitted wave, we need to project into that direction, that is, multiplying by the cosine of the angle these two lines form:

$$\upsilon = \upsilon_{max} \cos\varphi = -f_c \frac{v}{c} \cos\varphi$$

Notice that different MPCs can have different Doppler shifts, so the addition of all MPCs with different frequencies (due to Doppler shifts) leads to fading. Also notice that the rate of this variation is proportional to the speed of the MS. That is, at high speeds, there are more fading dips in the same period. Therefore, Doppler frequency gives us information on how fast the channel changes. In Fig. 5.5, a signal formed by two MPC components, with similar but not equal amplitudes, very similar frequencies (but not equal due to Doppler shift) and different but similar phase shifts (due to different paths) is depicted. Please notice the deep fadings.

Now let us generalize our model: Instead of two MPCs, we can have many, each one of them with a different frequency shift. This broadens the spectrum of the received signal. The minimum and the maximum values that we can have are:

$$f_c - f_c \frac{v}{c} \le f = f_c \left(1 - \frac{v}{c} \cos\varphi\right) = f_c - v \le f_c + f_c \frac{v}{c}$$

If we want to know how the power is distributed in the horizontal plane (that is, as a function of the incoming angle φ), weighted by the radiation pattern $G(\varphi)$ of the antenna, we have [100]:

$$S(\varphi) = \bar{\Omega}\left(pdf_{\varphi}(\varphi)G(\varphi) + pdf_{-\varphi}(-\varphi)G(-\varphi)\right)$$

where $\bar{\Omega}$ is the mean power in the receiver:

$$S(\varphi) = \bar{\Omega}\left(pdf_{\varphi}(\varphi)G(\varphi) + pdf_{-\varphi}(-\varphi)G(-\varphi)\right)\frac{1}{\sqrt{v_{max}^2 - v^2}}$$

If $pdf_{\varphi}(\varphi)$ is uniform and $G(\varphi)$ is constant, we have what we call Jakes spectrum (see Fig. 5.6), which in theory has singularities in the maximum absolute shifts. In practice, we would need infinite MPCs in the receiver to have a singularity. $S(\varphi)$ is called Doppler spectrum and describes frequency dispersion (that may hurt multicarrier modulations like OFDM—Orthogonal Frequency- Division Multiplexing) and gives an insight on the variability of the channel due to the speed of the MS.

However, even though Doppler spectrum is a complete characterization of the variations of the signal, it is often impractical. We use two other parameters instead: LCR and AFD.

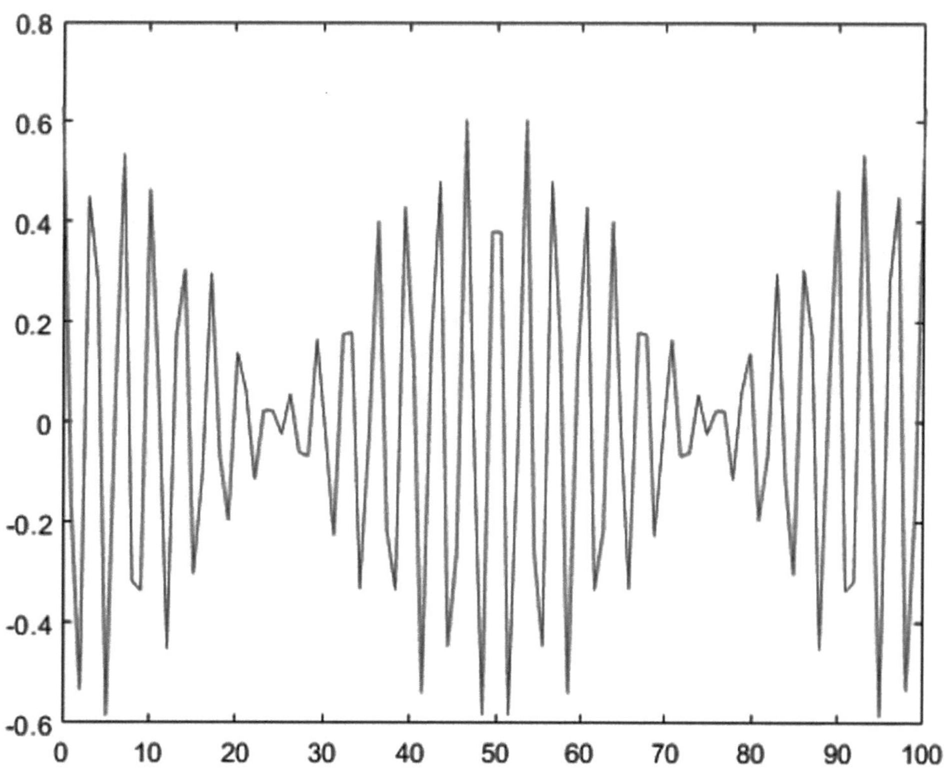

Fig. 5.5 Resultant signal from two MPC components with similar amplitude and Doppler shift

Fig. 5.6 Jakes spectrum

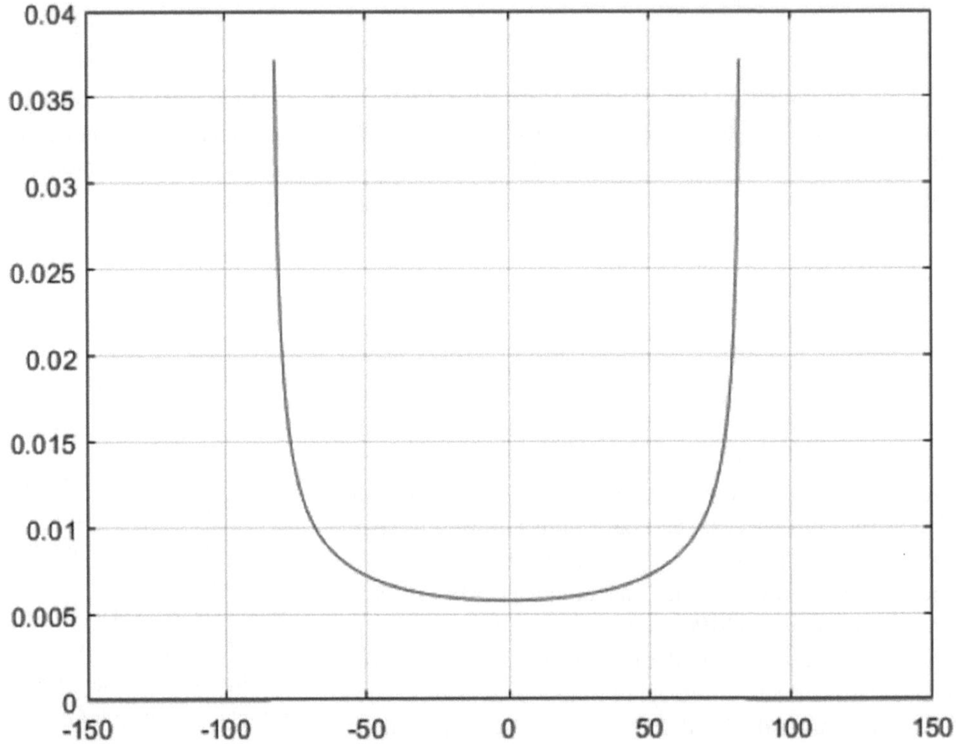

Level crossing rate (LCR) is the average number of times the signal amplitude crosses a certain threshold or level in the positive direction. Obviously, it depends on the threshold r we choose.

Average fading duration (AFD) is the ratio of time the signal is below the threshold and the total time.

Going back to the static scenario (no Doppler), we can have two different situations: to have a dominant MPC component, like in LOS links, or to have no dominant MPC at all. In either case, as we said before, when having a large number of MPC components, it is not practical to deal with them in a deterministic way. Therefore, instead of using a deterministic model, we choose a statistical one.

In the scenario with nondominant MPC, the amplitude follows a Rayleigh distribution, which basically implies that both the real and the imaginary parts follow a Gaussian distribution, both zero mean. The phase is assumed to be uniformly distributed between 0 and 2π. The probability density function (PDF) of the amplitude of the signal that follows a Rayleigh distribution is:

$$pdf_r(r) = \frac{r}{\sigma^2} e^{-\frac{r^2}{2\sigma^2}}, 0 \leq r < \infty$$

And the cumulative distribution function (CDF):

$$cdf(r) = 1 - e^{-r^2/2\sigma^2}$$

The peak of the PDF is when $r = \sigma$. The CDF is very useful to estimate outage probabilities, that is, how likely it is to have fadings of a certain depth.

Example

For a Rayleigh-distributed signal in the amplitude, estimate the probability of having fadings 20 dB below the mean power.

Instead of doing some (easy, by the way) math, in Fig. 5.7, the CDF vs fading depth in dB is depicted, so it is straightforward to compute the probability for a Rayleigh-distributed amplitude to be 20 dB below the mean power.

It is around 0.01 (1%). But notice that the probability for 10 dB is 10%, which is not negligible at all.

Notice that Rayleigh is a worst-case scenario (at least up to a certain bandwidth). Rayleigh distributions work very well in several NLOS mobile scenarios, but not in all of them (like ultrawide band modulations). For more details about the Rayleigh distribution, please check [100, 101].

On the other hand, if we have a dominant component, the fading is less severe. Assuming that in the LOS component we have $\varphi = 0$, we have two independent PDF for both the real and the imaginary parts of the amplitude:

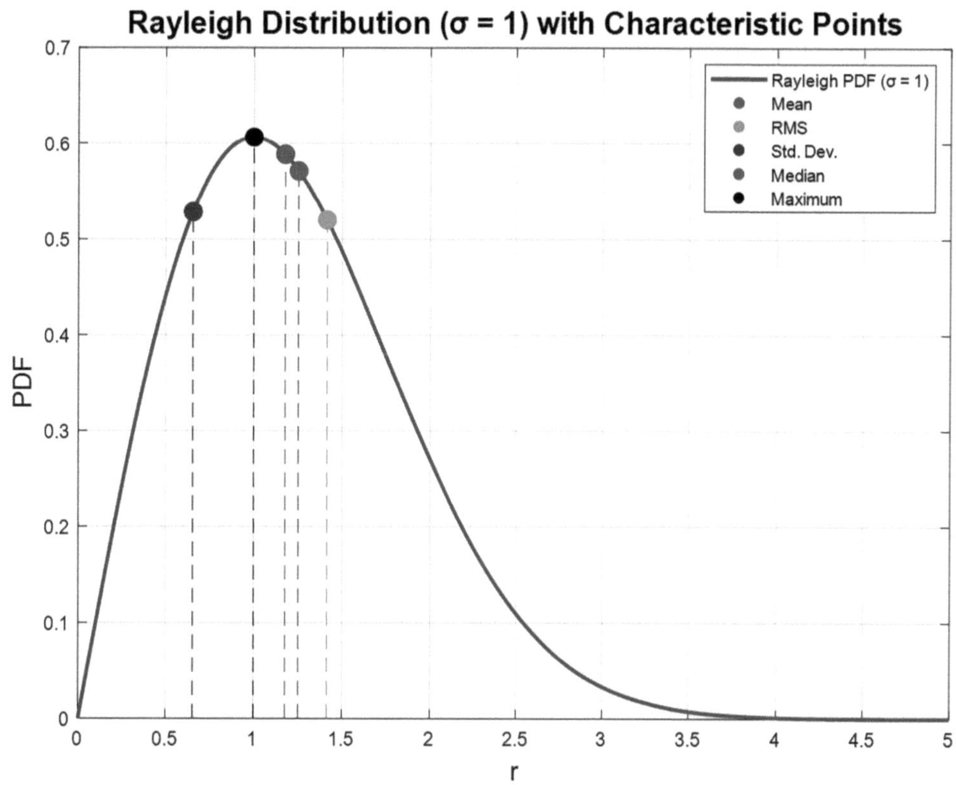

Fig. 5.7 PDF and main parameters for *Rayleigh distribution*

$$(E) \sim G(\neq 0, \sigma)$$
$$(E) \sim G(0, \sigma)$$

This is the same as in Rayleigh distributions, but here the mean of the imaginary part is not equal to zero. The PDF of a signal formed by several MPC that follow a Rician distribution is:

$$pdf_{r,\varphi}(r,\varphi) = \frac{r}{2\pi\sigma^2} e^{-\left[\frac{r^2 + A^2 - 2rA\cos\varphi}{\sigma^2}\right]}$$

Notice that, unlike Rayleigh, this distribution is not separable into phase and amplitude. A is the amplitude for the LOS component. A very useful parameter is the Rician factor $K_R = \frac{A^2}{2\sigma^2}$. If $K_R \to 0$, the PDF approximates Rayleigh (that is, there is no dominant MPC), but if $K_R \to \infty$, the distribution is Gaussian with average A.

Apart from Rayleigh and Rician, there are other distributions that can fit better in other scenarios, like wideband and ultrawide band [100] which are explained in the next section.

5.3.8 Statistical Channel Modeling: Wideband Channels

We have seen what happens on a narrowband channel, but most communications systems today are wideband ones, such as IEEE 802.11, LTE, etc. In this section, we explain the effects of multipath on wideband signals, which are mostly two [100]:

- Frequency selectivity: The channel transfer function (CTF) of the channel $H(f)$ is not constant for the whole bandwidth.
- Delay dispersion: $h(t) \neq \delta(t)$, where $\delta(t)$ is the Dirac delta, which means the received signal lasts longer in time than the transmitted one. This delay dispersion may lead to inter-symbol interference, a type of distortion where consecutive symbols overlap.

If we continue with the two-ray model, assuming static MS, the wideband channel is still linear time invariant (LTI) if the mobile is static, but could be delay dispersive, and perhaps, frequency selective. Let us see why.

We have two MPCs, each one of them has a different delay, say $\tau_1$and τ_2. Therefore, the channel impulse response (CIR) will be like this [100]:

$$h(t) = a_1 \cdot \delta(\tau - \tau_1) + a_2 \cdot \delta(\tau - \tau_2)$$

where $a_i \in C$. This means that the CIR has a duration longer than 0 (indeed, its duration is $\Delta\tau = |\tau_1 - \tau_2|$). To compute the frequency response, that is, the CTF, we only need to compute the Fourier Transform of the CIR:

$$H(f) = \int_{-\infty}^{\infty} h(\tau) e^{-j2\pi f\tau} d\tau = a_1 e^{-j2\pi f\tau_1} + a_2 e^{-j2\pi f\tau_2}$$

And the amplitude of the CTF is:

$$|H(f)| = \sqrt{|a_1|^2 + |a_2|^2 + 2|a_2||a_1|\cos(2\pi f\Delta\tau - \Delta\varphi)}$$

Example

We have two MPC on a wideband channel. Assume that the central frequency is 1.99 GHz, $\Delta\tau$=2 ms, $\Delta\varphi = \pi/8$ rad, and the amplitudes for each one of them are $a_1 = 0.8$ and $a_2 = 0.83$. If we plot the resulting, $|H(f)|$, we have (see Fig. 5.8):

Notice that $H(f)$ is not constant, meaning that the channel is frequency selective, or more precisely, that we have frequency-selective fading. The minimum values of the function are close to 0 and also follow a periodical pattern $\Delta f = 1/\Delta\tau$. At these minimum values, both MPCs are in counterphase (i.e., the phase difference is 180°). This is what we call destructive interference.

Actually, we deal with wideband channels in a more practical way, grouping all MPC into several equally spaced bins. This approach allows several MPCs coming from different angles and routes to be added together if they do not differ very much in delay terms. This can be explained more clearly looking at a power-delay profile (PDP) which is a representation of how the power of $H(f)$ is distributed in the delay domain. We'll see more on this very soon when we discuss "compact parameters."

To summarize, we can think in terms of Fourier analysis: if h(t) is a finite signal (as every real-world signal is), then $H(f)$ will be frequency selective.

From the mobility side, if neither the transmitter, receiver, nor the scatterers move, then the channel is LTI. That is, the CIR will be the same for every time t, and will depend only on the delay: $h(t, \tau) = h(\tau)$.

But if the MS or the scatterers (or, more unlikely, the BS) are in motion, the channel is linear time variant (LTV), which means that the CIR changes with time $h(t, \tau)$. Obviously, this is a typical scenario in vehicular channels. Remember that t represents absolute time and τ represents delay. However, some channels are "slow-LTV," that is, quasi-static, or "almost LTI" [100].

Starting from the general time-variant impulse response $h(t, \tau)$, we can take a Fourier transform on either t or τ. The results will be different. If we compute it on τ, the result will be $H(t,f)$, which is the time-variant transfer function. This is the usual CTF but time variant (i.e., changes with time). On the other hand, if we compute the Fourier transform on τ, we obtain $S(\upsilon, \tau)$, which is called the Doppler function impulse response. It provides information of the spreading of the signal in both the Doppler domain and the delay domain. Finally, if we calculate the Fourier transform on t for $H(t,f)$ or on τ for $S(\upsilon, \tau)$, the result will be the same: the Doppler-variant transfer function, which provides information of the signal in both the frequency domain and the Doppler domain, so it can be viewed as a generalized CTF. In Fig. 5.9, a diagram of all these functions and transforms can be seen. Notice that all this analysis is deterministic, as $h(t, \tau)$, $H(t,f)$, etc., are all deterministic functions. However, very soon we will again resort to statistics for some help.

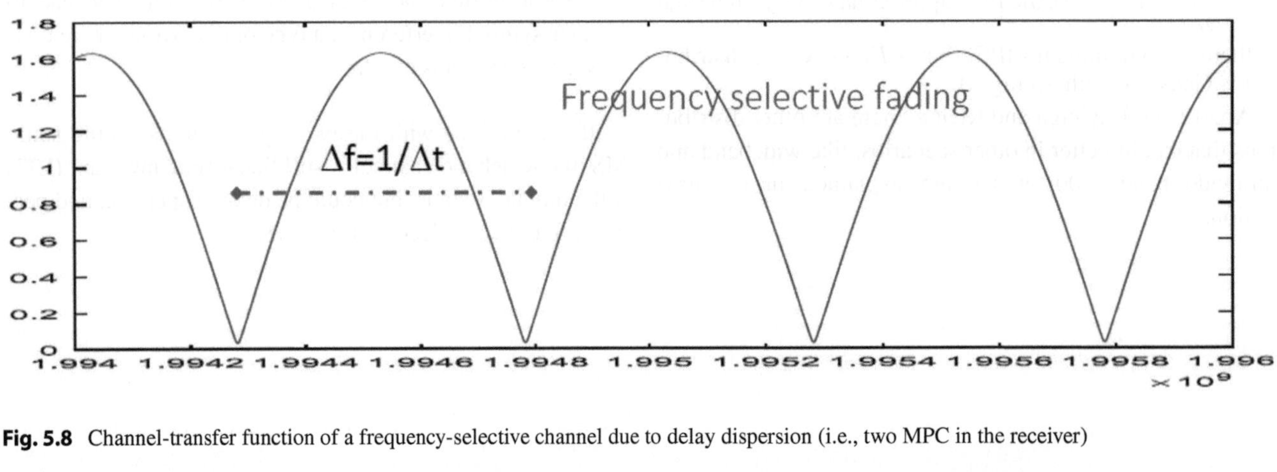

Fig. 5.8 Channel-transfer function of a frequency-selective channel due to delay dispersion (i.e., two MPC in the receiver)

Fig. 5.9 Channel functions adapted from

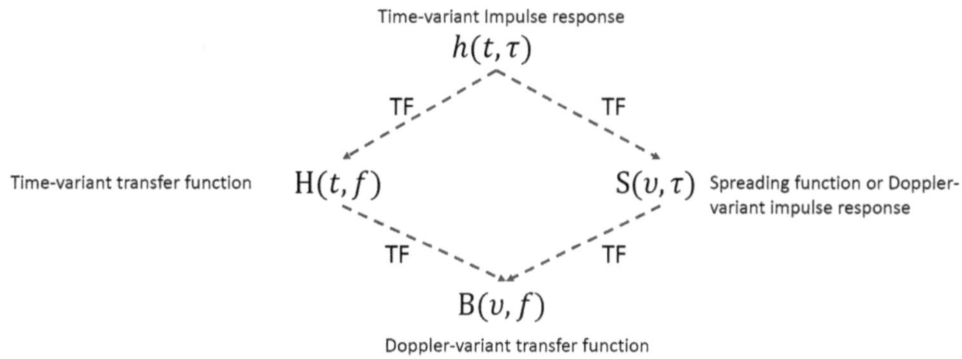

The problem is a frequent one: all these functions can be complicated to handle, so we use stochastic parameters instead. Under some circumstances, we can simplify these functions and use compact parameters instead. Let us see some of them:

Autocorrelation Function (ACF) A very common way to characterize a stochastic process, ACF describes the similarity of a function with a delayed copy of itself. Therefore, the ACF of the CIR is:

$$R_h(t,t',\tau,\tau') = E\{h(t,\tau)h(t',\tau')\}$$

where t and t' are time variables, and τ and τ' are delay variables [1], respectively.

Power Delay Profile (PDP) Denoted as $P_h(\tau)$ or delay power spectral density, which describes how power arrives to the receiver in the delay domain. It is a very important function:

$$P_h(\tau) = \int_{-\infty}^{\infty} P_s(\upsilon,\tau)\,d\upsilon = \lim_{T\to\infty}\frac{1}{2T}\int_{-\infty}^{\infty}|h(t,\tau)|^2\,dt$$

We can obtain the mean delay T_m as:

$$T_m = \frac{\int_{-\infty}^{\infty}\tau P_h(\tau)\,d\tau}{P_m}$$

where P_m is the total power of the signal:

$$P_m = \int_{-\infty}^{\infty} P_h(\tau)\,d\tau$$

Another one is S_τ, which is the root mean square (RMS) delay spread, also a very relevant parameter which can be obtained from the PDP as well (sometimes can provide enough information to replace the whole PDP):

$$S_\tau = \sqrt{\frac{\int_{-\infty}^{\infty}\tau^2 P_h(\tau)\,d\tau}{P_m} - T_m}$$

In some cases, S_τ is enough to know the bit error rate (BER) of a system, because the BER will be proportional to S_τ [1].

We can obtain other compact parameters like the rms Doppler spread, coherence bandwidth B_c, and the coherence time T_c.

The RMS Doppler spread is analogous to the RMS delay spread but obtained from the Doppler spectral density.

The coherence bandwidth is the bandwidth where the correlation function is below a threshold (usually 50% from the maximum value). It can be obtained from the Fourier transform of the PDP.

The coherence time is related to rms Doppler spread and measures how fast a channel changes, and a suitable estimation for this parameter is $T_c = 1/\nu_{max}$, where ν_{max} is the maximum Doppler shift of the signal, that is, T_c is the inverse of the maximum Doppler shift of the signal.

All these compact parameters come from a very common assumption: The WSSUS model. WSSUS stands for two assumptions: first, "wide-sense stationarity," which implies that properties of the channel do not change with time. Notice that this is not the same as a static channel, because amplitude may change, but not some essential parameters as total power or the Doppler spectrum. Second, "uncorrelated scatterers": that is, that different delays are uncorrelated. This is something generally true outdoor but not indoor, because reflections from tunnel walls are correlated.

When both WSS and US conditions are fulfilled, we are in WSSUS conditions. In the next section, we explain a common way to deal with WSSUS models: tapped delay line (TDL) models.

To sum up, multipath impacts severely the radio channel by at least three different ways: variability of the signal due to dynamic changes in the channel, variability of the signal due to movement of transmitter and/or receiver, and frequency selectivity.

5.3.9 Channel Models in Railway Environment

Once we learn about the math that helps us to explain mobile radio channels, in this section, we introduce channel models which are a more friendly engineering tool. This is the main objective of a channel model is to provide a mathematical description of the channel, balancing complexity and accuracy.

Channel modeling is a key topic for radio communications, with multiple research groups working on it, and still looks unending. It has several applications, such as channel emulation, input for planning software tools, and certification. Indeed, validation of radio communication systems in railways is costly because at the end, it usually requires on-site testing. However, with channel emulators of railway environments, these costs can be largely skipped.

Channel models can be of many different types. In this section, we do not intend to describe all types, but to allow the understanding of the most suitable ones for railway scenarios. Typically, channel models provide information about multipath, path loss, delay spread, and Doppler spectrum.

Furthermore, channel models can be based on several different sources; among them, we find data from measurements, data from simulations, or even mathematical and physical information. Regarding measurements, we call the techniques used to take those measurements "channel sounding," which comprise all the electronics and signal processing needed to retrieve either the CIR or the CTF of the radio channel [109].

In the case of simulations, the range of possible solutions is very large. However, in recent years, the dominant technology has been ray tracing (RT). RT is a technique that allows the computation of all the rays (MPCs in our notation) that go from the transmitter to the receiver. It is very useful to generate channel models. The main idea is to evaluate all paths to compute the amplitude, phase, angle of arrival, etc., for each one of the rays that gets to the receiver. This technique needs to have some geographical information of the environment and its materials. Sometimes modeling is done using 3D computer-aided design (CAD) software applications, and material characterization from laboratory measurements (for example, to estimate the reflection coefficient of the bricks that form a wall). It is a scalable technique, because the more reflections are considered, the more accurate will be the results (but more CPU time will be needed as well). But reflections are the easiest part, because other phenomena like scattering and diffraction are not so easy to model [9].

Path loss models: The simplest way we have to model the longitudinal attenuation of the signal is the following:

$$PL(dB) = L_0 + 10n\log_{10}(d) + \chi$$

where PL is the path loss in dB, L_0 is the attenuation at a fixed well-known distance (which makes the term $10n\log_{10}(d)= 0$), n is the path loss exponent (PLE), and χ is an extra term for the shadowing (which has a random nature). PL models are typically used in macrocells and are very helpful for frequency and coverage planning. Models can get more complex by introducing the impact of the antenna heights (for both BS and MS), the frequency (our basic model ignores the frequency of the signal, but we know that the frequency of the signal has impact on path loss), model the PL into more than one slopes, rather than only one, angles of arrival, etc. Examples of this are Winner II [110], 3GPP [111], Okumura-Hata [112], etc. For those interested in the prolific world of channel models in railways, our recommendation is EmulRadio4Rail project deliverable 1.3 [113], where a detailed state-of-the-art was compiled.

The tapped delay line (TDL) model is a type of nongeometric channel models (that is, it doesn't use or contain geometrical information of the environment) and represents the CIR of the channel. The CIR is depicted as a summation of a finite number of taps, each one of them with their own independent time-varying coefficients, like amplitude, delay, and Doppler spectrum:

$$h(\tau;t) = \sum_{k=1}^{K} c_k(\tau)\delta(\tau - \tau_k)$$

where $c_k(\tau)$ is the time-varying coefficient of the ith tap, τ_k is the delay of the kth tap, $\delta(\cdot)$ is the Dirac delta function, and K is the number of taps. For each tap, Doppler spectrum changes the value of $c_k(\tau)$. Therefore, each one of the K taps groups very close MPCs. Notice that it is possible to add a LOS MPC to the model, if needed.

We can implement WSSUS models with the tapped delay line (Fig. 5.10).

The (delay) resolution between two paths is limited by the bandwidth of the system, which is the inverse of the delay resolution: $\Delta\tau = 1/W$, where W is the bandwidth. For example, if we have a (remarkable) bandwidth of 1 GHz, we shall be able to distinguish MPCs separated 1 ns or more. If the delay between two MPCs is less than 1 ns, our setup will take both as only one.

TDL channel models are easy to implement in emulators and simulators because they resemble finite impulse response (FIR) filters.

In the remaining of this section, we show a TDL model for high-speed tunnels and some suitable references for other

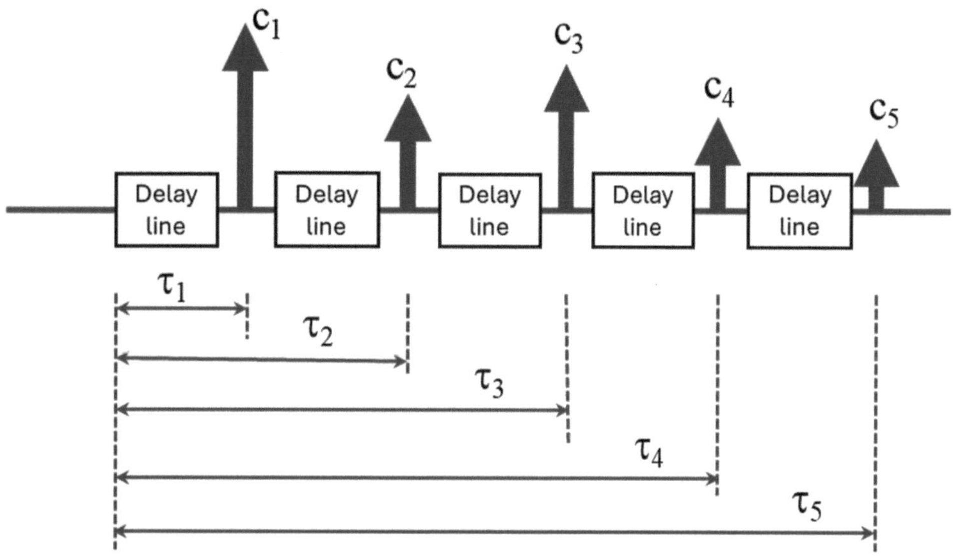

Fig. 5.10 Depiction of a tapped delay line (TDL) model, where all the power within each tap is grouped together

Table 5.3 Main parameters of the TDL channel model provided as an example for railway environments

Parameters	Value
Center frequency	2.4 GHz
Bandwidth	20 MHz
Speed	350 km/h (HSR)
Diversity	MIMO 2 × 2

TDL models related to other railway scenarios such as rural, viaduct, cutting, and hilly. Therefore, a set of channel models will be given, at least one for each scenario, with the most complete description possible in terms of:

1. Number of taps.
2. Delay associated to each tap.
3. Relative power associated to each tap (relative to the first one).
4. Doppler spectrum (i.e., distribution of the frequency shift in the Doppler domain).
5. Speed range considered.
6. Frequency range, as all channel models are defined for a finite frequency range.
7. Diversity scheme (i.e., single-input single-output (SISO), multiple-input multiple-output (MIMO), etc.). More on this in the next section.

Notice that the first four comprise the strict-sense TDL model definition, and the other three are metadata that allow us to determine the suitability of the TDL model. For example, if we want a TDL model for GSM-R, we should make sure that the model is defined for the 900 MHz band (see GSM-R section for more details), and that the Doppler spectrum at high speeds is considered as well. Otherwise, our model will probably fail to predict the real-world channel.

The details of the TDL model can be seen in Table 5.3. We have five taps, each spaced 8 ns, which means that all MPCs that arrive after 30 ns have very low impact and are not accounted for. For each tap, we have its power relative to the highest-power tap, the distribution of the fading (i.e., the statistical model for all MPCs that are grouped into that tap), and the characterization of the Doppler spectrum. In this case, the fading distribution is not Rayleigh or Rician, but a different one (Weibull), but the Doppler follows the classical Jakes spectrum that we explained in the previous section. More details of the modeling can be found in [114]. The 3D models for both the train and the tunnel are shown in Figs. 5.11 and 5.12 (Table 5.4).

For many different railway scenarios (cutting, viaduct, etc.), there are many TDL channel models available in the literature. A suitable landing paper is [115] which is a survey of many of TDL-based and non-TDL channel models for railways.

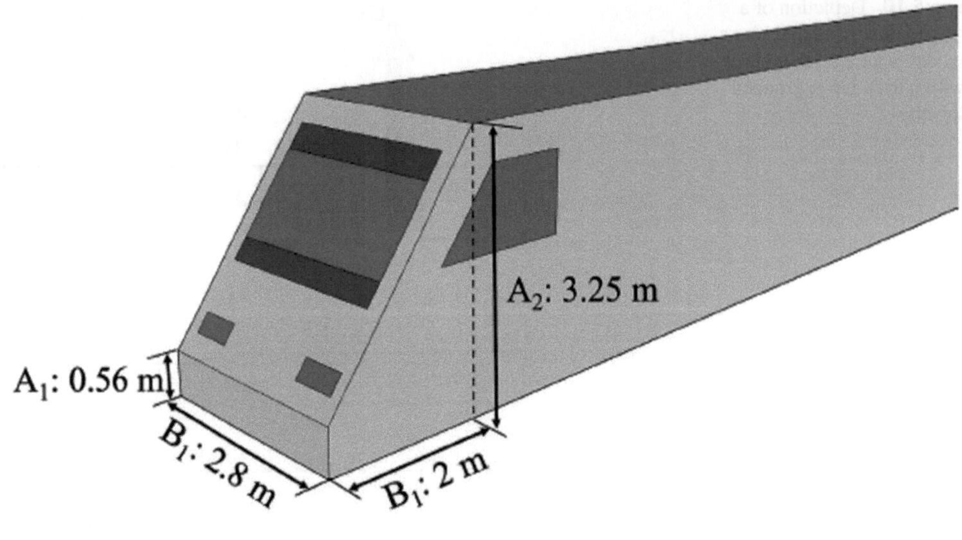

Fig. 5.11 3D model for a subway train used in

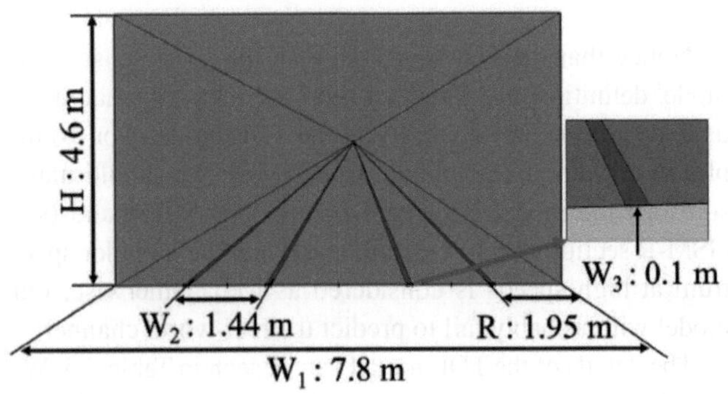

Fig. 5.12 Tunnel cross section used in

Table 5.4 TDL model for subway tunnel obtained in

Tap No.	Delay (ns)	Path Power		Fading Distribution		Doppler Spectrum				
		(lin.)	(dB)	Weibull β_ω	Weibull Ω_{weib}	Shape	Fitting Model of σ_{Jakes}	α	c	k
1	0	1	0	1.22	0.69	Jakes	Burr	0.6119	28.6175	0.2131
2	8	0.0377	-14.2312	0.68	0.44	Jakes	Burr	0.1568	39.1621	0.1061
3	32	0.0043	-23.6337	0.50	1.11	Jakes	None	σ_{Jakes} 0.1329		
4	48	0.0070	-21.5570	0.57	1.25	Jakes	None	0.2275		
5	64	0.0013	-28.9994	0.96	4.02	Jakes	None	0.1991		

5.4 Antennas

First, a formal definition "the antenna is a means for radiating or receiving radio waves" [IEEE Std 145-1983], and according to Balanis, in a classic textbook [116], it is a "transitional structure between free-space and a guiding device." Therefore, antennas[3] are a key element on every radio communications system.

In this section, we address the fundamental parameters of antennas and multiantenna technologies, which include both multiple-input multiple-output (MIMO), beamforming and general concepts about diversity, and, finally, integration of antennas onboard trains and on typical wayside scenarios.

5.4.1 Fundamental Parameters of Antennas

First, this subsection only intends to provide a handy summary of the essential antenna parameters. Consequently, for a more complete description of the radiation mechanism, types of antennas, or, perhaps more important, the EM formulation, please refer to some of the excellent books on this topic like Balanis [116] or Milligan [117].

[3]Or antennae, as we should call them to be more respectful with the Latin-speaking people who did not invent them (the romans.).

For our purposes, the essential antenna parameters are radiation pattern, beamwidth, directivity, gain, and polarization.

Radiation pattern is a depiction of how the antenna radiates/receives in spatial coordinates. Typically, this information is provided by two different graphs: the vertical and the horizontal planes (instead of a 3D depiction, which would be a more complete representation but certainly less useful). Indeed, these two planes are represented in dB, normalized in power (0 dB being the maximum value). Notice that working in dB implies a logarithmic plot, which distorts the actual power pattern from the linear depiction.

Looking at a radiation pattern (see Fig. 5.13), we can understand how the antenna works, if it is suitable for our application and retrieve even more information, as we will learn shortly.

The pattern may have lobes, which are the directions which represent the -local or absolute maxima for the pattern that could be classified as major, minor, side, and back lobes [116]. The major lobe is for the absolute maximum (in Fig. 5.13, it is in the 180° direction), the minors are for all the others, side lobes are lobes in different directions from the major one, and, finally, back lobes are those in the opposite direction to the major one. That is, both side and back lobes are minor lobes. Usually, minor lobes are undesired and may imply a source for interference (or more, precisely, an entry point for interference), but sometimes not.

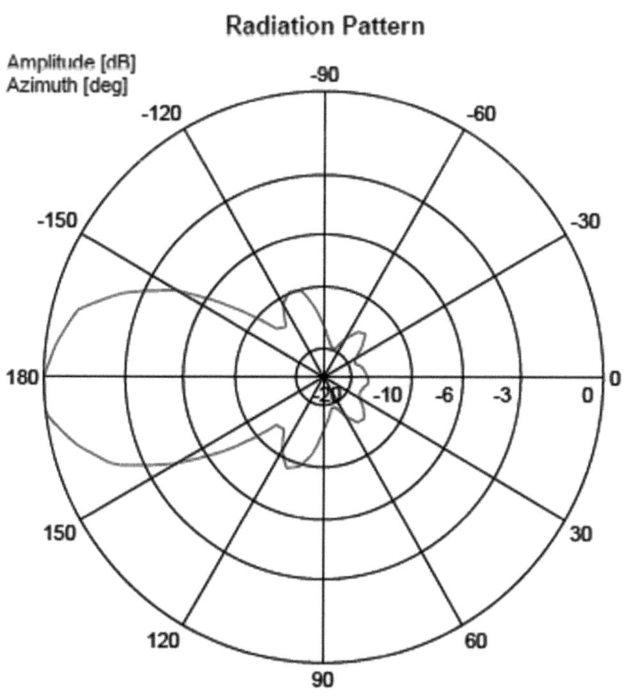

Fig. 5.13 Horizontal plane radiation pattern of the Huber-Suhner Sencity antenna at 2.4 GHz. This antenna is for railway use, to be mounted onboard trains

Typically, lobes are described by both the lobe level (i.e., the power level below the major one) and the direction that it points to. There are three essential radiation patterns:

1. Isotropic: Where the antenna radiates equally in every possible direction. This is a sphere in 3D space. This pattern is physically impossible, but it is very useful for theoretical analysis of antennas.
2. Directional: One direction in the pattern has more power than any of the others.
3. Omnidirectional: Nondirectional in the horizontal plane and directional in the vertical plane. This is a special case for directional antennas [116]. The planes are usually named "azimuth" and "elevation" plane, for the horizontal and vertical, respectively.

Beamwidth: Literally, the width of the major beam. The most common parameter to describe the beamwidth is the half-power beamwidth (HPBW), which is defined as the "angle between the two directions in which the radiation is one-half value of the beam" [116]. Something very important with HPBW and side lobe level (SLL) is that they are inversely proportional: that is, the smaller the HPBW (that is, the more directional the pattern is), the higher the SLL [116].

Directivity is the "ratio of the radiation intensity in a given direction from the antenna to the radiation intensity averaged over all directions" [116]. That is, how much more power is contained in the major lobe vs the average of all directions. It is usually expressed in dB, and there is a relationship between directivity and HPBW [116]:

$$D_0 = \frac{4\pi}{\Theta_1 \Theta_2}$$

where D_0 is the maximum directivity, and Θ_1, Θ_2 are the HPBW (in rad) in two planes separated 90°. This expression is not a general one but is a good approximation for many antennas.

Notice that isotropic antennas have a directivity equal to 0 dB, whereas omnidirectional antennas have a low directivity value but not equal to 0 dB.

Gain: Whereas directivity has to do with the radiation pattern only, gain is the ratio between the power in a specific direction compared to an isotropic radiator.

Polarization: The orientation of the electric field of the EM wave that departs the antenna. This can be linear (vertical or horizontal), circular, or elliptical.

Frequency range: The band where the antenna works.

Voltage standing wave ratio (VSWR) is a parameter that measures how well matched an antenna is to the transmission line that connects it to the transceiver. If it is unmatched, it means that less power is transmitted into the air. A value of 1 implies perfect matching, while values higher than 1 imply less than perfect matching, the higher the worse.

5.4.2 Multiantenna Technologies: MIMO and Beamforming

Modern communications systems use multiantenna technologies to increase both the capacity and the availability of the links. 3GPP LTE and IEEE 802.11 are good examples of this. These multiantenna technologies are of special interest in several railway scenarios, as we will see in this section.

For example, MIMO (multiple-input multiple-output) systems can increase data rates specially in harsh environments, with strong multipath, like railway tunnels. The reason for this is that tunnels tend to generate more multipath components than, for example, outdoor scenarios. This increases the diversity of paths and makes those paths more diverse, therefore increasing the capacity of the channel. We provide more details of this later.

However, in tunnels, an undesired effect can happen as well, decreasing the overall capacity of the channel: the keyhole effect [118]. This effect removes the diversity that comes from the existence of several subchannels on a MIMO system. This effect has been largely studied for a T2G link [119] in tunnels, and the results are straightforward: MIMO is beneficial for T2G in tunnels, but keyholes can reduce this advantage.

Let's back up for a moment: What is diversity? What is MIMO? Having more than one antenna element in the receiver or in the transmitter allows us to implement some diversity schemes. When we have more than one antenna in at least one end, we define subchannels, which are individual channels that go from one antenna element in the transmitter to another antenna element in the receiver. See Fig. 5.14. Once we have this diversity, there are three benefits or gains in this scenario: diversity gain, capacity gain, and array gain.

Diversity gain: When the result is the reduction of fading because fewer subchannels are affected.

Capacity gain or spatial multiplexing gain: Due to the spatial segmentation of the channel into independent subchannels, spatially multiplexing the information between them.

Array gain: When forming a classic antenna array, the radiation pattern of the array if different from that of the individual antennas alone. This new pattern can emphasize or attenuate some directions, therefore increasing the SNR. This is the basis of beamforming, which is addressed later in this section. More precisely, the signal-to-noise and interference ratio (SINR) as interferent signals can be suppressed by spatial filtering (that is, putting a "null" of the radiation pattern in the direction of the undesired interferent signal).

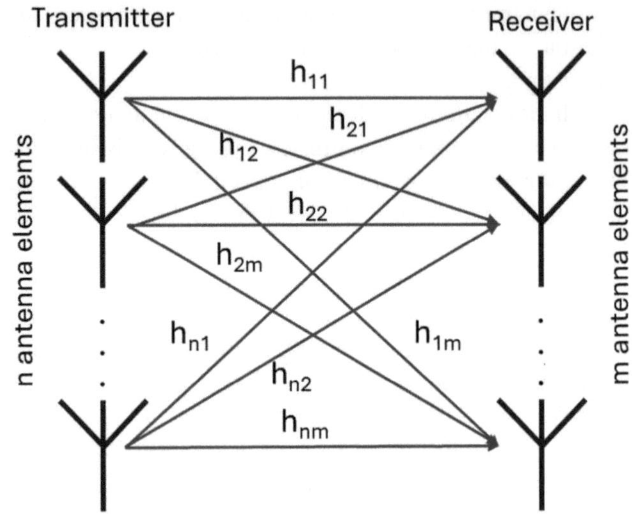

Fig. 5.14 MIMO channel schematic, with several antenna elements in the transmitter and several antenna elements in the receiver

Therefore, MIMO can convert something undesired and harmful for the capacity of the system (multipath and delay dispersion) into something advantageous.

So, let us explain further diversity and multiplexing. Diversity is about using more resources than strictly needed to deliver the information, to make transmission more robust against multipath. A good example from outside of the radio world: if instead of sending only one letter we deliver two copies of the same text in two different envelopes, the probability of the message arriving to destination is higher. But the complexity is also higher: both the transmitter and the receiver must deal with it. Back to radio communications, diversity can be temporal (sending the same information on two different time slots), frequency based (sending the same information on two different frequencies), polarization (sending the information on both vertical and horizontal polarization of the same link), or spatial (antenna elements within an array are separated at a certain distance, so the correlation between them is low). Typically, spatial diversity is the most attractive in terms of quality of service.

Regarding multiplexing, it consists of increasing the bitrate of the system sending independent data through each subchannel. These subchannels exist due to the different conditions among them (ideally, subchannels are independent). Therefore, multiplexing and diversity are opposite to each other, and it is not possible to achieve at the same time the maximum diversity gains and multiplexing gain [120].

There are two types of multiantenna systems: Based on diversity, like MIMO, where the signal that arrives to the receiver is not correlated, and beamforming based, where the signal is correlated. Beamforming is a technique that modifies the radiation pattern of the array to fulfill a certain requirement. For example, suppressing a certain direction from which an interferent signal arrives to the receiver or maximize the gain for a direction of interest (where a group of users are located, typically). The control of the radiation pattern is done by signal processing techniques which modify the input for each individual antenna element within the array.

MIMO systems are based on diversity, and we can find four potential scenarios:

1. MIMO (multiple-input multiple-output): That is, both the transmitter and the receiver have antenna arrays with more than one element. This means diversity can be present at both ends.
2. MISO (multiple-input single-output): Only the transmitter has antenna arrays with more than one element. The receiver has only one antenna.
3. SIMO (single-input multiple-output): Only the receiver has antenna arrays with more than one element. The transmitter has only one antenna.
4. SISO (single-input single-output): Neither the transmitter nor the receiver has more than one antenna element. There is no diversity here, and SISO systems are the classical communications links.

On a MIMO system, we have the advantages of a SIMO and a MISO system, that is, diversity in both the transmitter and the receiver. If we have, say, N_t antenna elements in the transmitter and N_r in the receiver, a priori, we can have $N_t \cdot N_r$ subchannels between the transmitter and the receiver. We say "a priori" because to have real (that is, useful) subchannels, they must be as independent as possible.

Typically, MIMO channels are formulated differently than SISO channels, but SISO systems can be seen as a particular case of MIMO systems (with $N_t = N_r = 1$). Therefore, the channel model for a MIMO system can be formulated as follows:

$$Y(t) = H(t) * s(t) + n(t)$$

where $H(t)$ is the channel matrix (size $N_t \cdot N_r$), $s(t)$ is the transmitted signal, $n(t)$ is the noise, and $y(t)$ is the received signal. On a SISO system, $H(t)$ is defined as $h(t)$, a scalar, instead of a matrix. The capacity of a system like this can be formulated according to Shannon [121], as the maximum bitrate that can be transmitted through the channel with an arbitrary low BER [122]:

$$C_{MIMO} = B \cdot \log_2 \left[\det \left(I_M + \frac{SNR}{N_t} HH^H \right) \right]$$

And for a SISO system, we have:

$$C_{SISO} = B \cdot \log_2 \left[1 + SNR |h|^2 \right]$$

where the capacity C is measured in bps (bits per second), B is for the bandwidth, N_t is the number of transmitting antennas, SNR is the signal-to-noise ratio, H is the channel matrix of size $N_r x N_t$, and I_M is the identity matrix of size $N_r \, x \, N_r$.

5.4.3 Wayside Antennas

Wayside antennas can be the same as in non-railway scenarios, but in certain cases, some differences may appear. For instance, in railways, it may be enough to have two sectors to cover the (linear) tracks, instead of having the typical three 120 degrees sectors.

In regional and high-speed lines, the antennas for the base stations are typically located on the top of tall towers. The height and location of the antenna must be carefully chosen based on the environment and the outcome of the channel models. Typically, towers are 20 or 30 m high, or even as high as 40 m, in some extraordinary cases. Depending on the height, the foundations of the tower must be calculated. Moreover, these towers must be able to withstand winds up to 150–200 km/h [123].

A common setup to provide connectivity inside short tunnels (less than 1–2 km long) is to point antennas (typically, Yagi antennas) external to the tunnel to the inside. If the length is higher, there are three ways to provide coverage:

Signal repeaters: The maximum distance is limited by the maximum admissible delay. They are easy to install, make use of directive antennas, and are usually connected by optic fiber links. However, the result is far from a uniform coverage inside the tunnel.

Leaky coaxial (LCX) is a particular type of antenna which is a coaxial cable, with periodic slots on its surface that radiate the signal into the air. Typically, have high coupling losses (60–90 dB), that is, outside of the LCX, but low propagation losses inside the LCX (1–3 dB/100 m, but these values are highly dependent on the type of cable, frequency, etc.). They have high acquisition and installation costs, but low maintenance figures. The coverage is uniform and can be easily reused for more than one radio carrier, using combiners to multiplex more than one signal into the same cable. Signal repeaters and leaky coaxial are not mutually exclusive.

Use normal base stations equipped with their own antennas inside the tunnel.

5.4.4 Onboard Antennas

On the other side, antennas must be installed onboard trains. The typical challenges here are three: first, lack of space, as there is almost no available space onboard trains, so placing antennas is difficult. Moreover, the cables to the radio transceiver must be as short as possible to avoid unnecessary attenuation. Second, gauge limitations. As we explained in Chap. 1, trains are largely limited by the maximum gauge. Therefore, antennas must be kept within the limitations derived from that maximum gauge, which forbids the installation of large antennas on the top of the train. Finally, aerodynamics has impact on the onboard (and wayside) antennas. At high speeds, the aerodynamic drag can be very relevant and antenna radomes can be torn off if not correctly bolted down. Therefore, antennas for railway vehicles to be installed on the top of the train usually have a streamlined design.

For T2G systems, onboard antennas usually have an omnidirectional pattern to be able to receive incoming signals no matter which direction they come from. On the other hand, satellite antennas may need some kind of mechanical steering and can be very large.

Another relevant aspect is the electromagnetic compatibility (EMC). This is explained in Chap. 6.

Regarding the testing, both wayside and onboard antennas are tested in terms of intermodulation, isolation, and VSWR. Depending on the specific case, some other tests may apply. For example, GSM-R equipment (not only antennas) is tested according to the EN 301515 standard [124].

Mobile Communications Generations

Since the 1980s, we have had the following generations of mobile communications systems:

1G: Analog systems, only for voice calls. Examples are Nordic Mobile Telephone (NMT), Advanced Mobile Phone System (AMPS), and Total Access Communication System (TACS). Very low data rates. Represented in railways by analog T2G systems.

2G: First digital generation and worldwide popular (1G had only marginal use, and none of its systems saw universal use, only regional), for both voice and data communications. 2G systems still have low data rates (around some kbps). The most relevant exponent of this generation is Global System for Mobile Communications (GSM), which is the basis for the GSM for Railways, a.k.a., Global System for Mobile Communications - Railway (GSM-R) (explained below).

3G: Universal Mobile Telecommunications System (UMTS) and Code Division Multiple Access 2000 (CDMA 2000) are the most relevant systems within this generation. This generation was the first wideband and enabled the introduction of more devices to mobile networks. It had no strict-sense railway use.

4G: LTE is the absolute protagonist of this generation. It is a full-IP technology, intended to last more time than its predecessors. This technology is explained later in this chapter because it is a relevant T2G for noncritical services.

5G: The fifth generation is an evolution of LTE which addresses more use cases, like massive machine-type mobile communications (mMTC), ultra-reliable low-latency communications (ULLC), and some others. It is also defined for high-frequency bands, which imply Gbps data rates. It has railway significance because FRMCS, the successor of GSM-R, is a 5G technology. More details of both 5G and FRMCS can be found in Chap. 8.

The industry and researchers are already working on the sixth one (6G). It will address many topics and use cases; among them is the Joint Communications and Sensing (JCAS), which could be useful for transportation systems like railways. Time will tell.

5.5 Radio Communications and Mobile Communications Systems in Railways

Once we have a sound knowledge of the underlying physics of radio communications in railway scenarios, in the following sections, we explain some of the most important train-to-ground communications systems that are employed in this field, like GSM-R and WiFi, among others.

5.5.1 GSM-R

Global System for Mobile Communications-Railway (GSM-R) is truly the most relevant train-to-ground system in railways so far. The history of this system talks about an absolute success and has starred (alongside with ETCS, the other half of ERTMS) in the communications chapter of high-speed railways and others.

It has been present since the very early days of this century and has represented a huge standardization in train-to-ground communications and has helped to do the same in a more important subject: signaling. In Europe, it is the standard T2G solution for high-speed lines but has been installed in regional and commuter lines as well. Outside Europe, it has also been widely deployed in countries like China, Australia, the United Kingdom (before and after Brexit), and Saudi Arabia.

GSM-R is, as its name suggests, an extension of GSM (from the second-generation mobile) for railway use. The idea was launched in 1993 by the International Union of Railways (UIC), following the same premises for ETCS: replacing all the national systems by an interoperable European alternative. Several parties were involved in GSM-R standardization, like UIC, European Telecommunications Standards Institute (ETSI), and European Union Agency for Railways (ERA). At that time, there were only two mobile communications systems that could be considered: 3GPP GSM and TETRA. The former was chosen probably for economic considerations, but not for technical reasons, as both systems are very similar[4] in terms of performance. Notice that both of them are narrowband communications systems.

GSM-R market is much larger than national T2G systems, but it is still a niche technology, not able to benefit from large economies of scale like GSM, LTE, WiFi, and many other mobile communications systems. Indeed, GSM-R (and more generally, ERTMS) is regarded as an expensive technology. Future evolutions may solve this issue, as can be seen in Chap. 8 of this book.

The purpose for GSM-R was twofold: voice and data communications, that is, both operational voice calls and data for ETCS L2 (and L3). It was also required to enable some railway-specific requirements which have strong implications in the T2G system.

Going back to the standard, two projects were launched: Mobile Radio for Railway Networks in Europe (MORANE) and European Integrated Railway Radio Enhanced Network (EIRENE).

MORANE (Mobile Radio for Railway Networks in Europe) [125] is about validation of GSM-R. Therefore, in this project, the main role is that of vendors, operators, infrastructure managers (IMs), and laboratories. Some of the most relevant outcomes of MORANE are:

- The sensitivity of the onboard radios shall be no worse than −102 dBm (−107 dBm for the base stations). Some IM usually require even better sensitivities.
- EIRP of the onboard radio: 39 dBm (45 dBm for the base stations).
- Received power (for planning): higher than −85 dBm (−70 dBm in tunnels).
- As usual in radio communications systems, requirements are given for a percentage closer to 100%, but not 100%. In this case, it is requested 95% of the time.

EIRENE (European Integrated Railway Radio Enhanced Network), a UIC project, is about the requirements for the GSM-based digital radio. These requirements are divided into FRS (functional) [126] and SRS (system level) [127].

Both EIRENE and MORANE results and specifications system are today part of the European TSI.

Moving into functional issues, GSM-R allows railway agents the following applications:

- ERTMS/ETCS data transmission for automatic train control.
- Voice calls. Often for communications between the OCC and the train driver, but many other alternative scenarios can be set up as well.
- Group calls.
- Broadcast calls.
- Emergency calls.
- Functional and regional addressing of calls.
- Train diagnosis data transmission.
- Half-duplex and full-duplex data communications.
- Caller identification.
- SMS (short message service), including predefined messages. To acknowledge the reception of SMS from the OCC, the radio cabin sends back a predefined message.
- Call hold and call waiting.
- Calls between drivers within the same train, which is not the same as direct mode, as these calls are routed through the network. This is necessary for those trains that do not have a wired line for audio calls (interphone), such as some old trains when they are coupled.
- Direct mode communications between terminals. This is an optional feature not implemented yet, but which can be very useful in some degraded modes.
- Push-to-talk. Starting calls with only pushing a button.
- Setting of different call priorities, such as "information," "operation," "emergency," etc. Depending on the priority, some of these calls are automatically answered, without human intervention.
- Voice communications among the crew of the same train (train engineer, conductors, catering staff, etc.).
- Inbound/outbound calls from out-of-the-railway-network numbers.
- Shunting, that is, operations related to starting the rail traffic, fixing failures, etc.
- There are many other minor requirements, like volume control, screen brightness adjustment, ergonomics, etc.

[4]Or perhaps even better, according to some tests done by Teltronic in Kazakhstan some years ago [239] and by Siemens in an undisclosed location [240]. This is very unlikely to be a real alternative for GSM-R, but it could be in those places where GSM-R is not mandatory (this is, some countries outside Europe). Anyway, the evolution of T2G systems is not related to these discussions.

5.5.1.1 GSM-R Architecture

The architecture of a GSM-R network is, like many other mobile technologies, composed of a core and a radio access network (RAN), plus the user equipment (UE) or terminals, and some other entities for management (Fig. 5.15). More, on a GSM-R network, we find several entities (see Fig. 5.16 for the complete architecture).

Base transceiver station (BTS): the base station itself, which interfaces with UE through the air (this is defined as the U_m interface) and with the BSC (A_{bis} interface). BTS can have several architectures, one of the most common of which is a separation between baseband units (BBU) and remote radio units (RRU), which act as the RF front end. The connection of BTS to the BSC usually follows an (optical fiber) ring topology, to leverage the risk of outage just in case a single wire fails. However, other topologies like star are also possible but are far less common than rings.

Base station controller (BSC): The main task of the BSC is to control BTS traffic and perform all the radio resource management (RRM) tasks. BSC control several BTS and interface with the Transcoder and Rate Adaptation Unit (TRAU). As in regular GSM networks, the protocol to interface with BTS is Link Access Protocol for D-channel (LAPD) and SS7 with the TRAU [128]. Both BTSs and BSC form the BSS (base station subsystem).

Transcoder and rate adaptation unit (TRAU): This entity adapts the 64 kbps channels from the core to the 16 kbps that the BTS can handle.

Core: Formed by several entities, like:

(a) Mobile Switching Center (MSC): Is the central unit of the GSM-R core, responsible of call-control, call switching, interconnection with other networks, handover management, etc.

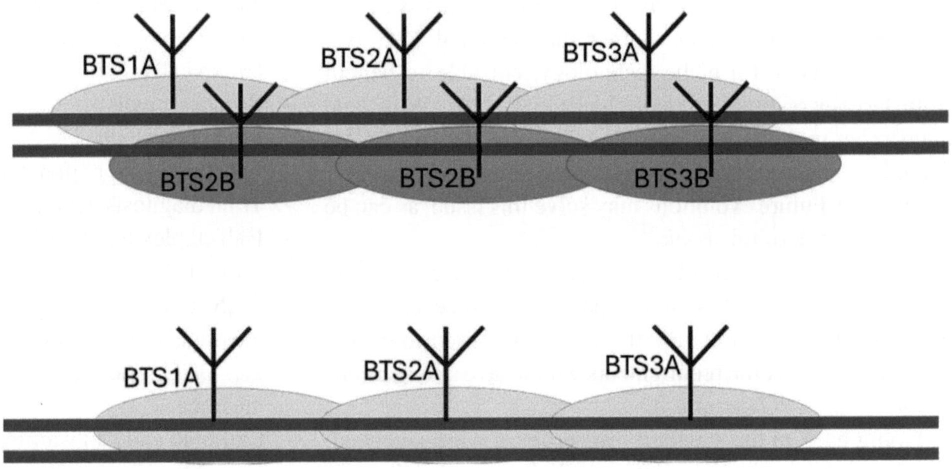

Fig. 5.15 Two-layer (top) and single-layer (bottom) cellular deployment for GSM-R coverage on a railway line. Blue and green layers behave independently, providing redundancy in case a single cell in one layer fails

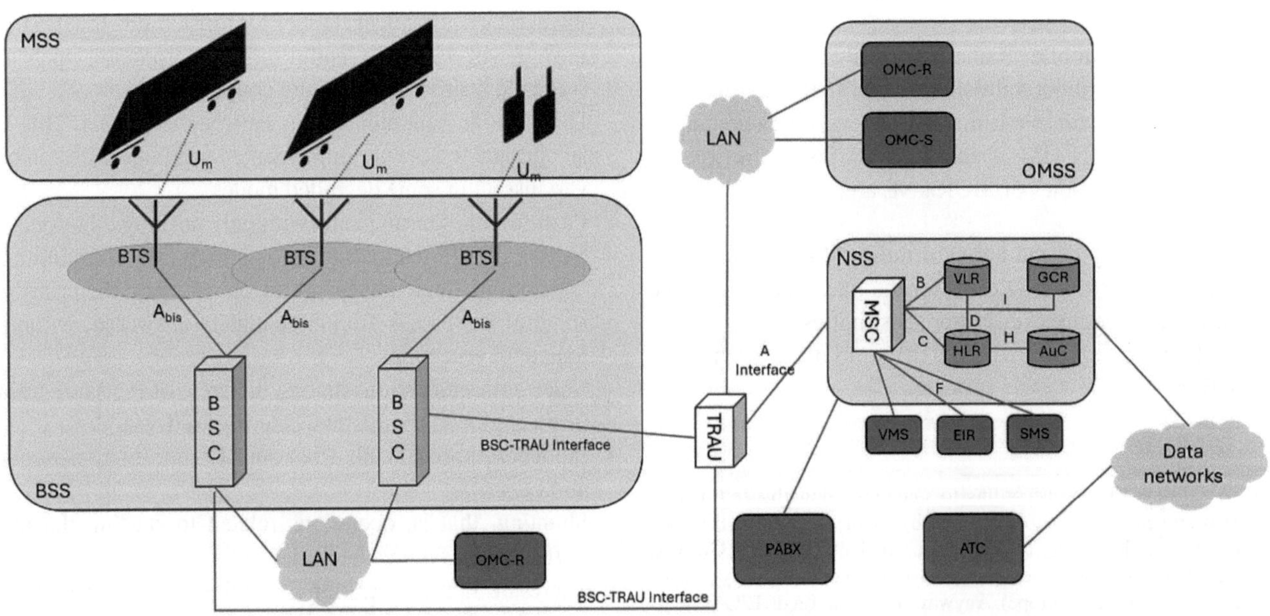

Fig. 5.16 GSM-R architecture. Adapted from

(b) Home Location Register (HLR): Database for the number of each MS registered in the network. Same as in GSM.
(c) Visitor Location Register (VLR): Same as HLR but for visitor MS (that is, roaming MS coming from other networks). Same as in GSM.
(d) Group Call Register (GCR): When a group call is defined, it is stored in this database. Notice that these groups can change depending on daily operational needs, that is, on the assignment of a train to a line or its removal from commercial service.
(e) Equipment Identity Register (EIR): This is the International Mobile Equipment Identity (IMEI) database, for every device registered in the network. That is, its role is more about not allowing blacklisted MS (stolen devices, among others) to register in the network. Same role as in GSM.
(f) Authentication Center (AuC): Database for authentication and ciphering of the data for each MS. In the SIM of each MS, there is a private key which is also stored in the AuC. Therefore, the role of the AuC is related to key management and encryption. Same role as in GSM.

User equipment (UE), or more precisely, mobile stations (MS): end user terminals, both handheld and onboard units (cab radios).

Operation and Management Subsystem (OMSS): For the management and supervision of the GSM-R network.

Transport network: It is a high-capacity network (1.2 Gbps), fully redundant, which provides redundant connectivity (two connections for every site). Typically, this network is formed by fiber loops of 565 Mbps every 100 km.

5.5.1.2 Handovers in GSM-R

Regarding mobility, handovers are much more frequent on a GSM-R network than on a regular PLMN (GSM, LTE, etc.) because onboard radios can travel at 300 km/h or faster. Therefore, the system must be designed accordingly. A very convenient (but costly) solution is oversize the cell overlap, up to 500–600 m, which will be covered by the train running at 300 km/h in 7 s. The maximum probability of a failed handover is 0.05%.

For availability purposes, another usual implementation of GSM-R is a dual-coverage, fully redundant deployment, that is, two independent layers of base stations, each one of them with its own optic fiber rings, different power supply, etc., and 100% overlapping. The topology is depicted in Fig. 5.15, with layer A base stations depicted in red and layer B in blue. This fully redundant deployment is at least twice more expensive than the single-layer coverage of PLMN deployments. However, the resulting availability is very high, because there are two independent base stations serving every single point of the tracks. Notice that this approach is protecting much more than a single point of failure: it covers the failure of a whole GSM-R layer of coverage.

Both layers of base stations shall be capable of handling the global traffic of the line, that is, to ensure no degradation should one of the layers fail.

Another interesting handover scenario is between two different networks (i.e., two countries). There are at least two different situations here:

1. When the train goes from one region to the other, the automatic network reselection must be activated in the cab radio of the train. To ensure no service disruption, an overlap zone between both networks is required.
2. When the train goes very close to another region but does not cross into it.[5] In this case, the MS must not change the network, so to avoid this change the automatic network reselection must be deactivated.

The reselection function can be triggered by either the train driver or automatically by reading a balise in the track. For more details on this, see [129].

5.5.1.3 GSM-R Radio Access Network

In this section, we cover the most relevant features of the GSM-R RAN. Many of them come from its parent technology (GSM), while some others do not. First, it is a narrowband technology, with a channel bandwidth of 200 kHz. As in GSM, the media access technology is Time Division Multiple Access (TDMA). This means that, within the time-frequency matrix resource for the radio channel, all the bandwidth (200 kHz) is allocated for a single user, but only for one time slot. In the next time slot, channel access will be granted to another user. In its frequency counterpart (FDMA), the division is in the frequency domain. There are many other ways to multiplex the channel, like CDMA, Orthogonal Frequency Division Multiple Access (OFDMA), etc.

The TDMA frame has eight slots to allocate voice and data traffic. Each of these slots lasts 0.577 ms, including guard intervals. Therefore, the eight-slot TDMA frame has a duration of 4.616 ms. As usual in TDMA, synchronization data is needed to manage the multiple access to the channel. In this case, the higher the number of traffic channels, the more synchronization channels are needed: for 7 data channels 1 synchronization channel is needed. For 14, two are needed.

The modulation used in GSM-R is Gaussian minimum shift keying (GMSK), the same as in GSM. It is a low-order modulation with one bit per symbol.

For the frequency planning, to avoid co-channel interference, cells (i.e., BTS) using the same channel must have the

[5] In Europe this can happen frequently, like on the East and West shores of the Rhine River.

Table 5.5 Reusage pattern for GSM-R channels

Pair of channels	Channel
A	955
B	957
C	959
D	961
E	963
F	965
G	967
H	969
I	971

largest possible separation. Let's see what this means: for either the downlink or the uplink, we have 4 MHz, but we must leave a guard band of 100 kHz at each side, which leaves only 3.8 MHz. If the defined channel bandwidth is 200 kHz, this implies that we have only 19 channels to plan our coverage across the tracks. Then we need to know how to allocate these channels to each BTS, that is, the conversion between channel ID and frequency. This is done using the following two equations:

$$f_{ul}(n) = 876.2 + 0.2 \cdot (n - 955), 955 < n < 974$$

$$f_{dl}(n) = 45 + f_{ul}(n), 955 < n < 974$$

(Both equations are valid only for the European GSM-R band).

That is, the 19 channels are defined from 955 to 974, and each one of them is related to a central frequency of a 200 kHz channel that can be obtained using both formulae (one for the DL and another for the UL). Regarding the pattern for consecutive cells, it is repeated after nine cells. This is shown in Table 5.5. Only the primary channel is depicted in the table, but there is also a backup channel which is the next one for each case (i.e., 958 is the backup for 957, which is the B pair).

As usual in many mobile communications systems, the signal bandwidth is higher than the nominal channelization (200 kHz in this case); therefore, some adjacent channel interference is generated.

Interference

In every mobile communications system, interference levels must be carefully considered. There are two types of interference:

- Co-channel interference: This is the interference generated by another transmitter which is working in the same frequency. Given that channel frequencies will be reused sooner or later, the objective here is to separate these two transmitters working on the same frequency as much as possible.
- Adjacent channel interference: In real-world transmissions, the actual bandwidth is not limited to the nominal one, which means that some power is radiated outside of the defined band (therefore interfering neighboring channels). This type of interference can come either from other channels of the system (i.e., other GSM-R BTS) or from other communications systems (i.e., in the GSM case, Public Land Mobile Network (PLMN) base stations). This is very frequent: frequencies near the GSM-R bands are used for PLNM, which sometimes do not have the required filters to avoid interfering GSM-R.

Interference from other sources rather than the communications systems (i.e., Bluetooth devices in 2.4 GHz IEEE 802.11) can be considered either co-channel or adjacent channel interferences, depending on the frequency.

Regarding the BTS, the antennas are usually placed in towers, very near to the track at a height of 20 m (sometimes up to 40 m). Given that the track is linear, so is the needed coverage. Therefore, instead of the common PLMN, three 120° sectors for each BTS, in GSM-R, we have two sectors with a narrow 30° horizontal beam (in some scenarios like curves, the width is 60°). Typically, BTS that belong to the same layer are separated 5–6 km, but this strongly depends on the environment (and BTS from redundant layers are separated half that distance.). The equalizer can work up to 500 km/h, which is an improvement from 120 km/h in GSM.

5.5.1.4 GSM-R Drawbacks and Limitations

Even though GSM-R has been a success, it has several limitations that cause many problems for both IMs and operators for many reasons, the most important being that it has a very limited number of channels (19), which is not enough for some scenarios, like shunting yards, depots, and large stations. This was partially solved in some countries which allocated 3 extra MHz.

Moreover, GSM-R, like GSM, is a circuit-switched technology, which implies that the path (i.e., the circuit) between two communication ends is assigned only to them. This is a solution suitable for voice communications but not for data and leads to very low usage rates (less than 1% on average).

Indeed, it has a low tolerance to interference coming from neighboring PLMN base stations. On the other hand, there are many cybersecurity vulnerabilities that have not been fixed (so far).

Another important drawback is to deal with the costs: As GSM-R is a niche technology, it is expensive. Moreover, GSM-R is not IP based, which makes more expensive any system integration that must be carried out.

Something very relevant for the sector in general terms is the fact that it will be supported no longer than 2030. This is perhaps the most important drawback of GSM-R.

For all these reasons, there is a replacement coming, called Future Railways Mobile Communications System (FRMCS), which is described in Chap. 8, as is the migration problem from GSM-R.

5.5.2 IEEE 802.11 (WiFi)

IEEE 802.11 is the official name for a worldwide famous technology: WiFi. In this text, we use both terms plus WLAN (wireless LAN). In railways, WiFi is mostly for the following applications:

1. Train-to-ground system in metros and, to a lesser extent, tramways. Not only for Communications-Based Train Control (CBTC) but also as a general-purpose T2G technology. Remember that the spectrum allocation for general purpose T2G (that is, not Communications-Based Train Control CBTC and not GSM-R) is very scarce.
2. To provide connectivity onboard trains for passengers and railwaymen.
3. The same in stations and other railway facilities (such as depots, etc.). An advantage of 802.11 technology in this scenario is to use real-time location services (RTLS) capabilities and to provide related services to passengers, like pedestrian routes and other useful information.
4. As a high-capacity train-to-ground system, to offload massive data files to the ground. This kind of system is usually installed at depots or terminal stations, to allow fast transmission when the train is stopped.

But before we start talking about technical details of WiFi, we make two remarks: WiFi is (most of the time) very quick and easy to deploy. In simple and small setups like homes, small offices, etc., only a WiFi router and the backhaul are needed. There is no need at all for network planning, core management, troubleshooting, etc. Second, this is not true at all for large deployments, not to mention ultra-dense environments (explained later in this section). In railway scenarios, WiFi deployments are very often large networks, with many access points (AP), which is how base stations are called in this standard. Therefore, let us not make the mistake of thinking that WiFi is a commodity technology.

5.5.2.1 Spectrum for IEEE 802.11

Unlike GSM-R or 3GPP LTE, IEEE 802.11 works on three unlicensed spectrum bands:

1. 2.400–2.483 GHz: In 2.4 GHz, the spectrum is divided into 14 overlapping 20 MHz-wide channels. Actually, we can also define this in terms of nonoverlapping channels, but the number decreases to 3. Fig. 5.17 depicts how all these channels are defined. The usual way to do the frequency planning of a WiFi network is to allocate only nonoverlapping channels, like numbers 1, 6, and 11, leaving the other channels for special scenarios.

   This band is also called ISM, for industrial, scientific, and medical, that is, use cases outside of the communications world. See "unlicensed spectrum" box for more details.
2. 5 GHz bands: Indeed, this band is not only one but four (UNII-1, UNII-2, UNII-2-extended, and UNIII-3). In Fig. 5.18, the details can be seen, where 20 MHz, 40 MHz, 80 MHz, and even 160 MHz channels are defined. Moreover, the number of available channels is much larger than 14. These 5 GHz bands are as follows:
   (a) 5150–5350 MHz
   (b) 5470–5875 MHz

   These bands are much less crowded than the 2.4 GHz one (as of 2023), but the frequency is more than twice higher which means free-space attenuations four times larger. This fact has a strong impact on planning.
3. More recently, the 6 GHz band has been allocated for this use. The so-called WiFi 6E (more details on the "WiFi evolution" subsection in this chapter) enabled 1 more GHz from 5.925 GHz up to 7.125 GHz: Dealing with wireless communications systems requires knowledge of spectrum regulations.

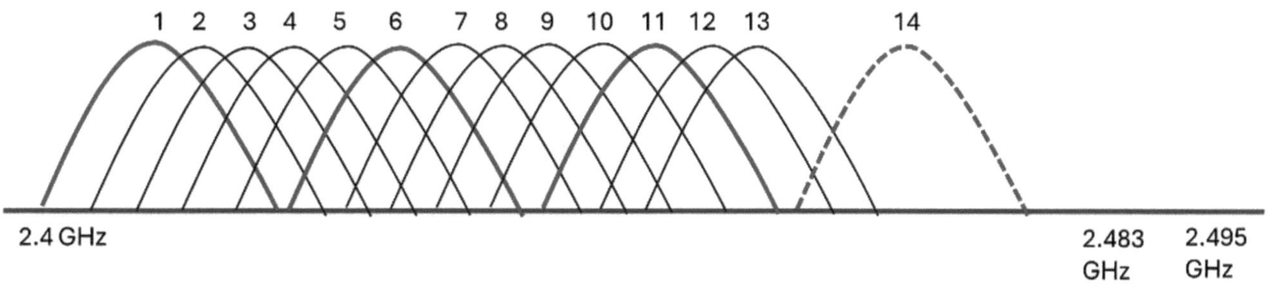

Fig. 5.17 WiFi channels in the 2.4 GHz band. Channels 1, 6, and 11 have been highlighted to emphasize that for the first 13 channels, only three are nonoverlapping. Channel 14 is not worldwide available

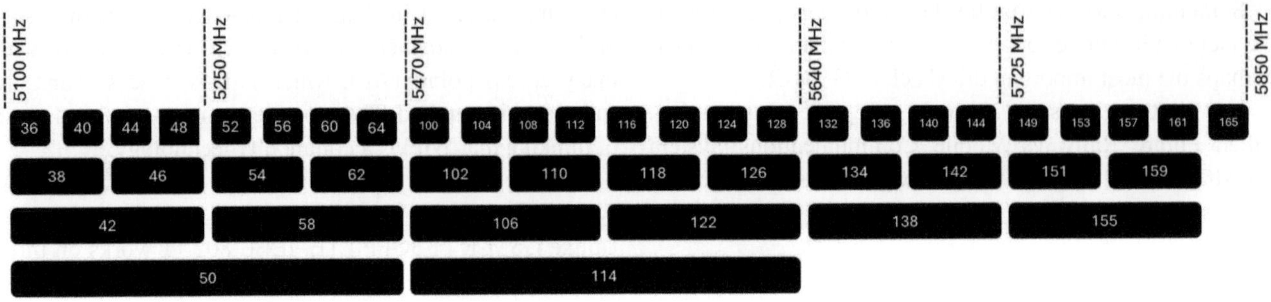

Fig. 5.18 WiFi channels defined in the 5 GHz band

Another important issue is that there are subtle differences across countries in spectrum for WiFi. That is, the channels that can be used in a particular country or region differ based on the regulations.[6] That is why in the initial WiFi setup, it is required to introduce the country, to avoid illegal transmissions. This will prove more relevant when we introduce radars and dynamic frequency selection (DFS) in the 5 GHz band.

5.5.2.2 Dynamic Frequency Selection

There is a part of the 5 GHz spectrum which is not allocated primarily for WiFi but for radars. See Fig. 5.18 to identify those channels in the 5 GHz band that might interfere radars. These radars can be of many types, like weather or military, so a specific method must be implemented in WiFi access points to avoid this interference. This is called DFS and is a function that enables WLANs to use 5 GHz frequencies that are generally reserved for radars. The procedure is as follows: WiFi APs detect when they are interfering nearby radars, and then move to a different channel. This shall be done when the AP boots, so if the AP detects that a radar is using a particular DFS channel, the AP will exclude that channel from the list of available channels. This state will last for 30 min so when this time expires, the AP will check again. This task can take up to 10 min (although the normal time is much shorter). This is why DFS channels are not immediately available when an AP boots. This is what DFS is about and not implementing it in 5 GHz APs is illegal and can be grounds for fines.

Another important point is that DFS can force channels to change (in case a radar is detected), which can have impact on network performance. For all these reasons, some engineers do not use DFS channels on their deployments, whether there are radars in the vicinity or not. Therefore, DFS should be considered in the design phase.

[6]For example, in the USA, there is more spectrum allocated for unlicensed use than in Europe.

5.5.2.3 IEEE 802.11 Architectures

There are three main elements in a WiFi network: first, access points (AP): network entities which have at least one radio module allow wireless connections from clients and are connected to the wired network (through a switch or a router). Second, all the clients, that is, the end devices that are connected to the APs of the WiFi network. Finally, wireless LAN controllers (WLC) are the entities that manage the wireless LAN (i.e., the network).

The most important functions of WLCs are authentication and security, that is, both encryption and client registration in the network; QoS control, mobility management, i.e., to ease roaming between APs. Remember that the term "roaming" in WiFi is (more or less) the same as "handover" in 3GPP mobile communication systems. Remember that "roaming" in 3GPP notation is about being a visitor on a different network than your home network.

More WLC functions: Radio resource management (RRM), that is, AP channel assignment, transmission power, and load balancing: forcing clients to change AP if their home AP is overloaded; configuration management, such as firmware updates, cybersecurity functions, like detection of rogue APs, and real-time monitoring and analytics reporting.

Compared with some other wireless technologies, IEEE 802.11 is very flexible as it can be deployed in several different ways, such as:

- WLC based: This alternative is the most centralized one. A WLC is needed. This is very common in large deployments, to ease management and help them to scale.
- Autonomous AP: In this topology, APs assume functions from the WLC (which may be no longer needed). This is common in small deployments to save costs. WLCs are expensive and take time from the O&M teams, so they are not justified in small deployments. Centralized management pays off when there are many APs and clients in the network, not when there are few APs.
- Direct mode (ad hoc): A direct connection between clients without an AP.

– Mesh: On a mesh network, APs are interconnected, instead of being independently connected to the network. The resulting network can be more resilient against failures and be deployed more flexibly (for example, to avoid wiring in some difficult scenarios).
– Workgroup bridge (WGB): In this topology, a device (the WGB) connects a wired LAN to a wireless AP. That is, the WGB acts as a collective wireless client for the wired network. This is very useful for moving networks or vehicular scenarios, for example, to connect an onboard LAN to a WLAN infrastructure on the ground. It is possible to have a car or a train with many onboard devices and only one connection to the ground. This technology is not standard but rather a Cisco extension of the standard. Figure 5.19 represents a typical network topology of a WGB.

Step-by-Step Guide to Choose a Suitable WiFi AP for My Network

1. Make sure your AP supports your design. For example, not all APs can work on mesh networks, or they may have only one radio module (instead of two) or they may not support WGB, etc.
2. Consider the WiFi generation: 802.11n, ac, WiFi 6, etc. Sometimes are backward compatible, but some others are not. Some of them work only in 2.4 GHz (legacy APs) and some others only in 5 GHz (like 802.11 ac).
3. Consider carefully operation and maintenance (O&M) issues. Whether the management system is CLI based or not, cybersecurity management, etc. This will impact largely the needed skills for your maintenance team, training plans, etc.
4. Price (as usual).
5. Size to install them in your railway facilities and trains.
6. External or internal antennas. Having external antennas can improve the quality of the signal, but this also means that one needs to maintain antennas and coaxial wires (antennas can be stolen or vandalized more easily).
7. Obsolescence and evolution. If the network is a large one, it is very likely to be upgraded in several phases so a plan is needed to do this efficiently, which requires advance knowledge of when the APs will be no longer supported by the vendor.
8. Is this AP already in the library of your favorite troubleshooting software? This allows to save a lot of time dealing with it. Note that having a suitable tool for both design and troubleshooting is pretty much needed. More details about this are in the further reading section.
9. Is this AP suitable for the environment? This means outdoor/indoor, certified for railway environments, fire protection, etc. Checking the nominal MTBF for a device can help to figure out the needed resources to maintain a WiFi network (and of any other kind, for that matter).

To do this, the technical details of the APs must be retrieved and also opinions from other sources must be checked. But one will often have almost no choice and have to deal with a particular brand because a WLC from a vendor only works with APs from that very vendor.

5.5.2.4 Mobility and Handovers in IEEE 802.11

Like in every mobile communications system, when the quality on a cell degrades, the client may switch to another one. This is called roaming in IEEE 802.11 (and handover in 3GPP), but the procedure to decide when to move to another cell is not defined by the 802.11 standard. Typically, it is based on signal strength measurements but could be trig-

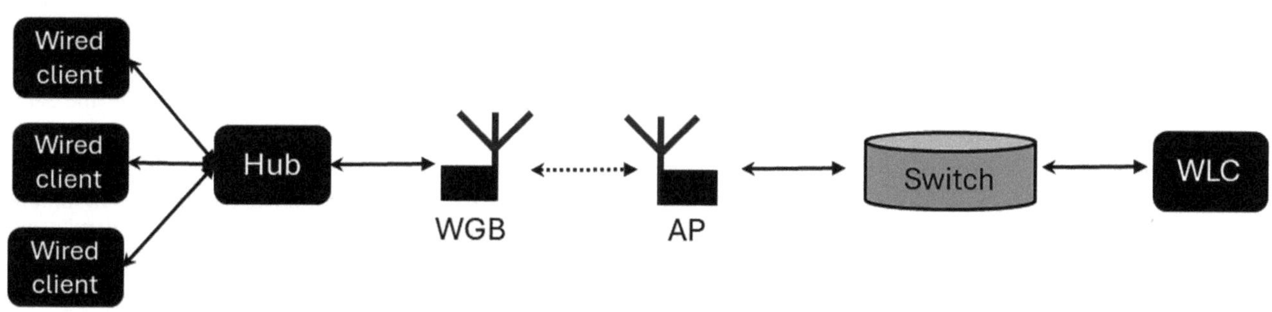

Fig. 5.19 Workgroup bridge mode for IEEE 802.11. Several clients (on the left) are wirelessly connected to the network through the WGB-Access point link. Adapted from

gered by other parameters like bit error rate (BER), SNR, or even network load. Roaming has a large impact on network performance, with latency-sensitive services like voice calls being especially vulnerable.

The basics are very similar to any other wireless system, but there is an issue worth mentioning. It is called the "sticky client problem" and comes from the origins of 802.11 as a nomadic technology (instead of a pure mobile one). WiFi is a very resilient technology because it works even in bad channel conditions (i.e., SNR close or even below 0 dB). Therefore, it is very hard to have a client detach from the AP even if the radio link is very degraded. This is both good and bad news because clients are very conservative, that is, they tend to stay connected to an AP even if the connection is poor and, what is worse, even if there is a better AP available. This is the "sticky client problem" because the client gets stuck to the AP. This is a problem for both the client and the network, because the client experiences low QoS conditions, and the network allocates resources for a client that does not transmit and receive data.

The solution for this is very easy to say: Let the AP force the disconnection of clients with poor connection. But there are two main problems here:

What to do if there are no other APs available?

Real-time and latency-sensitive services can have their performance severely degraded in case the AP disconnects the client and the client needs to begin the discovery procedure of a new one.

Unlike GSM-R, WiFi is not a technology designed for railway use. Indeed, in the 802.11 amendments, the number of mentions to railways is zero. But this does not mean that it is not suitable for this environment. More in particular, in the early days of the standard, it was not even intended for mobility (only for nomadic users, that is, to provide wireless access to laptops in offices and homes, for people that move from one room to another, at walking speeds). But very soon, some amendments came to the rescue: more in particular, three 802.11 amendments are essential to understand mobility for WiFi:

- IEEE 802.11r: fast roaming using pre-authentication, that is, pre-saving client credentials to accelerate roaming. Authentication is done before the client moves to the new AP, by client to AP or even AP-AP forwarding of the credentials.
- IEEE 802.11 k: An AP knows its neighbors. When a client requests a list of AP candidates for roaming, such a list is sent by the AP the client is connected to. This procedure saves both time and energy.
- IEEE 802.11v: Eases transition toward the best AP (that is, a faster trigger of the roaming procedure). Allows detaching clients from overloaded APs. Also allows detaching clients which are unfit or slow or which fall below a given signal strength threshold.

5.5.2.5 IEEE 802.11 Radio Access

The physical layer for 802.11 is not homogeneous among all the generations of the standard (b, g, n, ac, etc.) but since 802.11 g, orthogonal frequency division multiplexing (OFDM) is the media access scheme. This technique is very popular in broadband mobile communications systems like LTE and 5G, so we will stop to examine how OFDM works. The main idea is to divide the bandwidth into narrowband subcarriers, which are orthogonal and behave as parallel data streams. Each subcarrier is modulated by symbol of a specific modulation, like QPSK, 16-QAM, etc. Therefore, OFDM can be seen as the aggregation of several narrowband subcarriers that form a single wideband carrier. A guard interval between symbols is introduced to remove (or attenuate) interference between consecutive symbols. This interference is called ISI, and in strict sense, it is a distortion rather than an interference. Multipath can lead to ISI as well.

Multipath has two undesired effects over an OFDM signal: inter-symbol interference (ISI) and inter-carrier interference (ICI), which removes, albeit not necessarily in a complete way, the subcarrier orthogonality. Cyclic prefixes (CP) come to the rescue for both ISI and ICI. CP is based on prefixing each symbol sent with the end of the same symbol, thus allowing the receiver to differentiate a symbol from the next in an environment where copies of the same symbol may be received at different times because of multipath propagation. Therefore, the CP extend the duration of the signal but remove the undesired effects mentioned before. Moreover, it is a simple technique that makes time-domain equalizers unnecessary at the expense of reducing transmission efficiency. This is a remarkable thing, because UMTS, from the previous mobile generation, required equalizers which mostly raised costs. Notice that only equalizers in the time domain are removed, but not the frequency-domain ones, which are very much needed in OFDM.

An important parameter is the subcarrier spacing, usually denoted as Δf. In general terms for WiFi, it is 312.5 kHz, but recent generations of the standard lower this value to 78.5 kHz. The narrower, the more subcarriers in the same bandwidth, which means higher data rates (if the symbol rate does not change), but a drawback is that the interference between subcarriers can be higher (especially in mobile channels, where the Doppler shift is present). As a reference, in 3GPP LTE, this value is 15 kHz.

In 802.11, depending on the WiFi generation, we have a different number of subcarriers for data, pilots (for channel estimation), and nulls. Each 20 MHz channel is composed of 64 subcarriers, spaced 312.5 kHz. 802.11 a/g use 48 for data,

4 for pilots, and 12 as null subcarriers. 802.11 n and ac use 52 subcarriers for data, 4 for pilot, and 8 as null subcarriers.

Regarding the data rate, it is proportional to:

- Number of data subcarriers in the defined bandwidth (20, 40, 80, or 160 MHz, depending on the 802.11 generation).
- Number of bits per symbol for each subcarrier: i.e., 2 bits for QPSK, 4 bits for 16-QAM, 8 bits for 256-QAM, etc.
- The number of transmitted symbols per second.
- The channel coding.

This is straightforward, but it is not always "the bigger the better," because wider channels also imply higher noise power (indeed, doubling the bandwidth, for example, from 20 to 40 MHz means +3 dB of extra noise power) plus twice the number of available subcarriers to transmit data. This is called "channel bonding" and is possible in WiFi networks since its introduction in IEEE 802.11n. It is considered a good policy to avoid using 80 and 160 MHz channels, on every deployment (not only in railways). Perhaps in railways, it could be justified for some particular use cases, where interference is nothing to worry about, like one-cell deployments for fast data transmissions in depots.

5.5.2.6 Some Considerations About Interference in IEEE 802.11

Interference is often a key issue in mobile networks. Given that WiFi works only in unlicensed bands, one can take for granted a high interference level (both co-channel and adjacent channel). Therefore, 802.11 networks must be designed around this fact. Indeed, RF engineers identify interference as the main "capacity killer" in WiFi. It is very common to fix a degraded 802.11 network by turning off an AP that interferes with the others.

Typically, there are many other interference sources that may affect your 802.11 deployment, not only neighboring APs. For example, in homes, microwave ovens are the usual suspect (they work on 2.4 GHz too), as are Bluetooth devices (something very relevant when dealing with large crowds), etc. A good policy for WiFi maintenance is to schedule periodic site surveys to detect problems as soon as possible.

5.5.2.7 IEEE 802.11 Evolution

Since the very beginning of WiFi, the technology has been in constant evolution. Amendments and new versions of the standard were released every year (sometimes even more often than yearly!). Older versions (802.11a, b and g) of the standard have been obsolete for a long time, so we don't cover them here. We start with 802.11n which is still used in many places.

IEEE 802.11n was ratified in 2009, so the first commercial products came out around 1 or 1.5 years after. It is the first time the standard defined two bands: both 2.4 and 5 GHz. It is backward compatible with 802.11b and g. But 802.11n also introduced other novel concepts in WiFi, like MIMO, beamforming, spatial multiplexing, and channel bonding. The nominal data rate that could be achieved is 633 Mbps, but in practice, this value is significantly lower.

IEEE 802.11 ac (also called WiFi 5, but the "ac" brand is much more used), released in 2013, brought a significant step forward in several aspects, but one was very relevant: the decision to forget the 2.4 GHz band. This band was very crowded, and the assumption was to consider it the legacy band for legacy devices. The future would be 5 GHz only. Multiuser MIMO (MU-MIMO) was introduced, plus channel bonding to form 80 and 160 MHz channels, 256-QAM modulation and more spatial streams than in 802.11n. This means a nominal data rate higher than 3 Gbps (only in ideal conditions, but still a remarkable improvement over 802.11n).

WiFi 6 (or 802.11ax), released in 2019, introduced many interesting new features, like high order modulations, like an impressive 1024-QAM (10 bits per symbol!). It is unrealistic in 99.9% of the time and scenarios, but nominal data rates are above 9 Gbps.

It is greener than former generations, by the implementation of the target wake time (TWT) mechanism, which allows devices to determine when to wake up to transmit and receive data. This is very important for battery-powered devices which are no longer forced to send and transmit data constantly. Moreover, the access to media is much more deterministic than in former WiFi generations, and the spectral efficiency is improved. Actually, TWT was first introduced in WiFi HaLow 2 years before WiFi 6, but the impact on wide-sense IoT networks was really felt with the advent of WiFi 6.

OFDMA, which means allocating sets of subcarriers to different users, instead of giving the whole channel to a single user. This is very relevant in crowded scenarios, with a large number of devices contending for air time. In previous versions of the standard, all subcarriers are assigned to the same device each time. Therefore, OFDMA allows a more efficient resource allocation, with less fragmentation.

Another useful feature for dense environments is basic service set (BSS) coloring, which allows the identification of data packets sent from different networks even if they work in the same channel.

A larger guard interval is more suitable for outdoor scenarios (where the delay times are longer than in indoor scenarios). In WiFi 6, it is possible to have more subcarriers in the same bandwidth than in former generations. This can be achieved by decreasing the subcarrier spacing. In WiFi 6, this value is 78.125 kHz instead of 312.5 kHz.

The 2.4 GHz band is no longer regarded as a thing from the past, so WiFi 6 is defined in both 2.4 and 5 GHz bands.

An intermediate step between WiFi 6 and 7 is WiFi 6E. The most important variation for this technology is the band where it is defined: 5945–6425 MHz in the EU, but this is tricky because in Europe, this band has been allocated for unlicensed use, but vendors do not manufacture devices working in this band yet, and 5945–7125 MHz in the USA. Notice that in the USA, the band is significantly larger.

WiFi 7 standard has not been released at the time of this writing, so it is briefly explained in Chap. 8, like other technologies that will impact the mid-term future of railways.

IEEE Standardization

There is no technical need to understand the "behind-the-scenes" activities of the IEEE 802.11 committees, but it may be worth spending some time there, so we leave a couple of references for this.

https://www.ieee802.org/11/Reports/802.11_Timelines.htm

https://www.ieee802.org/11/

There are three main bodies involved: Wifi Alliance: in charge of product certification, that is, "seals of approval"; IEEE: creates standards, designing protocols and all the related documentation and CE and CEPT (European Conference of Postal and Telecommunications, that is, for Europe only) and its counterparts in other regions to set local regulations.

5.5.2.8 Very High-Density Environments (VHD)

WiFi was not designed for large crowds, but the need to provide connectivity in such a scenario is very frequent. Sometimes this comes from mobile operators who are unable to offer service to large crowds of people (even less in the UL). The challenge is remarkable, because in large gatherings PLMN very often do not work properly (at least, up to 4G networks, but much less for 5G), so WiFi networks are usually the only connectivity alternative. This problem will be fixed in 5G deployments, while 2G through 4G are usually inadequate.

Let us begin with a definition: We face VHD conditions if we have more than 100 clients in an area of 250 m2. There are some classical examples, such as stadiums, rallies, subway stations, large squares (like St. Pietro in the Vatican City, where over one million people can gather), and even airports.

VHD environments are a special scenario where the design, to work properly, should not only follow the same rules, but also consider some additional aspects. In railways, VHD may apply in some stations and onboard metro trains.

There are some considerations for the design of a VHD WiFi network:

Divide and conquer: That is, plan the coverage as an ensemble of several small cells, planned to allocate no more than 25–35 clients.

Only use 5 GHz channels, because many channels are needed, and interference levels would otherwise skyrocket. This means that the 2.4 GHz band is not to be considered at all in the design.

Use special VHD APs instead of regular ones. These APs can have more active clients and registered clients than the normal ones.

Human bodies attenuate the signal. The impact of one or two people is negligible but having ten or more people between your device and the AP may imply up to 20–25 dB of extra attenuation. There are many channel models for this, but the extra attenuation can be up to 5–10 dB per human body.

AP installation is a very sensitive issue for two reasons: the number of APs to be installed and the propagation impact. For example, in stadiums, it is common to install APs underneath the seats.

The key concept to understand VHD is the "collision domain" where all clients compete for the "airtime." The more clients, the more likely to have collisions, and the more clients to share the resources of the AP. That is why the most important premise in VHD is to design for small cells. But this is very expensive.

Carry out site surveys before the installation (design phase) and after. This is good advice in every situation, but even more here. There are professional tools which are very helpful in the design phase and also for troubleshooting. Indeed, in WiFi, it is strongly recommended to do periodical site surveys to prevent degradation of the overall quality of the network.

Pay attention to VLAN sizing: Avoid having a VLAN at 99% of its capacity and others at 10%. This is a relevant point in every WiFi deployment, but especially so in VHD networks with many connected clients.

Avoid bottlenecks in the backhaul, by estimating the traffic figures for that part of the path. Very often, the bottleneck is not on the AP side but on the network side.

For more details about VHD scenarios, there are guides for VHD design edited by Aruba (now part of HP), which are more than recommendable [132, 133].

Real-Time Location Services (RTLS)

The most common use case for a WiFi network is connectivity, but sometimes it is requested to locate people or assets in real time. Given that GNSS does not work properly in indoor areas (like large stations), and PLMN-based location is not accurate enough for some applications, setting up RTLS over a WiFi network could help. Shopping malls, railway stations, and airports are common scenarios for RTLS.

There are two scenarios:

1. It is needed to retrieve analytics about which areas or places are more crowded, so the WiFi beacons sent by the client (received by the APs) can be used to do these estimations. It is not needed to have the client connected to the AP (if this is the case, even more data can be gathered from it, but it is no longer needed). A centralized platform is needed to put everything together and generate the aggregated analytics.
2. In case the objective is to help passengers to navigate inside a station and arrive to their destination platform, locate themselves and wait by the train door nearest to their seats. This can be done using an app installed in the passenger device, which receives WiFi beacons from the APs distributed across the station and estimates the location of the device, providing the user extra information about the route to be followed.

A final remark: The more accuracy is needed, the more specific RTLS requirements must be considered, which means that sometimes it is necessary to design the WiFi network specifically for RTLS, but it can be used for connectivity as well.

5.5.3 Tetra

Terrestrial Trunked Radio (TETRA) is a European standard [134] formulated by ETSI, to be used by law enforcement, government agencies, civil protection, first responders, railway applications, etc. It is a very robust technology and is the most used professional mobile radio (PMR) system in the entire world.

Allows point-to-point and point-to-multipoint communications, for both data and voice. It is a narrowband technology since channels have a bandwidth of only 25 kHz. It is usually allocated in the 380-470 MHz band, which makes it more efficient than GSM-R (which works in the 900 MHz band), needing fewer base stations for the same coverage area. Moreover, TETRA requires 25 kHz for four channels and GSM-R 200 kHz for eight channels [135].

TETRA can work in three modes: direct mode or walkie-talkie mode, where the infrastructure is not needed; normal mode, i.e., using the infrastructure and gateway mode, where a terminal can act as a relay for other terminals not connected to the network, therefore extending the coverage.

It may be viewed as the GSM-R for subways, sharing many characteristics but not the same, since TETRA does not carry critical data (i.e., signaling). Otherwise, TETRA has the same limitations as GSM-R.

The typical railway-specific applications (group calls, functional addressing, geographical addressing, push-to-talk, etc.) are supported by TETRA.

5.5.4 3GPP LTE (Long-Term Evolution)

3GPP LTE is relevant in railways for two reasons: first, it is widely used in its PLMN form in several railway use cases, mostly for operational services, like digital train, remote maintenance, CCTV, etc.; second, there are many LTE private networks deployed in railways. There is also an intermediate solution between these two, which are private access point names (APNs). Private APNs are, in the strict sense, PLMNs, not a private LTE network. In this section, we introduce all of them, but first we explain LTE technical details.

5.5.4.1 LTE Architecture

LTE is the most relevant fourth generation mobile communications technology. Its main feature is the flexibility it offers on many different levels, for example, its architecture (see Fig. 5.20) which is deliberately simple and flat. Compare it with GSM-R: in LTE, there are far fewer entities, and the functionalities are more grouped, which means less equipment to maintain and less latencies. We see that, as usual, there are two main networks:

- RAN, called E-UTRAN (Evolved Universal Terrestrial Radio Access Network), which includes MS, base stations (here called extended Node B, or eNB) and the related interfaces.
- Core or Extended Packet Core (EPC).

In the first one, we find the eNB (which concentrates all the functions that belong to the RAN domain) and the defined interfaces with MS, core, and other eNBs: U_u, S1, and X2, respectively (see Fig. 5.20). More precisely, the S1 interface is twofold: S1-U for the user plane and S1-MME for the control plane. eNBs perform critical tasks such as radio admission control, mobility management, and scheduling. Let's get into more detail.

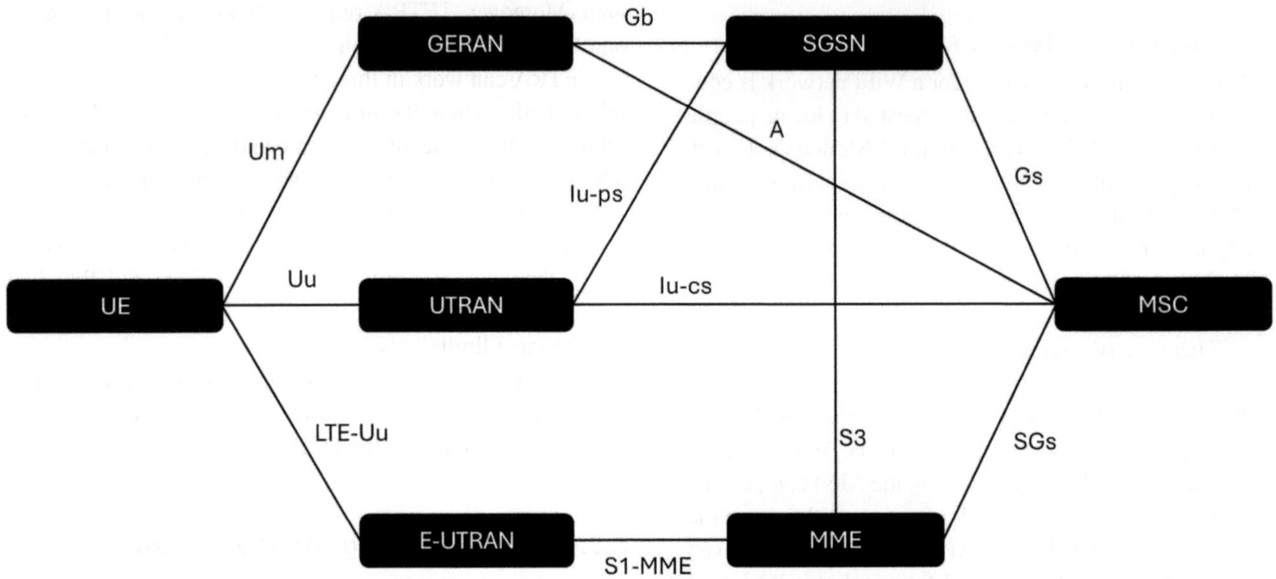

Fig. 5.20 LTE architecture with all the interfaces between entities

Radio admission control (RAC): This function is about accepting or not a new UEUser equipment (UE) in the network. This has a strong impact on network performance and security as well.

Mobility management includes handovers, cell selection and reselection. It is an essential aspect in mobile networks, for obvious reasons. For example, handover has a strong impact on the overall performance and can affect the availability of the communications services.

Scheduling, is the allocation of user data requirements in the time-frequency grid, plus setting the modulation and coding scheme to achieve efficient transmission and reception of the data. This is a very important task, even more so if the 3GPP does not standardize algorithms for any of these functions.

Interference control is to allow more efficient transmissions in the cell edge (which is the area most vulnerable to interference). Within a cell, the eNB keeps different users from transmitting at the same time on the same frequency, but the problem comes when they are in different cells. Therefore, ICIC enables a coordination technique between neighboring cells that can allow them to use the same channels. When a UE detects high-interference conditions, it raises a flag to let eNBs know. Then, eNBs implement the coordination in terms of transmission power control and an alignment in the transmission (to avoid simultaneous use of those subcarriers that were subjected to heavy interference). Notice that this alignment happens only in the frequency domain, because doing this in the time domain could imply loss of eNB-UE synchronization. This approach is a disruption from previous mobile generations, where the available spectrum was divided into neighboring cells (therefore, decreasing the overall capacity), and channel assignment was carefully planned to minimize interference.

Channel measurement and estimation is done to allocate resources efficiently and maximize the overall performance of the system.

Plus, the interfacing already mentioned (X1, S1, and U_u).

Finally, an eNB can be connected to more than one EPC and one or more cells can be managed. All radio related protocols terminate in the eNB.

We cannot state that LTE RAN is an evolution of Universal Mobile Telecommunications System (UMTS), a 3G mobile communications standard), as they have many differences from each other, but the LTE core is an evolution of General Packet Radio Service (GPRS), 2.5 generation). The main task for the core is to provide IP connectivity and to connect the LTE network to other networks (both IP and non-IP). Here we find three entities [136]:

Mobility management entity (MME): It offers control-plane functions, which include authentication and security, bearer and QoS management, mobility management of idle users (i.e., paging), and intra-LTE mobility management (that is, handovers between eNBs). It is also responsible for the network attachment of new users in the network. Bear in mind that idle users are non-active users, which still need to be located to send them data or transfer calls to them if needed. This management of idle users is called paging.

Packet Data Network Gateway (P-GW): This is the gateway between the LTE network and other packet-based networks, like the Internet or private networks. Among its functions are billing (for data usage), IP address, and QoS management. An MS can be connected to one or more than one P-GW.

Serving Gateway (S-GW): The gateway between the RAN and the core, with functions like IP routing, packet buffering when the MS is out of coverage, and it enforces MS connectivity during handovers. An MS only can be connected to one S-GW at a time. LTE handovers, unlike UMTS, are "hard handovers." That is, when moving to a different cell, the connection with the cell-to-be-left is broken, and then the connection with the new cell is established. This procedure is called "break-before-make." There is a lot of literature on handovers in LTE (see Further Reading section).

5.5.4.2 Multiantenna Transmission in LTE

Since the very first LTE release (release 8, December 2008), LTE included support for multiantenna technologies, like MIMO (both SU-MIMO and MU-MIMO, that is, single user and multi-user MIMO, respectively) and beamforming (release 12 (March 2015) addressed massive MIMO as well). All of them are much needed to achieve the desired data rates and, more importantly, the desired spectrum efficiency (measured in bps/Hz, which is an essential KPI in mobile and wireless communications systems). The trend regarding multiantenna technologies is to manufacture larger and larger arrays.

We do not explain any further these multiantenna technologies as more details can be found on a previous section of this chapter.

5.5.4.3 Multiple Access Technologies

The multiple access technology for LTE is OFDMA, based on OFDM, and the same one we explained in the IEEE 802.11 section. We have explained OFDM in a previous section of this chapter, so here we address only the differences between 3GPP LTE and IEEE 802.11 in this aspect.

OFDMA is the first of these differences: OFDMA is used in LTE since its first version, in both links. Strictly speaking, we can say that multiple access is only used in the UL, and DL transmissions are multiplexed instead. Therefore, one could say that OFDM is used in the DL and OFDMA in the UL. On the other hand, more generally, we may say that OFDMA is used in both links. IEEE 802.11 did the same starting with WiFi 6. This is a recent development, not yet widely deployed in railways, but it will be. OFDMA allows the partial allocation of subcarriers to different users, instead of assigning all of them to the same user. In Fig. 5.21, we can see a visual representation of this.

The advantages of OFDMA include multiuser diversity, robustness against multipath, and resource allocation granularity, all of which imply a higher spectral efficiency. As for disadvantages, perhaps the most important one comes from amplifier nonlinearities. In OFDM, the peak-to-average power ratio (PAPR) can be high, so amplifiers need to work with a wide range of input amplitudes. Otherwise, the signal can be distorted by the nonlinearity of the amplifier, which will amplify differently some subcarriers from the others. Another drawback is that OFDM is very sensitive to frequency errors.

High PAPR has a special impact on MS (UE in LTE), as they are much more power limited than BS (eNB). Hence, the decision is to use SC-FDMA instead of OFDM in the UL. SC-FDMA is an adaptation of OFDM that removes abrupt amplitude differences between neighboring subcarriers. This is achieved by a discrete Fourier transform (DFT) precoding of the signal, before the OFDM IDFT generation (see Fig. 5.22).

Another difference between multiple access in LTE and IEEE 802.11 is that in LTE the subcarrier spacing is 15 kHz, both for the UL and the DL. Therefore, as LTE channels are defined for several bandwidths (1.4, 3, 5, 10, 15, and 20 MHz), the number of subcarriers also varies (73, 181, 301, 601, 901, and 1201, respectively). The allowed modulation type for each subcarrier varies among 802.11 versions and LTE releases, namely QPSK, 16-QAM, and other higher-order QAM-based modulations (i.e., 64-QAM, 256-QAM, etc.).

Regarding the time-domain resource allocation in LTE, the main concept is the resource block (RB), which is the minimal resource that can be assigned by the eNB to a user. Indeed, it is a bidimensional resource, defined both in frequency and in time. In LTE, a RB lasts 0.5 ms, and its bandwidth is 180 kHz (that is, 6 or 7 OFDM symbols, depending on the CP duration) or 12 subcarriers (spaced 15 kHz). Typically, in large macrocells with corresponding longer delays, CP tends to be larger as well, so in this case we have six symbols only.

In FDD, we have 10 ms frames, divided into 0.5 ms slots, but the minimal number that can be transmitted is 2 (i.e., 1 ms). As said, each one of them is divided into six to seven

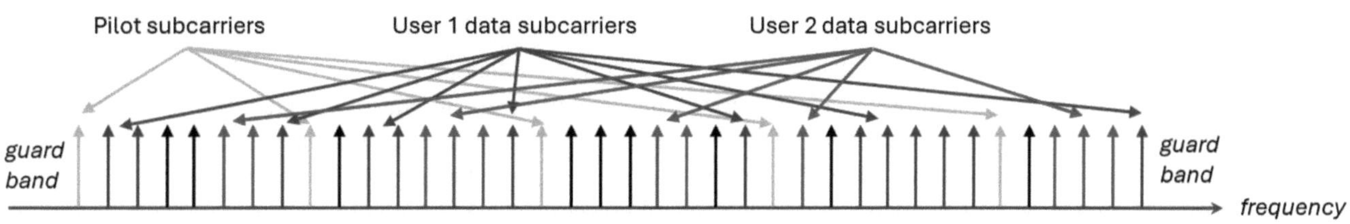

Fig. 5.21 Orthogonal Frequency Division Multiple Access (OFDMA) with subcarriers assigned to different users

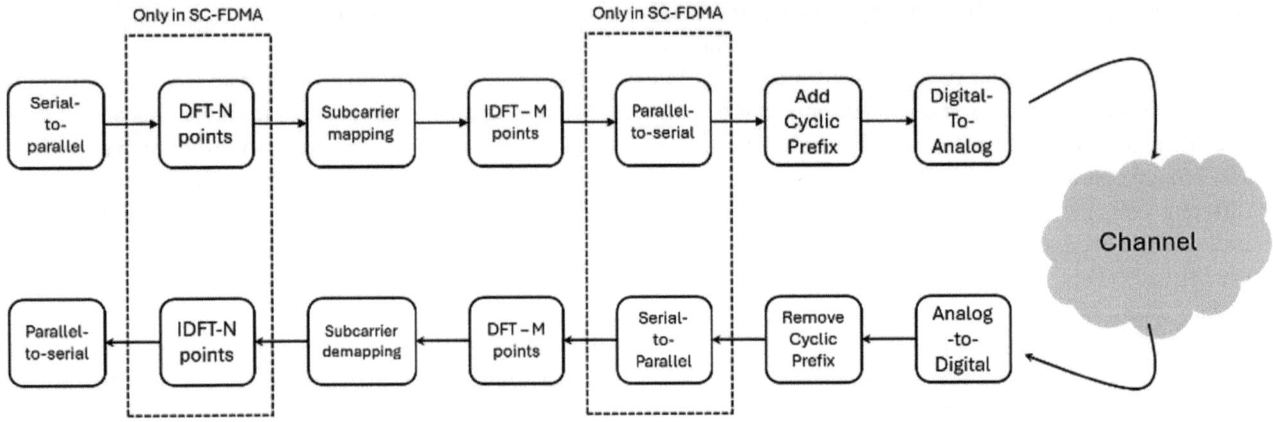

Fig. 5.22 SC-FDMA block diagram for the transmitter

symbols, each one of them with 66.7 μs length. The CP will take 16.67 μs if we have six symbols and 4.7 μs if we are on a seven-symbol configuration.

In TDD, we have 10 ms frames, too, but the minimal division is 1 ms (instead of 0.5 ms). These subframes can be either UL data, DL data, or pilots to sound the channel. In terms of spectrum usage, TDD is more flexible than FDD in that TDD allows for seven different frame configurations, each offering a different UL/DL split, whereas in FDD, the use of separate paired bands for UL and DL may be wasteful in the case of asymmetrical UL/DL requirements. Depending on your needs, which may change largely as not all railway operators have the same ones at the same time, a more efficient LTE system can be implemented.

Finally, some remarks about LTE: As explained, it is very flexible in many terms, like spectrum (is defined for both TDD and FDD on many bands), channel bandwidth, etc. Moreover, LTE has an outstanding spectral efficiency, much larger than 802.11. Please notice that a technology working in unlicensed bands is not comparable to another one that does not, as there are many differences between them (the most relevant one is the media access). Furthermore, there is LTE-U, where "U" stands for unlicensed, which is an adaptation of LTE technology to work in 5 GHz band (the same as 802.11). The performance of LTE-U is very similar to IEEE 802.11 in terms of spectral efficiency (that is, bits per second and Hertz).

LTE is applicable on private rollouts, not only PLMNs, but spectrum availability remains a problem in many countries, while others have allocated bands for private deployments. The need for a core makes LTE technology more expensive, but there are micro-core solutions that solve this. The 3GPP defined some use cases for railways, like the moving relay [137] but it received no attention from LTE vendors. 3GPP is standard, which means that the RAN and the core can be from a vendor, and all the onboard equipment from a different one.

5.5.4.4 LTE-r

LTE-R stands for LTE for railways, but it is not an international standard as GSM-R is. Indeed, it is an LTE-based, trunking system developed for the Chinese market by some of its flagship vendors, like Huawei and ZTE.

It includes several railway functionalities such as group calls, PTT, voice and video calls, and others like PIS or CCTV [138]. Some of these functionalities, like PTT or voice calls, are not supported by the 3GPP standardization of LTE.

Among its advantages, we find that it is closely related to future developments that will converge on the successor of GSM-R (see Chap. 8 for more details about this new technology).

5.5.5 Legacy Analog T2G

GSM-R was conceived to replace all legacy T2G systems in Europe with an interoperable solution. For example, before GSM-R, Germany had eight different T2G systems in operation. It is worthless to get into much details about legacy systems, so in this brief subsection, we explain only one example of a legacy T2G still in use in Spain, the so-called Tren-Tierra (literally, "train-to-ground" in Spanish). Some others across the world are similar and some others may be not, but the differences between more modern T2Gs are remarkable.

The "*Tren-Tierra*" is an analog system that has been in use since 1982, in the 447.400–448.65 and 457.4-458.6 MHz bands. For the BS, the EIRP is 20 W, according to the Spanish spectrum regulation. It is a narrowband technology (25 kHz) and is still in use in almost the complete Spanish rail network. It works mostly as a backup system for GSM-R, but in some lines, it is the only T2G available. It has a high degree of obsolescence, but it is not very costly to maintain for both the IM and operators.

The users of this system are OCC staff, train engineers, and other operational and maintenance staff.

The architecture is as follows [139]:

- A central radio system (PCR or "central radio system" in Spanish): Usually located in control centers, is the core of the "*Tren-Tierra*" system.
- Base stations (PFR, for "*Puesto Fijo de Radio*," or fixed radio site in Spanish), with antennas and radios distributed along the tracks.
- Mobile stations in train cabins (PMR, for "*Puestos Moviles de Radio*," which means mobile radio stations). The train desk has a PTT telephone plus a keyboard and a display, to show the train number, communications mode, system errors, and other relevant information. The keyboard is used to send text messages from the PMR to the PCR.
- Portable radios or walkie-talkies, for other railwaymen rather than train engineers.

Finally, regarding the functionalities, this system can provide voice communications, short, predefined messages, originated in the train or the OCC, group calls, and remote public announcement from the OCC to trains.

5.5.6 LoRa

5.5.6.1 Introduction to LoRa and LoRaWAN

We have seen some of the most relevant mobile and wireless communications systems in railways, so before we get to the end of this chapter, we must introduce one more family of technologies: short-range wireless communications (SRWC) systems. In this section, we explain the most relevant SRWC we have today. Satellite systems for railways are explained in Chap. 8, as they are not very common in this field yet (but they probably will be in the future).

LoRa stands for "long range" (nice name for a short-range technology), and, in the strict sense, it is a low-power wide area network (LPWAN), very much used for IoT applications, like sensor networks. It can work in two radically different modes: public and private networks. In the public one, typically, an operator provides the service to its customers, while the other one is a self-provisioned service over a privately held network.

As its name implies, LoRa has a range of several km but is mostly used in shorter distances. LoRa devices have very low power consumption figures, typically 25 mA in transmission mode and 100 nA in sleep mode. As we said before, it is meant for IoT use cases, so the needed data rates are consequently low, between 300 bps and 50 kbps. It is a narrowband technology, with bandwidths of 125 kHz, 250 kHz, and 500 kHz.

LoRa Alliance, a nonprofit organization, stands behind all the interoperability certification for all devices involved.

Regarding the protocol stack, we have LoRaWAN (long-range wide area network) and LoRa. The difference between them is straightforward: LoRaWAN defines the MAC and the network layer, over LoRa. Moreover, the system architecture is defined by LoraWAN, and LoRa defines the physical layer only. Therefore, it is completely agnostic on the upper layers of the stack.

5.5.6.2 LoRa Physical Layer

We introduce the LoRaWAN architecture later in this section. Let us address LoRa PHY first: It is based on spread spectrum, more on chirp spread spectrum (CSS), a spread spectrum technique that, unlike direct sequence spread spectrum (DSSS) or frequency-hopping spread spectrum (FHSS), has no pseudo-random elements.

Spread spectrum (SS) is a technique that increases the bandwidth of a signal by either multiplying it by a pseudo-noise sequence (case for DSSS, used in IEEE 802.11b) or rapidly changing the carrier frequency of the signal (FHSS, in Bluetooth). SS techniques provide more immunity against intentional interferences (i.e., "jamming") and provide a multiple access mechanism, giving each user a specific code to de-spread the received signal.

In CSS-modulated signals, the modulation technique consists of increasing/decreasing the frequency of a sinusoidal signal (the modulated signal here receives the name of "chirp") accordingly to the variations in the amplitude of the modulating signal. The result is a wideband signal. Chirps are present in some radar and sonar systems. CSS allows LoRa receivers to have a very high sensitivity (around −130 dBm or even −140 dBm) which is key to achieve long distances.

Regarding the spectrum, it is defined for different ISM bands, depending on the country or region: 863-870 MHz and 873 MHz in Europe, 915–928 MHz in South America and Asia, 902–928 MHz in the USA and Canada, and 2.4 GHz everywhere.

5.5.6.3 LoRaWAN Architecture

The typical architecture is depicted in Fig. 5.23, with three different entities: sensors (also called nodes), gateways (or concentrators), and the server.

We have the sensors connected to the gateways by the LoRa protocol. Notice that a sensor can be connected to more than one gateway, forming a star-like topology (which resembles more a "star of stars" topology). It is a robust architecture since multiple gateways receive the same packets from the sensors.

Gateways act as transparent bridges, converting RF LoRa frames to IP packets. Gateways also communicate with the server, via a TCP/IP backhaul (i.e., 4G, Ethernet, WiFi, etc.).

Server has two functions: network server, for data persistence, and the application server for data processing and display of results.

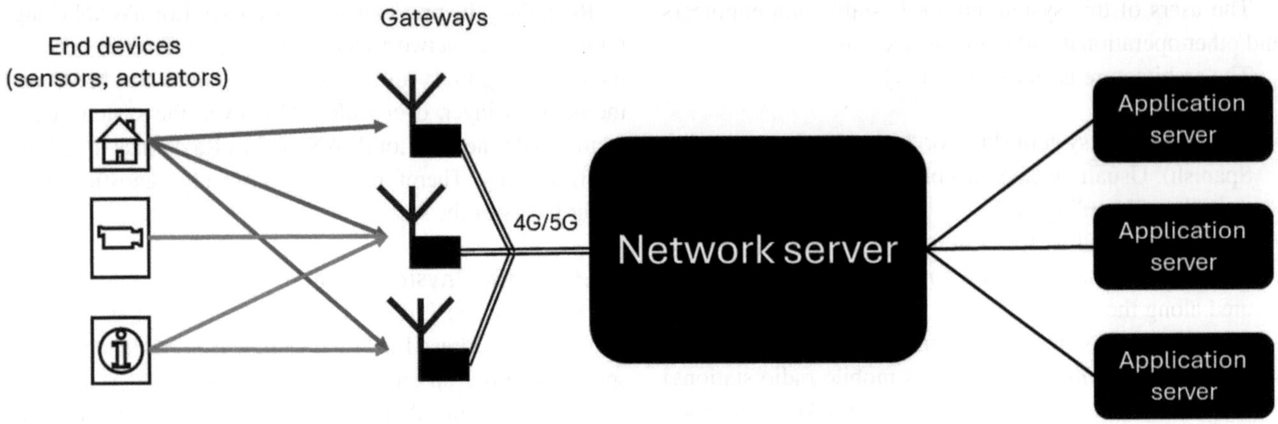

Fig. 5.23 LoraWAN architecture

Regarding the nodes, we find three types depending on the communication profile:

- Class A (or "Aloha"), which must be supported by all devices. It is meant for bidirectional communications. It is intended for battery-powered devices that spend most of the time in sleep mode, and only wake up from time to time to transmit and receive data. Therefore, the latency is very high, but they are very energy efficient. This class is suitable for actuators that have no stringent latency requirements or sensors which need to save power.
- Class B ("Beacon"): In this class, there are scheduled windows intervals for downlink reception of data. This implies a balance between latency and power consumption and is appropriate for battery-powered actuators.
- Class C is defined for sensors and actuators that are not powered by batteries, so devices always listen for downlink messages unless they are transmitting. This class is suitable mostly for actuators and can achieve very low latencies.

Finally, some key points for LoRa/LoRaWAN: perhaps the most advantageous feature is that it requires little or no planning to add nodes and gateways to the network. It is also a robust system since multiple copies of the sensor data can be received by different gateways. Moreover, LoRa/LoRaWAN is a low-cost technology due to hardware expenditures and unlicensed spectrum. It has good security features and is well established in the IoT market (with more than 10,000 private networks around the world). Therefore, LoRa/LoRaWAN is a suitable technology for low data rates use cases (like sensor networks), no matter the range is either short or long. Perhaps the only dark clouds on the horizon are the fact that 5G targets all Lora/LoraWAN use cases (more on this in Chap. 8).

Regarding sensor networks in railways, we find several examples of use cases, most of them related to maintenance and operation, but not only. Typically, the servers are in the public cloud (but can be on-premises as well). A non-exhaustive list of these railway-related applications for SRWC technologies like LoRa/LoRaWAN is the following:

- Monitoring electric currents in rail switch engines.
- CO_2 meters onboard trains.
- SynchronisingSynchronizing station clocks.
- Offload passenger counting data.
- Monitoring of spare parts inventory.
- Monitoring energy consumption in stations and trains.
- All of them have in common two things: need for ubiquitous connectivity and low data rates.

IEC 61375-2-6 Standard

There is a standard for train-to-ground communications: the IEC 61375-2-6 "Electronic railway equipment – Train communication network (TCN) – Part 2–6: On-board to ground communication" [140].

This document provides requirements for the selection of the T2G alternative depending on the QoS required by the application and to allow onboard applications, to have access to ground-based applications and vice versa. Cybersecurity aspects are addressed as well. This standard belongs to the train communications network series of standards, so the perspective is rolling stock centered.

It provides general requirements for both the onboard equipment (called Mobile Communications Gateway (MCG) and Ground Communications Gateway (GCG). Four different data classes are specified: process, message, stream data, and best effort data (that is, the same as for TCN networks. See Chap. 6 for more details).

Summary of the Chapter

This chapter has two differentiated parts: the first one is an introduction to the basics of radio communications services, including specific details about the mobile radio channel, which is essential to understand the complexity of radio communications. To design radio communications systems, it is not necessary to have a deep understanding of the physical-layer complexities, like statistical modeling or channel sounding, but it is very helpful when things get ugly. The same is true for antenna design, as many times is more than enough to find a suitable one looking at the data sheet only. Anyway, in this chapter, we have introduced many mobile radio issues which are relevant in railways (and many other scenarios) like waveguiding, diffraction, multipath, etc. The essential aspects of mobile communication-like cell, coverage, etc. are presented on a comprehensive manner.

The second part of the chapter is about real-world T2G systems like GSM-R (perhaps the most relevant today in railways), TETRA, WiFi, LTE, etc. Some of them are general-purpose systems like WiFi or LTE, whereas some others are railway specific (GSM-R). All have its mission or historic context. For example, GSM-R has played an essential role in railway interoperability since the early days of the twentieth century, but now it faces its end to be replaced by FRMCS.

Exercises

1. What is EM spectrum?
2. What is the radiation pattern of an antenna?
3. In which bands does GSM-R work?
4. How many different channel bandwidths are defined in IEEE 802.11?
5. What is the Doppler effect?
6. Calculate the maximum Doppler shift of a signal transmitted at 2.4 GHz from a train running at 300 km/h. Is this higher or lower than the subcarrier spacing in IEEE 802.11 or LTE?
7. What can happen if an onboard antenna is out of the gauge envelope defined for a train?
8. What is the directivity of an antenna?
9. What is the difference between a Rician and a Rayleigh fading?
10. What is LOS?
11. Do you think that the prevalence of LOS or NLOS conditions has an impact on the antennas to be placed onboard the train?
12. Explain the differences between MIMO, SISO, SIMO, and MISO.

Further Reading

There are plenty of books and papers for all the topics covered in this chapter. For antennas, the classic book by Prof. Balanis is essential [116] and for the mobile radio channel, there are some fundamental texts as well, like Molisch's [100].

Regarding the communications systems, for LTE, a very good reference is the book by Dahlman [136], and for WiFi, there are many excellent books but is more common to go to online resources from Cisco, Aruba [132], or Extreme. These guides are very interesting to read, and you'll learn a lot about WiFi. If the reader only has time to read one document from this topic, please go to Aruba's theory guide [132] or the engineering guide [133].

Another relevant way to go deep on a topic is to explore strategies and marketing issues (that is, apart from the technical part). This analysis is about how Cisco (a very large networking company) sees its competitors [141].

Sometimes, GSM-R gets severely interfered by PLMN. This is explained in this report by the ERA [142].

Communications Systems Onboard Trains

6

6.1 Train Communication Network

Networks onboard trains is a topic that is largely ignored by the literature but, unlike wayside networks, it is railway-specific and, more importantly, has experienced a very relevant evolution in recent years. Time will tell if wireless standards and time-sensitive networks will define the future of train networks. See Chap. 8 for more details.

This chapter is about several topics closely related to communications onboard trains:

- First, some definitions about trains, types of trains, etc.
- The most relevant one: train communication network Train Communication Network (TCN), for the train control and management, with all its evolutions and technologies.
- Hardware onboard trains must be designed considering some details of the environment, like vibrations.
- Apart from vibrations, one of the specific aspects to consider is the electromagnetic compatibility (EMC). We include a section in this chapter about this.
- The connectivity onboard trains is usually divided into two different layers: Train Control and Management System (TCMS) and Onboard Multimedia and Telematic Services (OMTS). In this chapter we explain how to deal with both as well as their differences.
- Finally, we explain some aspects like sensor networks onboard trains, some nonstandard technologies, train couplers, etc.

Regarding the references, the most relevant one in this topic is the IEC 61375 family of standards, which is the main source of the TCN part of this chapter.

Supplementary Information The online version contains supplementary material available at https://doi.org/10.1007/978-3-031-82247-6_6.

6.2 IEC 61375 Standard for TCN: Definitions and Types of Trains

Before we learn about the different technologies for TCN, we need to have a clear mind about some terms. All these definitions come from the IEC 61375-1 standard [143].

6.2.1 Types of Trains According to IEC 61375

First, the concept of train: Given that it can have many different forms and even change during the day, we introduce the term "composition," which is the "*number and characteristics of the vehicles forming a train*." The term "vehicle" is not addressed by the standard, but is the basic rolling stock unit, like a car or a wagon (remember that cars are for passengers and wagons are for freight).

The second more relevant definition is for the term "consist" (which is not very well known): "single vehicle or a group of vehicles which are not separated during normal operation, and which contains no, one, or several consist networks." Consists are also called train sets.

Therefore, the train composition of one or a set of consists, which can be operated as an autonomous unit, and at least one driver's cab. Trains can be categorized into open trains, closed trains, and multiple unit trains, as depicted in Fig. 6.1.

- Open train: "Train composed of one or a set of consists, where the configuration may change during operation, as, for instance, locomotive hauled International Union of Railways (Union internationale des chemins de fer, (UIC)) trains." In Fig. 6.1, the open train is formed by a six-unit train.
- Closed train: "Train composed of one or a set of consists, where the composition does not change during normal operation, for instance, metro, suburban train, or high-speed train units." In Fig. 6.1, the depicted closed train is formed by a six-car train.

J. Moreno García-Loygorri, *Telecommunication Systems for Modern Railways*, Textbooks in Telecommunication Engineering,
https://doi.org/10.1007/978-3-031-82247-6_6

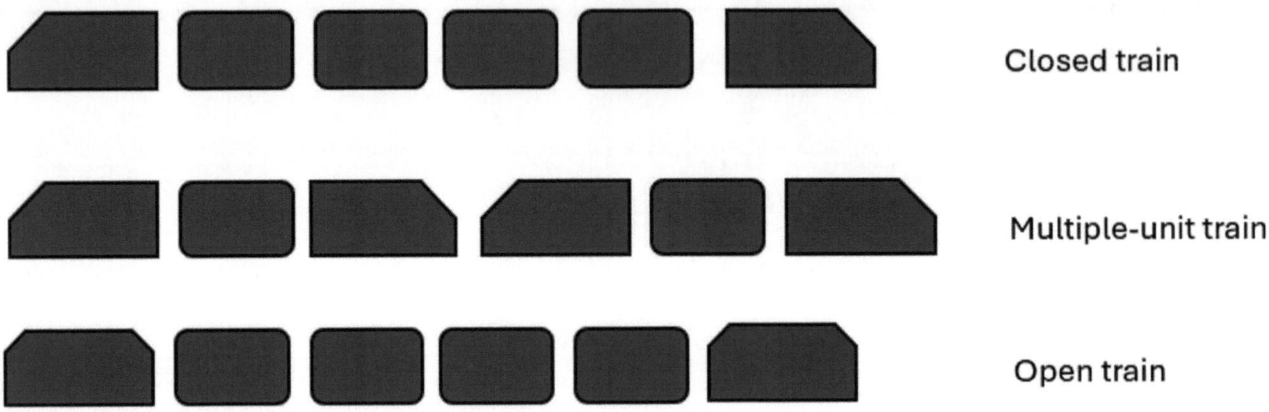

Fig. 6.1 Closed train, multiple-unit train, and open train

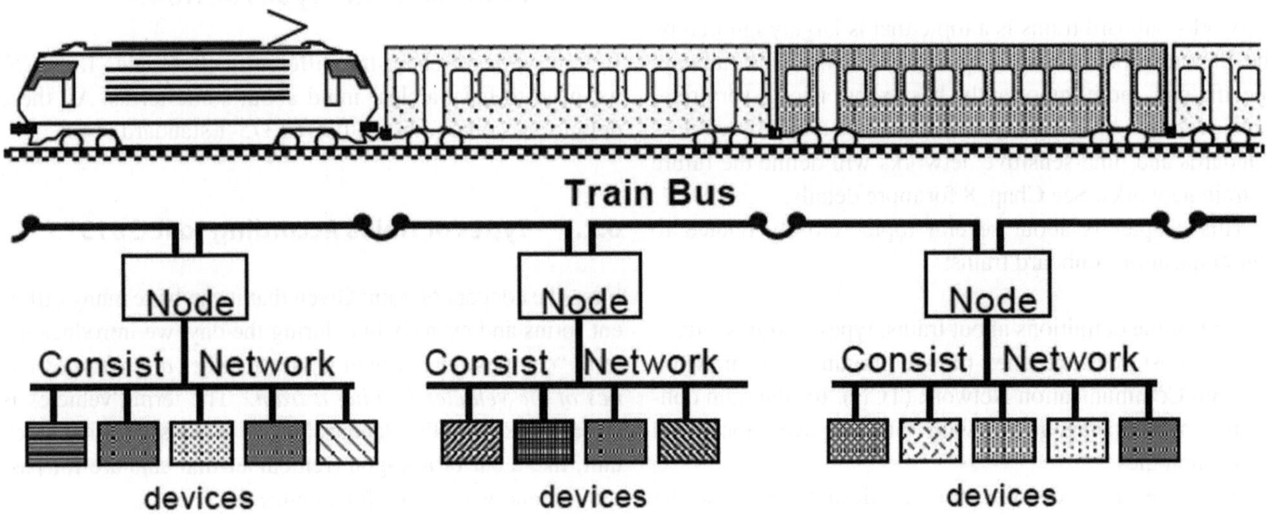

Fig. 6.2 Train backbone and consist network [144]

- Multiple-unit train: "A train consisting of a set of closed trains, where the composition of the set may change during normal operation." In Fig. 6.1, the multiple-unit train is formed by two coupled units.

6.2.2 Train Backbone and Consist Network

Once we know what a train is, let's go for the networks. The train communication network (TCN) is a "data communication network for connecting programmable electronic equipment onboard rail vehicles." Here we find two basic domains: consist networks and train backbone (network), as depicted in Fig. 6.2.

- The consist network is a "communication network interconnecting communication devices in one consist."
- The train backbone is a "bus connecting the vehicles of a train and which conforms to the TCN protocols."

A definition of bus is "a communication medium which broadcasts the same information to all attached participants at nearly the same time, allowing all devices to obtain the same sight of its state, at least for the purpose of arbitration." Finally, two more definitions:

1. Gateway: "Connection between different communication technologies."
2. Message: "Data item transmitted in one or several packets."

Depending on the desired level of redundancy, we can find three different topologies for consist networks (see Figs. 6.3, 6.4, and 6.5):

- Linear, with no redundancy.
- Ring.
- Ladder.

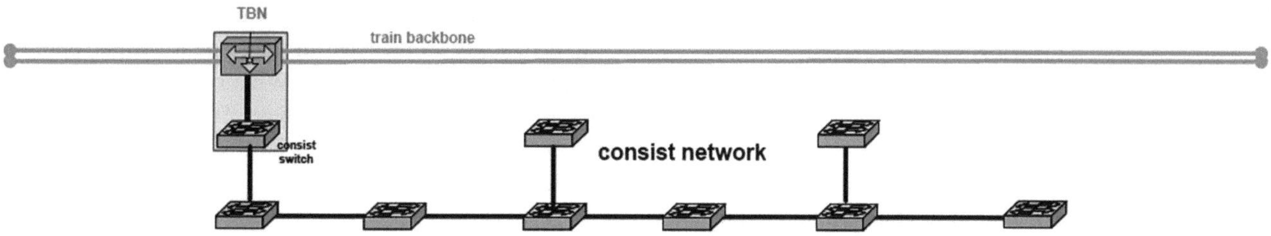

Fig. 6.3 Linear topology for a consist network, according to IEC 61375-1 [143]

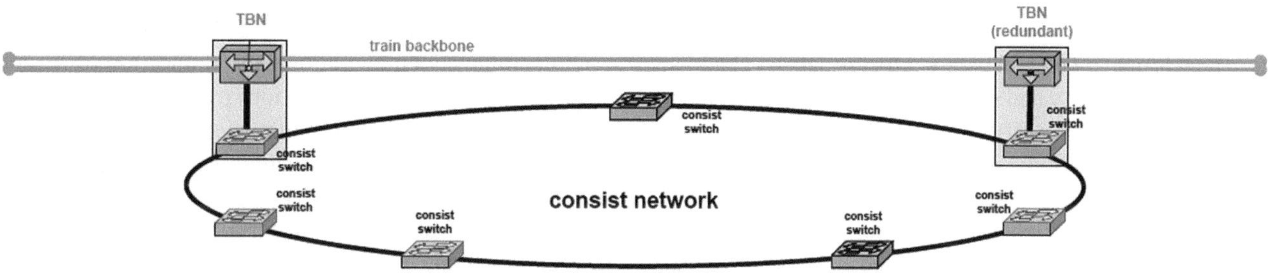

Fig. 6.4 Ring topology for a consist network, according to IEC 61375-1 [143]

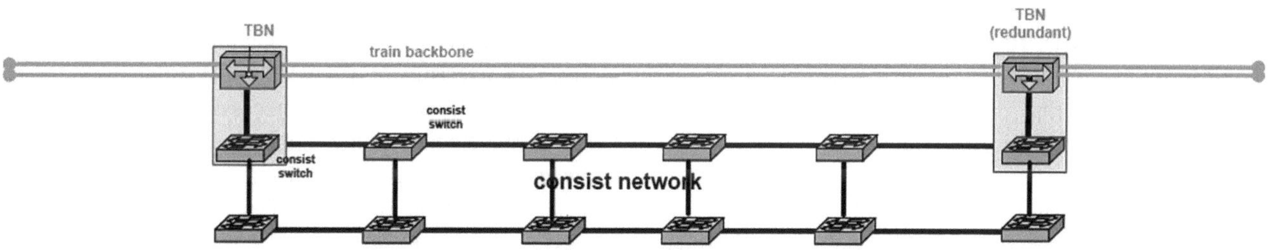

Fig. 6.5 Ladder topology for a consist network, according to IEC 61375-1 [143]

As happens in L2 networks in communications, a protocol must be defined to avoid network floodings due to loops (like the Spanning Tree Protocol). This is needed in all topologies that have physical loops that may lead into logical loops, which are all topologies, but the linear one.

Redundancy can be achieved either by network topology or connecting a device to two different nodes of the consist network. Moreover, the consist network can have subnetworks and can also be heterogeneous, that is, a consist network can be formed by a segment of a technology (i.e., multifunction vehicle bus (MVB)) and another segment of another technology (i.e., CAN). The heterogeneous approach is common when the train manufacturer is not able to have all train subsystems on a single technology. This fragmentation means that gateways are needed, and the topology gets more complicated.

TCMS

Not being part of the TCN itself, the Train Control Management System (TCMS) is the main service that works over the TCN. This means that takes control of the train control functions (but not signaling) and requires a dedicated and usually redundant device to process all the data and provide all the required functions. This device is called central control unit (CCU).

Another relevant device (generally redundant as the CCU) is the input/output (IO) which connects all the wired parts of the train to the TCN, enabling a higher degree of reliability (in case the TCN fails) and also allowing the control of many onboard devices (like buttons, analog sensors, other actuators, etc.).

6.2.3 Types of Data in TCN

Regarding the type of data to be transmitted over the TCN, there are five classes defined:

1. Supervisory data: Non-application data, related to network functions, like safe train inauguration, redundancy management, etc.
2. Process data: Real-time data for train control and monitoring. This is cyclic data where high integrity and safety is required, plus very low latency.
3. Message data: Data for train control and monitoring, which is not "hard" real time as process data is. The focus here is on integrity and safety, rather than latency (but can be demanding as well) or data rate. Typically, message data has not a periodic nature, so it is transmitted only on demand.
4. Stream data: Data packets for voice or video services, which means high data rates required and low latency and jitter.
5. Best effort data: Low-priority data transmissions.

6.3 TCN Technologies

For the TCN, we find different technologies, which are explained in the next sections:

- RS-485 (recommended standard 485):
- CAN (controller area network): A serial protocol primarily designed for automotive (where it is widely used) extended to other sectors and applications, like industry and railway vehicles, where it is for consist networks. CAN is sometimes present in legacy TCN and also in accelerometers to sense axle boxes' condition of high-speed trains bogies. It is described in the IEC 61375-3-3 standard [145].
- MVB (multifunction vehicle bus): It is a technology for the consist domain. It is described in the IEC 61375-3-1 standard [146].
- WTB (wire train bus): It is a technology for the train backbone. It is described in the IEC 61375-2-1 standard [144].
- ECN (Ethernet consist network): An Ethernet-based standard for the consist network of a train. It is described in the IEC 61375-3-4 standard [147].
- ETB (Ethernet train backbone): The backbone counterpart of the ECN, that is, an Ethernet-based standard for the backbone of the train. It is described in the IEC 61375-2-5 standard [148].

There are other parts of the IEC 61375 standard which apply here, but not for the definition of the technologies and functions, but for testing. We encourage interested readers to study all the parts of the standard which apply to the work they are focused on. What comes next is a summary of the main aspects of these TCN technologies but does not replace the whole standard with all its details and complexities.

6.4 TCN: RS-485

RS[1]-485 is much more than a technology for TCN, because it is widely used in many industrial applications. It is for serial communication and allows the connection of several devices (multipoint communication) over a single bus. The main advantages of digital communications are simplicity and cost. For the current standards, it is not a high-speed solution but can achieve tens of kbps for train-like deployments (that is, hundreds of meters). There is a trade-off between range and data rate, so larger distances mean less data rate and vice versa.

The typical topology is linear, that is, a succession of point-to-point links, connecting nodes. At the end of the linear bus, a terminator is needed to avoid data corruption due to reflections. Typically, the terminator is a resistance, equal to the characteristic impedance of the cable used in the RS-485 bus (typically, but not necessarily always, 120 Ω). Other arrangements rather than the linear topology are allowed, but assuming shorter distances and slower transmission speeds than the linear one. Typically, low speed in this context means 9600 bps, which is more than enough for some TCNs, especially for legacy trains.

RS-485 buses are implemented by two wires: the positive one and the negative one (as it is a balanced system, where the signal is transmitted over a wire and the opposite of the signal is transmitted over the second wire). This way allows longer wires as the immunity against noise is higher. However, two-wire setups are half-duplex as four wires are needed for a full-duplex RS-485 bus (see Fig. 6.6).

Regarding the wires, a couple of wires are usually twisted together (and consequently called "twisted pairs"), to reduce noise or, more precisely, the impact of electromagnetic interference (EMI).More details on this can be found later in this chapter.

All nodes in the bus receive the information published by other nodes, and an arbitration mechanism is needed to regulate the timing of the access to media for nodes that write data in the bus. In Fig. 6.7, a two-receiver, one-transmitter RS-485 network is depicted.

How many nodes can an RS-485 network have? The most common answer is 32, but this is not actually true, as it

[1] Actually, it is EIA-485, but the informal "RS" notation is way more popular, so we decided to keep that one instead.

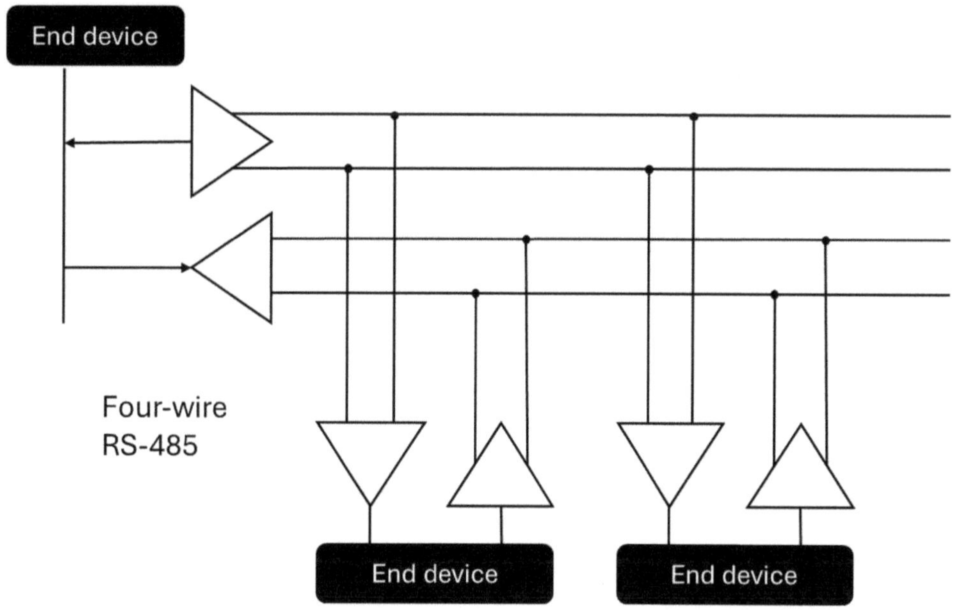

Fig. 6.6 Four-wire (full-duplex) RS-485

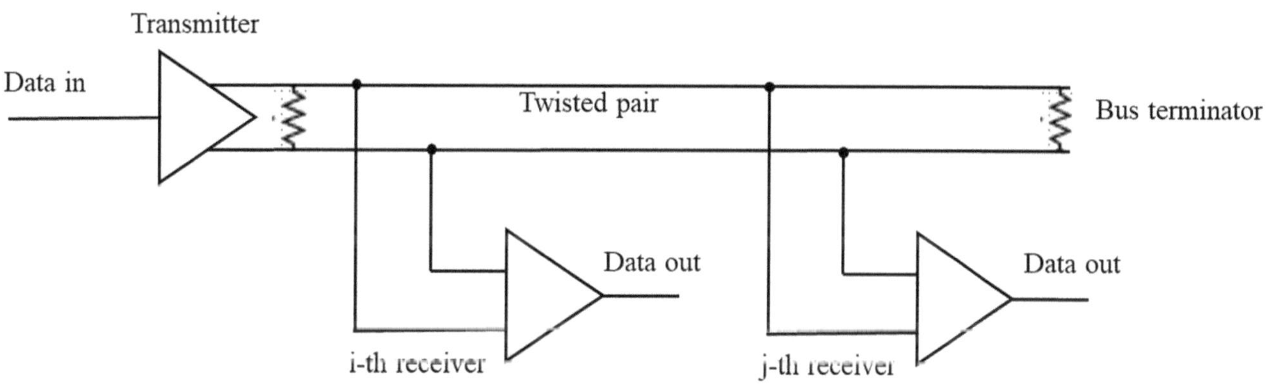

Fig. 6.7 A two-receiver, one-transmitter RS-485 network

depends on how much the devices low down the system [149]. In the data sheet of the device, the reader may find this information expressed in unit loads and make the calculations to make sure that your network does not exceed 32-unit loads (which can mean more than 32 devices). Typically, RS-485 buses that connect modern transceivers can have up to 256 devices, but reaching this number is not recommended, as the overall performance of the network will be degraded. TCNs are very often far from that figure.

In strict sense, it is not a communications protocol, as it only specifies the electric requirements or, more precisely, the interface. This means that signal levels, formats, connectors, and the protocol itself of an RS-485 bus can be very different from one example to another. This is both an advantage (more freedom for developers) and an inconvenience (no standardization at all, so more development costs). Moreover, the RS-485 guidelines [149] are very useful but not exhaustive; so, many important details are left open.

It is widely seen as a legacy technology in railways, but still present in old trains (as the first nonstandard TCMS implementations were done in different technologies, one widely used to be RS-485) and some slow connections between peripherals and a control unit (like LED displays on a passenger information system). Anyway, RS-485 is being replaced by IP-based or Ethernet-based solutions which can provide better performance.

On the other hand, RS-232 shares some features with RS-485, but the former can provide lower data rates and work for shorter distances.

6.5 TCN: Controller Area Network

A controller area network (CAN) is a standard that allows the communication between nodes typically on a vehicular scenario. A message-driven protocol was developed in the 1980s by the automotive industry, which was looking for weight and complexity savings. The very first car equipped with a CAN bus was the Mercedes Benz W140. The purpose in automation is the same we already know about TCN in railways: connecting sensors, actuators, and control units to safely operate the vehicle. As in railways, the initial approach was multiple point-to-point wirings, which are costly to install, hard to maintain, and heavy to transport. Since the 1973 oil crisis, all industries, but especially those dependents on oil like automation, were desperate for energy-efficient solutions. In this crisis, the oil tripled its value, shaking the world economy severely.

In railways, CAN buses are for consist networks and use a specific implementation of CANopen, which is a standard solution for CAN buses in industrial applications. The reference for this is "Electronic Railway Equipment – Train communication network (TCN) – Part 3-3: CANopen Consist Network (CCN)" [145]. The remaining of this section is largely based on this standard.

6.5.1 CAN Architecture

The typical architecture of a CAN network is shown in Fig. 6.8. We see two different elements:

1. Several nodes (called "virtual devices" by the standard) which represent train subsystems, like brakes, PIS, lighting, etc.
2. A gateway to the train network, that is, the backbone. The role of the gateway is the same as usual ("connection between different communication technologies").

The topology (as depicted in Fig. 6.8) is linear, which means that, as in RS-485 buses, terminators are needed at both ends. The total length of the bus must not exceed 450 m for a 125-kbps transmission speed.

Devices are connected to the bus with short stubs. These individual stubs must not be larger than 22 m, and the aggregated stub length must be shorter than 110 m (for 125 kbps speeds). Like in automotive, it is recommended to have the shortest possible stubs in a CAN network.

To address CAN devices, each one has a unique Node-ID defined from the hexadecimal 01 to 7F.

6.5.2 Physical Layer

Regarding the physical layer, as in RS-485, twisted pairs are used, and DB-9 (actually, the real name for these connectors is D-sub9; see Fig. 6.9) connectors are recommended and largely used (as in other TCNs like MVB). Another alternative is the M12 family of connectors (Fig. 6.10). Only five pins are needed for the following signals:

- CAN_L bus line (dominant low).
- CAN ground.
- Optional CAN shield.
- CAN_H bus line (dominant high).
- Optional CAN external positive supply, just in case optocouplers must be powered here.

There are several transmission speeds than can be defined (1 Mbps, 800 kbps, 500 kbps, 250 kbps, 125 kbps, 50 kbps,

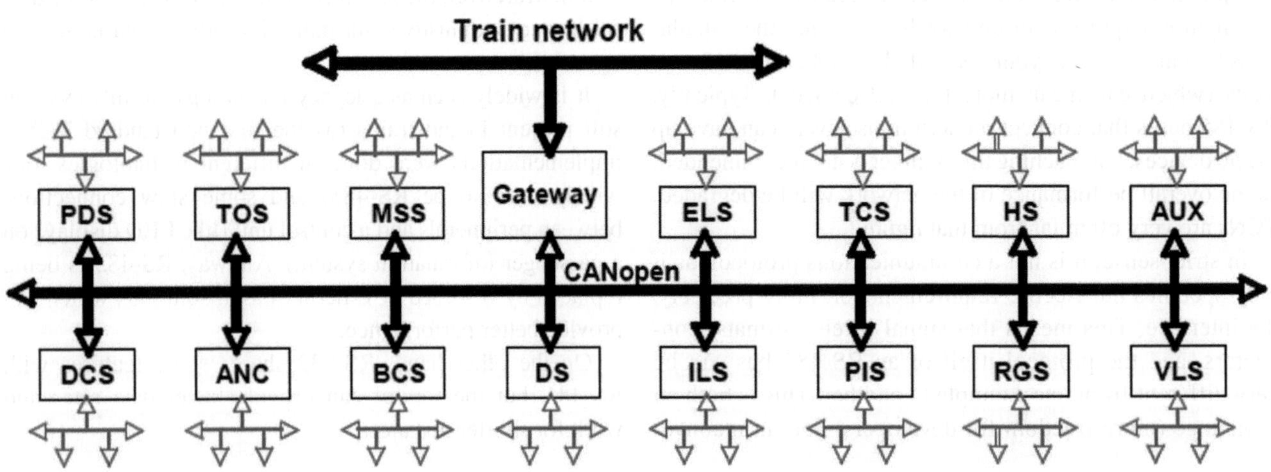

Fig. 6.8 Logical architecture for a CAN network of the consist network [145]

Fig. 6.9 DB-9 connector

Fig. 6.10 4-pin M12 male connector

20 kbps, and 10 kbps), but only 125 kbps is mandatory. All CAN devices must support this data rate to comply with the IEC 61375 standard. The encoding and synchronization must be defined according to ISO 11898-1 [150], which is part of the CAN standard, but for automotive (IEC 61375-3-3 [145] describes only railway-specific aspects).

6.5.3 CAN Interfaces and Importance on Railway Vehicles

Another relevant aspect of CAN buses in railways is the interface with the backbone network, that is, the gateway function. Recall that CAN in railways is defined only for the consist domain, so the backbone will be a different technology, like WTB. The standard also provides some guidance about how this interface shall be carried out.

By the time of writing this book, CAN buses remained a legacy technology but still popular in tramways. For other types of trains (high-speed, regional, metro, etc.), the most common solution for the consist network is MVB and, more recently, ECN. MVB is addressed in the next section and ECN is explained later in this chapter.

6.6 TCN: Multifunction Vehicle Bus

MVB stands for multifunction vehicle bus, and the reference for the standard is "Electronic railway equipment—Train communication network (TCN)—Part 3–1: Multifunction Vehicle Bus (MVB) IEC 61375-3-1:2012 [146]. This section is largely based on this part of the standard.

This technology is, by the time of writing the book, the most popular one in consist networks, but is being replaced by Ethernet consist network (ECN) TCNs. It is the first TCN standard specifically designed for railways, unlike CAN (which was for automotive) or RS-485 (for industrial applications).

ECN slowly replaces MVB as trains are manufactured to last at least 40 years or more. New trains are not fully ECN as some systems are only compatible with MVB or CAN. However, the clear trend is to have MVB and CAN replaced by ECN in the long term.

6.6.1 MVB Architectures and Types of Devices and Data Classes

Depending on the capabilities of the device itself, the standard defines six classes of MVB devices. The capabilities are the following:

- Device status: The device can send its "Device Status" upon request. This is mandatory to all classes of devices.
- Process data: The device can send and transmit process data (that is, real-time data) when needed.
- Message data: The device can send and transmit message data when needed.
- User programmable: These devices can run user-defined applications. Implies message-data capabilities.

- Bus administrator: The device can act as a master, which means message data, process data, and device status capabilities.
- TCN gateway: The device can access another bus, which can or cannot be MVB.

There are six classes:

- Class 0: These devices do not support any capability. Repeaters are a typical example of class 0 devices.
- Class 1: These devices must support "device status" and "process data" capabilities.
- Class 2: These devices must support "device status," "process data," and "message data" capabilities, being not programmable devices.
- Class 3: The same as class 2 plus "user programmable" capability.
- Class 4: Class 3 plus "bus administrator" capability, being "user programmable" optional.
- Class 5: Same as class 4 plus "TCN gateway."

The transceiver interface between the MVB bus controller and the transceiver of a particular device is depicted in Fig. 6.11. This interface defines three signals:

- RxS: receiver signal.
- TxS: transmitter signal.
- TxE: transmitter enable, set to "1" to enable the transmitter.

6.6.2 MVB Physical Layer

MVB is a technology more flexible and capable than CAN, for example, in terms of the physical layer. An MVB network (or, as it is more often called, "MVB bus") is formed by one or more than one bus segments of these three types:

1. Electrical short distance (ESD).
2. Electrical medium distance (EMD).
3. Optical glass fiber (OGF).

Therefore, an MVB network can be formed by several segments, each one with either the same or different technologies as the others. Moreover, for the sake of redundancy, segments can be duplicated. All these segments are interconnected by couplers, which can be either a repeater (typical approach) or a star coupler (only for OGF).

ESD medium: Segments of up to 32 nodes and up to 20 m are allowed in ESD, without galvanic isolation. In case galvanic isolation is present, it is denoted as ESD+, and distances of 200 m can be achieved. Four wires are present: VCC (typically, 5 V), DATA_N, DATA_P, and ground. Like other buses, a terminator is required (120 Ω). For the MVB connector, a device connected to two redundant MVB lines (depicted as Line_A and Line_B) is depicted in Fig. 6.12. The standard for ESD MVB provides a lot of details, like the minimum and maximum current for a device in the connector (20 mA and 300 mA, respectively), whose pins shall be female or male within the DB-9 connector, how to arrange

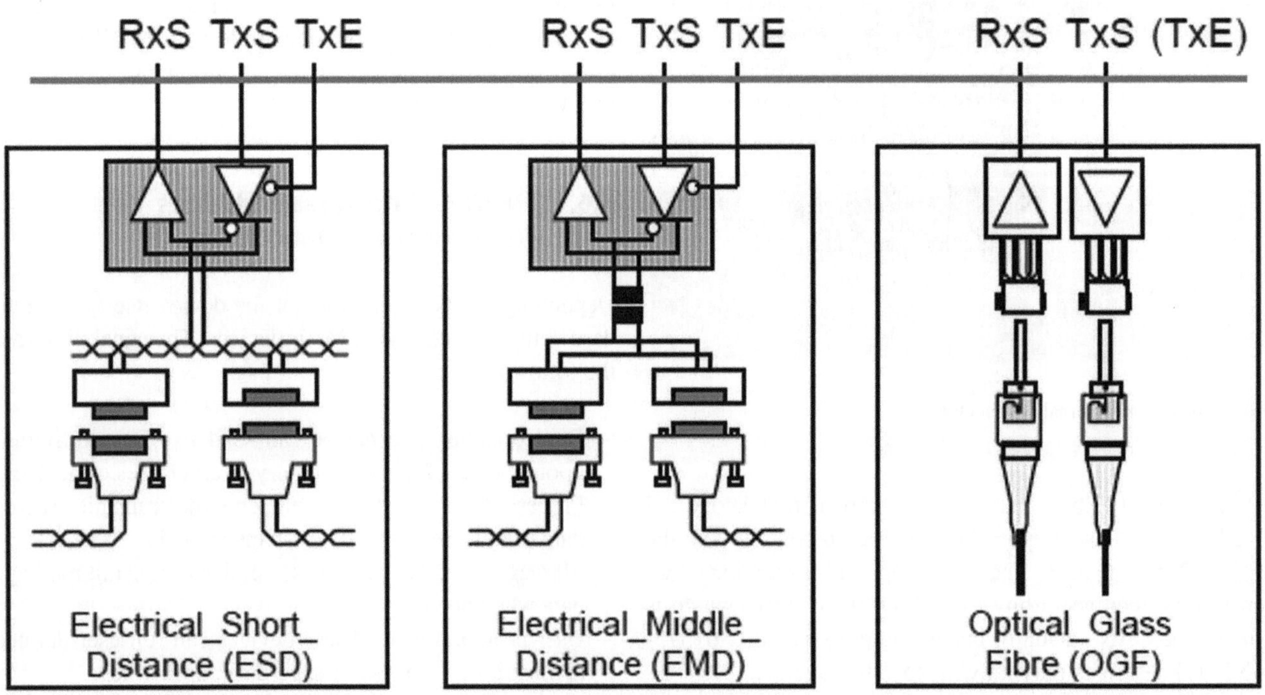

Fig. 6.11 Transceiver interfaces for MVB networks [146]

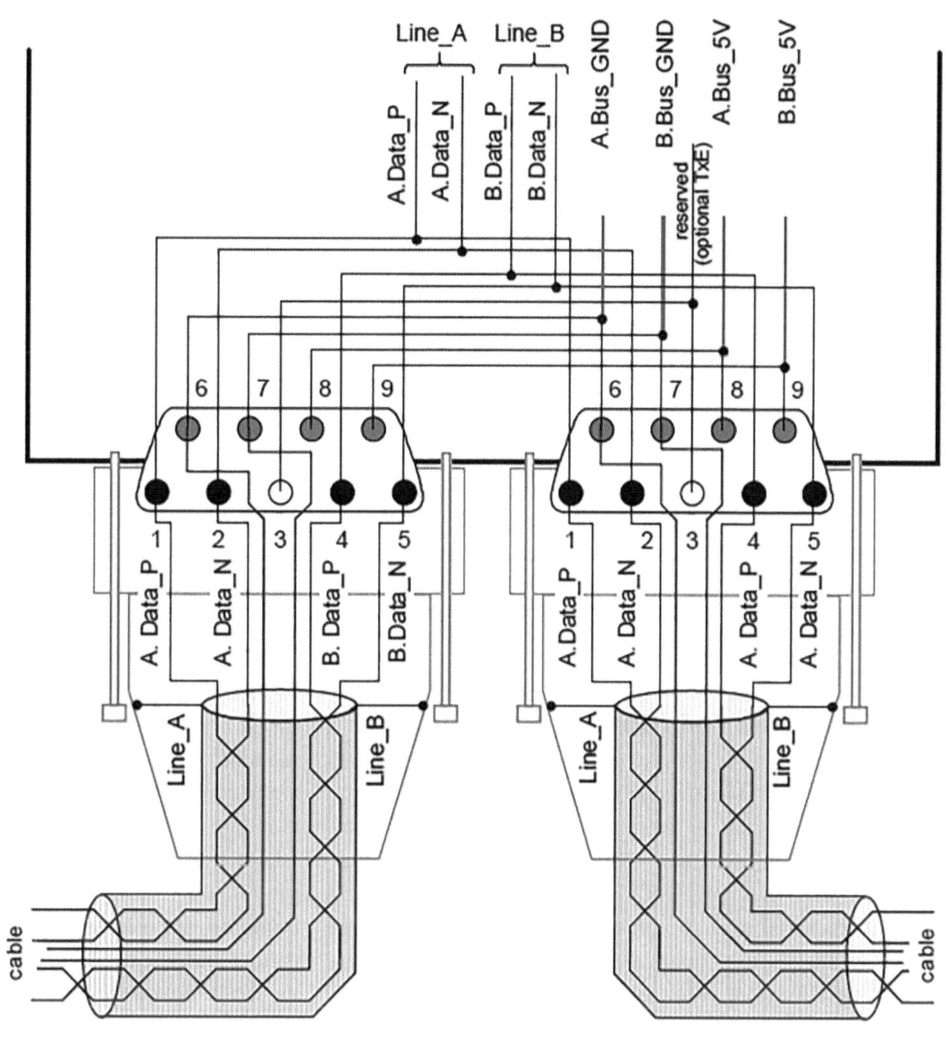

Fig. 6.12 ESD connectors within a device for two redundant MVB lines [146]

the connectors in the terminator, etc. regarding the insertion losses, what we explained about RS-485 buses remain valid.

EMD medium: It is very similar to ESD, but with galvanic isolation and transformer coupling, which allows distances up to 200 m and more than 32 nodes in the segment. However, both EMD and ESD busses are limited by interference and noise rather than by a "theoretical" bound. EMD is typical in modern tramways and subways, but ESD+ busses can be easily found in subways as well.

OGF medium: This employs fibers rather than twisted pairs. Therefore, distances of up to 2000 m between star couplers are achieved. The topology is much simpler than EMD or ESD, as it consists of a pair of fibers that form a full-duplex link. The standard is very detailed on the specification of the fiber, the needed certifications, maximum attenuation, etc. OGF is often present in locomotives and other environments where EMI levels are high.

All three media (ESD, EMD, and OGF) work at 1.5 Mbps; up to 4096 devices can be connected to an MVB bus, and the arbitration of the bus is based on a deterministic polling (for process data), where every device is requested periodically to send data (the master frame is sent by the master, and the slave frame contains the data pulled by the end device). The typical response time for the slave frame to be sent is 4 μs (but the maximum allowed is 43 μs). Large delays are for long links, typically OGF. ESD and EMD are usually closer to 4–5 μs rather than 40 μs. Although many devices can act as MVB masters, only one can do at the same time.

Both the master frame and the slave frames are broadcasted. All frames are read by every device in the bus, but only the end device addressed in the master frame replies to it. This approach enables digitalization applications which only read data from the bus and offload them into IP networks to process the data for operational and maintenance use cases. This feature is common to other TCNs and is very relevant for operators which want to develop their own predictive maintenance algorithms or other related applications.

6.6.3 MVB Interfaces and Importance on Railway Vehicles

Typically, when MVB is present in the consist level, WTB (see Sect. 6.7) is the backbone technology. However, this is not mandatory at all, and many scenarios can be drawn. On the other hand, typically MVB network segments coexist with other MVB segments, RS-485, and CAN networks. If the coupling between consists is required in normal operation (that is, not for emergency mode), more train functions are required at the backbone level and more capable this network should be.

The IEC 61375 standard also provides requirements for conformance testing of MVB networks. This way, interoperability is assured. The tests for the conformance are resistance test, inductance test, measurement of insertion loss, and measurement of signal waveform during transmission [151].

6.7 TCN: Wire Train Bus

Wire train bus (WTB) is a technology for train backbone networks. The reference is the standard "Electronic railway equipment—Train communication network (TCN)—Part 2-1: Wire Train Bus (WTB)" [144]. As RS-485, CAN, and MVB, it is a serial data communications bus. As we already know, the backbone network is for connecting consist networks, that is, a train configuration which may change frequently. These types of trains that may change its configuration are called open trains.

There are many alternatives for a TCN, such as MVB in the consist domain and WTB in the backbone, but this one is not the only possibility we have. It is common to see other types of consist networks, such as ECN or CAN, interconnected by a WTB. Moreover, the maturity of ECN systems is much more than ETB's, so this means that ECNs are usually deployed together with WTB rather than ETB. Only time will tell if ETB networks replace WTB as the flagship technology for train backbone networks.

As in MVB, the WTB standard [144] is flexible in several aspects, like the allowed configurations. We can have a WTB without consist networks at all, or with MVB consist networks or even with another technologies. For example, the norm addresses the coupling of several open trains, allowing the interconnection of up to 32 nodes and up to 15 consist networks in the complete train. An extreme example allows having an MVB bus as both consist and train backbone networks. This is frequent in closed trains which are not coupled and decoupled frequently.

As a reference, typically, it is assumed that the cable length per vehicle is 150% of the vehicle length. The standard estimates that the maximum distance for the cable (that is, between repeaters) is 860 m.

6.7.1 WTB Physical Layer

For the WTB physical layer, we find only twisted pairs (not fibers as in MVB OGF) at 1 Mbps. Between any two nodes, the maximum propagation delay is 60 μs.

The WTB bus is formed by nodes interconnected by bus sections that can have three different forms:

1. Trunk wires that go from one end of the consist to the other.
2. Jumper wires that connect consist trunks.
3. Extension cables from a trunk to a node.

The standard provides a detailed specification for the cables, so they must be a twisted, shielded, and jacketed cable, with no less than 12 twists per meter, and the recommended cross section of the cable is also given: 0.75 mm^2 for trunk cables and 1.34 mm^2 for jumper cables. The characteristic impedance of the cable is 120 Ω, and values for capacitances, cable attenuation, and other electrical aspects are provided as well [144]. The objective for such detailed requirements was twofold: enforcing the quality of the TCN rollout and ensuring interoperability.

We find two types of nodes: intermediate nodes, located in non-extreme cars, that connect two bus sections, and end nodes, located in extreme cars, and terminate two bus sections. See Fig. 6.13 for more details.

As in MVB, WTB lines can be redundant, being the nodes connected to two WTB lines. For each line, only two pins are needed, as shown in Fig. 6.14, where the WTB connector (DB-9) is depicted.

Device's addresses in WTB use 8 bits, where "all ones," i.e., 255 is for broadcast messages. Frame data is formatted as specified by high-level data link control (HDLC) protocol [152]. The bus arbitration is like MVB: each segment has a unique master and several slave nodes, which transmit data only when requested by the master. The transmission unit within the WTB bus is the telegram, which is a pair of master frame and slave frame.

6.7.2 WTB Functional Aspects

The WTB standard also considers several specific functions, like train inauguration, which is to be launched in case of a

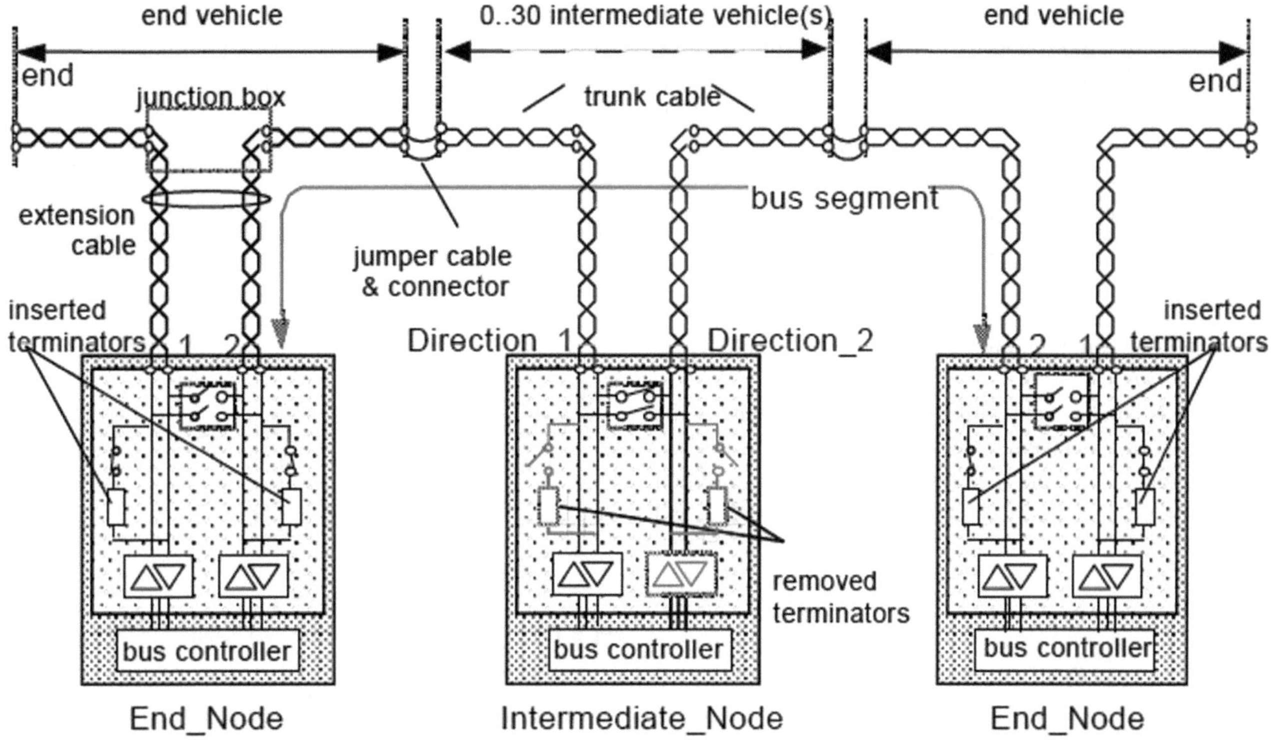

Fig. 6.13 WTB intermediate nodes and WTB end nodes [144]

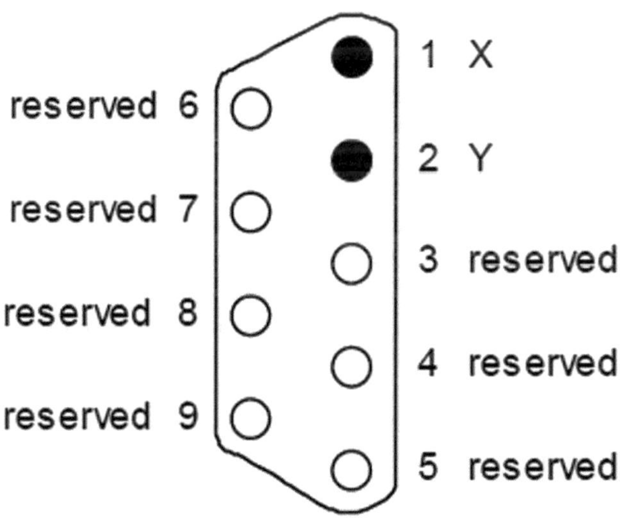

Fig. 6.14 WTB connector pinout [144]

change in the train composition. In this procedure, every WTB node receives its address from the master, the orientation, and the description of all other nodes within the bus. After the inauguration, the WTB starts normal operation.

Finally, another special feature of WTB is the "fritting" to overcome oxidation of electric contacts. At train setup, a pulse is transmitted to remove the oxide in connectors. Fritting is an optional feature but, in case it is implemented, it means that 48 V is applied between two pins of a connector, and a current no higher than 80 mA is applied as well.

6.8 TCN: Ethernet Consist Network

Ethernet consist network (ECN) is a more recent standard for train consist networks. As its own name indicates, it relies on Ethernet technology, which is the most used solution for local area network (LAN) connectivity. This move denotes an intention of convergence toward open solutions, avoiding niche markets, same as in other railway fields, like train-to-ground and many others.

The reference for the standard is "Electronic railway equipment—Train communication network (TCN)—Part 3-4: Ethernet Consist Network (ECN)" [147]. As in previous sections, this one is based on the essential contents of this standard.

6.8.1 ECN Architecture

As in MVB, ECN can be used not only for consist networks but for closed trains and multiple-unit trains as well. The architecture for the network is the same as in previous cases: two layers (backbone and consist) and a node which interconnects both of them.

ECN is formed by Ethernet switches (IEEE 802.3) and the end devices (ED) connected to them. This is the essential fact about ECN. Optionally, the network can be more complex and includes subnets, routers, etc. Still optionally, EDs can be connected to one or more networks, in case redundancy is needed. In case a device is connected to more than one network, we call this "dual homing," and it is also considered in both IEC 61375-1 [143] and IEC 61375-3-4 [147].

The gateway function toward the backbone must be either WTB, ETB, or other ECNs. That is, the interconnection of different ECNs could be done by a router or by an application-layer gateway.

The ECN can have any of the topologies explained in the first section of this chapter: tree, ring, or ladder. Moreover, ED can make use of the dual homing or not. The resultant topologies are depicted in Figs. 6.15, 6.16, 6.17, 6.18, and 6.19.

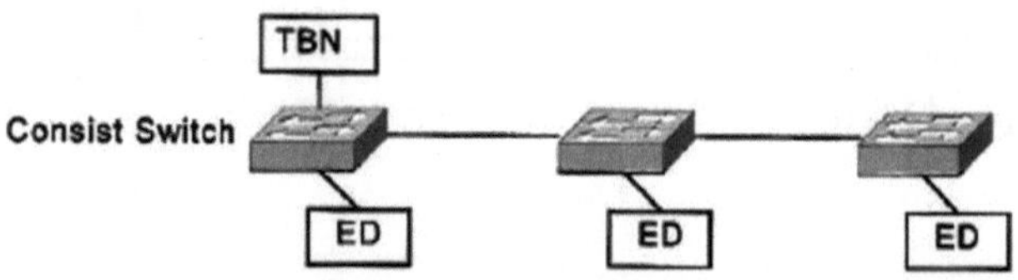

Fig. 6.15 ECN with linear topology [147]

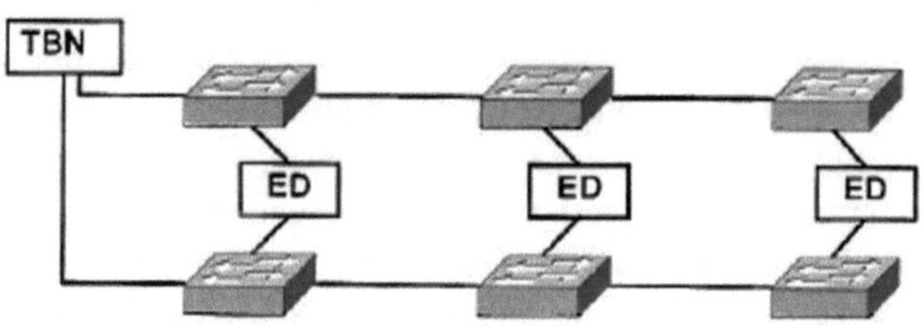

Fig. 6.16 ECN with linear topology and dual homing [147]

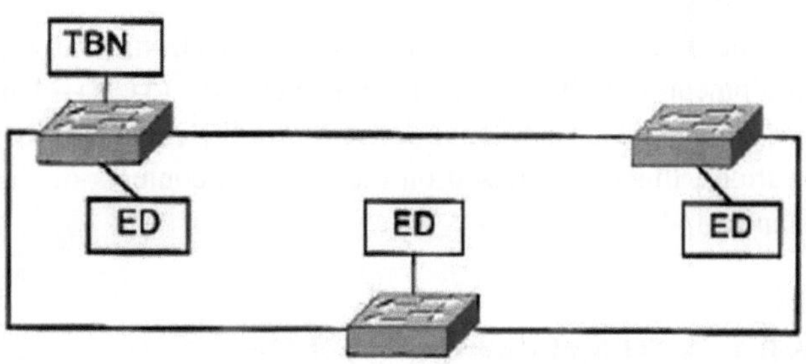

Fig. 6.17 ECN with ring topology [147]

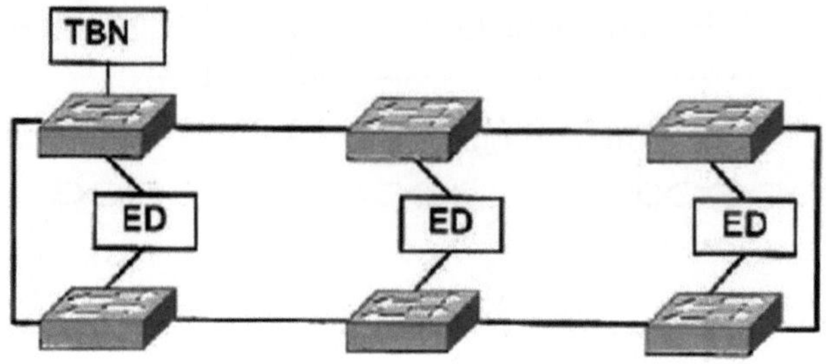

Fig. 6.18 ECN with ring topology and dual homing [147]

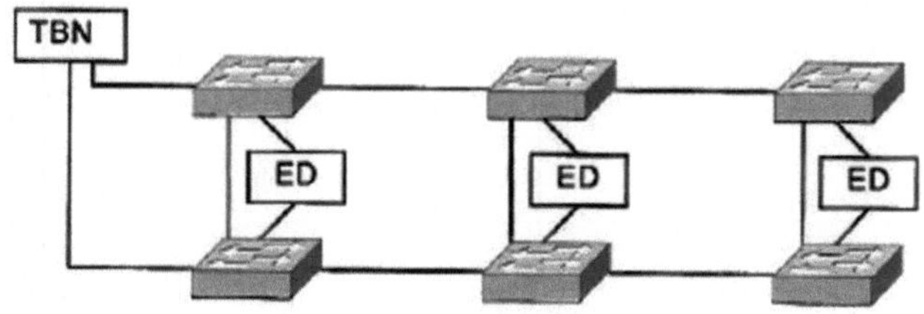

Fig. 6.19 ECN ladder topology with dual homing [147]

The network devices we can find on an ECN are of three types:

- Consist switches: This is the main type of ECN devices. It performs switching functions (that is, relaying link-layer frames) and can be either managed or unmanaged. Unmanaged switches are simpler devices, not able to set up virtual LANs (VLANs) or manageable at all.
- Repeaters: If needed, they extend the physical layer (that is, the distance between switches) and are transparent for upper layers.
- Routers: Routers are network devices with at least two network layer interfaces (i.e., IP interfaces). However, not all network-layer functions must be implemented on routers, as they can run on EDs. These functions are Dynamic Host Configuration Protocol (DHCP) for the IP address assignment; Network Time Protocol (NTP), for time synchronization; or Domain Name System (DNS), among others.

6.8.2 ECN Data Classes

For the data classes specified by the IEC 61375-1 standard [143], explained in the first section of this chapter (process, message, stream, best effort and supervisory data), this standard provides guidance (that is, not a hard requirement, but a list of typical values for some network parameters, like

Table 6.1 Types of data as defined in IEC 61375-1 [143]

Data class	Parameter	Typical value
Process data	Minimum cycle time	20 ms
	Maximum data size	1500 bytes
	Maximum latency	10 ms
	Maximum jitter	10 ms
Message data	Minimum cycle time	–
	Maximum data size	1500 bytes
	Maximum latency	–
	Maximum jitter	–
Stream data	Minimum cycle time	–
	Maximum data size	1500 bytes
	Maximum latency	125 ms
	Maximum jitter	25 ms
Best effort data	Minimum cycle time	–
	Maximum data size	1500 bytes
	Maximum latency	–
	Maximum jitter	–
Supervisory data	Minimum cycle time	10 ms
	Maximum data size	1500 bytes
	Maximum latency	10 ms
	Maximum jitter	10 ms

latency, jitter, etc.). These values are depicted in Table 6.1 but must not be taken as a reference because the network shall be designed to fulfill the requirements from applications.

Data size is the link-layer payload, that is, the number of bytes of a frame. The cycle time is between two consecutive cyclic frames; the latency is measured between two EDs, and the jitter is the variation of the latency.

6.8.3 ECN Functional Aspects

For the functions that an ECN shall provide, we find the following:

- Frame relaying: Relaying MAC frames originated at an ED toward the destination ED. The reference for this is IEEE 802.3D.
- VLAN functions as defined in IEEE 802.1Q.
- Management of redundancies: In case it is needed. For example, in the case of link failure, this situation must be detected; the failing route must switch to the redundant one and the network must reconfigure accordingly to the new situation. All these steps must be carried out faster than the time it takes for the applications to stop working. Please notice that redundancy can be either managed at network level or at ED level (by dual homing).
- QoS management: To guarantee application requirements, implementing traffic prioritization policies. IEEE 802.3 standard considers eight different priority levels. Consist switches must support at least two levels (more precisely, two different queues, grouping several levels into a queue). The most priority data class is supervisory, the second process data, then message and stream data, and, finally, best effort data.
- Gateway functions: toward the backbone.
- Train network management and name resolution (optional) between IP addresses and names: This function can be either performed by network devices or by EDs. DNS servers must be redundant, and it is recommended to place them at backbone level (in case there is a backbone). Same policy on redundancy in general for all functions that may apply, like DHCP for IP address management, etc. In case there is a backbone network, all servers to provide network functions must be placed at backbone level. On the other hand, hostnames must be unique at ECN level.
- Dynamic IP addressing (optional): There are two networks to be considered here: ECN addresses and train addresses (that is, when ECNs are connected to the backbone). The latter can change on every train inauguration.
  - IP addressing on an ECN can be either dynamic or static. All ED must have at least a unique IPv4 address within the ECN. IP addresses within the ECN are to be assigned within a subrange of class A addresses, from 10.0.0.0 to 10.127.255.255 (10.0/9). The host bits for the addressing can be divided into subnetworks or not.
  - Train addressing on the train network. Not all EDs must be addressed in both train and consist domain, but ECN addressing is mandatory. Both train and ECN addresses can be the same, but in case they are not, ECN must implement a procedure to map one addressing into the other. Typically, this mapping can be done by using[2] network Address Translation (NAT). As in ECN, train network addressing is IPv4 based and are defined in the following way:

    00001010.1bbxssss.sshhhhhh.hhhhhhhh/18,

    where "b" is the identification of the ETB, "x" must be set to 0, "s" is the ECN id assigned at inauguration time (s = 0 is reserved to the backbone), and "h" is for the host (which means that we have 14 bits for host addressing, that is, 16,382 hosts per consist network and 18 bits for network addressing). 2^{14} = 16,384 minus 1 for the network (the first one) and minus 1 for broadcast address. Result: 16382 hosts.

Group addresses (that is, IP multicast addresses) can be defined, that is, several ED may belong to a group, either defined at train or consist level. Membership is usually static but may change at inauguration.

[2]There is a railway-specific protocol for this, called railway network address translation (R-NAT).

Train Real-Time Data Protocol

When TCNs embraced the IP world, it was clear that an on-purpose protocol was needed. The answer was Train Real-Time Data Protocol (TRDP), which is an evolution of the previous one developed by Bombardier Transportation (now part of Alstom). This protocol is now a standard and was developed by TCNOpen, a joint initiative from several train and TCN manufacturers. It is based on User Datagram Protocol (UDP), which means connectionless, is open source, and the configuration is done with XML files.

TRDP is optional within consists but mandatory for the train backbone (for interoperability reasons). TRDP uses UDP port 17225.

6.8.4 ECN Interfaces

Regarding network device interfaces, several are mandatory, others are optional, and some are not. Depending on whether the switch is managed or unmanaged, some technologies are mandatory or not, like 100Base-TX, 100Base-T, PoE, RJ45, and some others. For example, for ED connection, 100Base-TX full-duplex links are mandatory,[3] and PoE links are optional (but very helpful many times: intercoms, CCTV cameras, speakers, WiFi AP, etc.; see Chap. 4). For network-layer protocols, the situation is the same: some mandatory and some optional. For example, Internet Control Message Protocol (ICMP), which is the basis of the famous *ping* function, the best friend of network engineers and technicians, is mandatory for managed switches, and DNS, as we already know, is optional, like NTP.

For the connectors, M12 D-coded are mandatory for permanent EDs, so RJ-45 connectors are allowed for temporary EDs. Furthermore, the pinning of the M12 connectors is also specified by the standard.

For EDs, the standard also has similar requirements in terms of interfaces, and link and network layer protocols.

6.9 TCN: Ethernet Train Backbone

Finally, the last is TCN technology. Ethernet Train Backbone (ETB) is the backbone counterpart for ECN. However, it is less extended than ECN. The reference standard is "Electronic railway equipment—Train communication network (TCN)—Part 2-5: Ethernet train backbone" [148]. When discussing ECN, we have introduced several ETB concepts, so we don't repeat them in this section.

[3] 100Base-T is optional just in case it is needed due to higher EMC levels.

The standard divides the physical layer into three regions:

- Intra car: Cables and networks inside cars.
- Inter car: Mostly cables that connect cars, but potentially inter car networks as well.
- Inter consist: Cables that connect consists, that is, couplers.

There are several Ethernet technologies considered for these three scenarios.

Regarding the physical architecture, redundancies are allowed, typically a dual Ethernet trunk line between the ETBN (ETB Nodes).

6.10 Onboard Multimedia and Telematic Services

Typically, TCN networks are for Train Control and Management Systems (TCMS), but there is another group of functions and services that must be considered. They are grouped under the Onboard Multimedia and Telematic Services (OMTS) umbrella . They comprise some services which are less critical but more demanding of data rate, like CCTV, PIS, and some others.

A typical approach is to separate both domains (TCMS and OMTS) into two separate networks, as the requirements for each one of them are far from equal. However, there is no standard to consider this separation, and some operators prefer to implement different VLANs for OMTS and TCMS, whereas the most of them use completely different networks, connected, if needed, by firewalls.

The reference standard for OMTS is the IEC 61580, which has some parts in progress.

6.11 Onboard Electronics: EN 50155

We have discussed many aspects of onboard networks, but we have not introduced the specific requirements for the onboard equipment (network or not). The reference standard for this is the "EN 50155:2021 Railway applications—Rolling stock—Electronic equipment" [153], or simply, "50155 certification" as it is usually addressed. This standard covers several aspects related to onboard electronic equipment that we discuss in this section, including design, manufacture, documentation, and testing. In general terms, this standard supports a certification process, for electronic hardware onboard trains. That is, several requirements for the hardware are introduced (temperature range, vibrations, etc.) and how to test them. If a device passes all the required tests, we say it is "EN 50155 certified" or, less precisely, "railway certified." Notice that this standard only addresses rolling stock equipment and not wayside. However, as wayside is in general terms less demanding than onboard, 50155-certified devices can work in the wayside as well.

Railway operators usually require this certification for all onboard electronic equipment, with some possible exceptions, which need an agreement between the operator and the manufacturer.

As the reader may notice when it learns about the contents of this standard, this certification makes electronic devices more expensive, as many modifications are needed to pass. The extra cost will depend on many factors (as economies of scale and the type of device), but the increase is undeniable. Moreover, it also impacts market time, as these devices need to pass the certification process, which takes some weeks or even months, delaying the time to market.

6.11.1 Aspects to Be Considered for Onboard Electronics

EN 50155 has three aspects of special importance:

- Temperature: It has a strong impact on the aging of electronics. Therefore, six classes are defined, depending on the operating temperature range. See Table 6.2 for more information. The default one is OT3. This means that the certification process depends heavily on the housings of the electronic equipment (see Fig. 6.20).
- Vibration and shock: The testing for vibrations and shock is done according to the EN 61373 [154] standard.
- EMC: It can be a major issue onboard trains, so electronic equipment must be designed around this. The reference standard is EN 50121-3-1 [155]. This requirement is twofold: both EM perturbations from the device under test to the other electronic equipment and the impact of external perturbations into the device under test must be minimized.

There are many more requirements apart from these three. Here we will discuss only the most relevant ones, but we encourage the reader to study EN 50155 standards, as it includes many important aspects related to onboard electronics.

6.11.2 Other Aspects Addressed by the EN 50155 Standard

A very relevant one is the voltage supply: the standard identifies several DC voltage levels (24, 28, 36, 48, 72, 96, and 110 V), but only three of them are recommended: 24, 72, and 110 V. It is very common to find different voltages onboard trains, but the most common are these three. Notice that these three are not the typical ones for consumer electronics (12 VDC, etc.). So very frequently, power supplies must be

Table 6.2 Temperature classes as defined in EN 50155 [153]

Class	Temperature (°C)	Comments
OT1	−25 to +55	Driver's cabin and passenger's area cabinets
OT2	−40 to +55	Driver's cabin and passenger's area cabinets
OT3	−25 to +70	Lockers
OT4	−40 to +70	Lockers
OT5	−25 to +85	Not recommended in general. Only for boards inside exterior cubicles, which are much more exposed to extreme temperatures
OT6	−40 to +85	Same as OT6

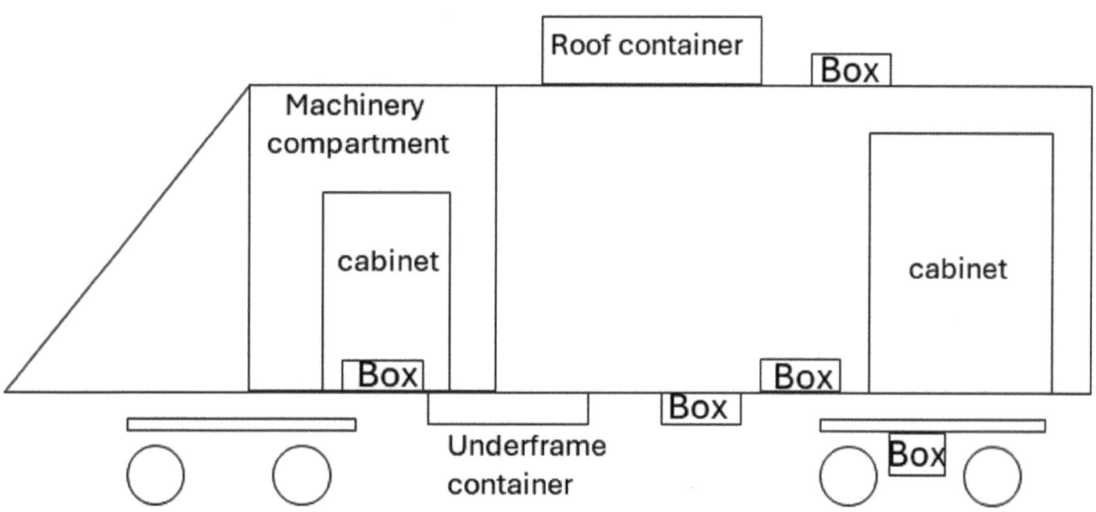

Fig. 6.20 Types of enclosures according to EN 50155 [153]. Figure adapted from [153]

introduced when redesigning an electronic device for railway use.

The standard also provides an upper and lower margin for voltage supplies, as they may change inside the train. Therefore, power supplies must be designed to work with voltages −30% and +25% of the nominal one. For example, electronic trains with the pantograph down (that is, powered by batteries only) provide a significantly lower voltage than when the pantograph is up (for example, 120 VDC vs 95 VDC, for a nominal value of 110 VDC).

There are other requirements like transitory over voltages (+40% for a second or less).

The lifespan of an electronic device depends on several aspects, like the type of device. The EN 50155 standard specifies five classes:

- L1: 5 years.
- L2: 10 years.
- L3: 15 years.
- L4: 20 years.
- LX: agreement between both parties.

This is a very relevant issue for operators, as it has a strong impact on costs and availability of the rolling stock. Therefore, this information must be very clear in the specifications of the equipment, and in case it is needed, some testing must be done in advance to certificate the proper lifespan class of the device. Notice that the expected lifespan of a train is around 40 years, so it is assumed that onboard electronics last shorter and must be replaced.

Large renewals are called retrofits and are a hard thing to do. However, they are much necessary to have trains properly working until the end of the expected lifespan. Typically, a "mid-life retrofit" is carried out by operators and usually include new TCN, door control units, etc., plus all the elements that are replaced on a regular basis.

On a low-level basis, the standard also tackles aspects about connectors, soldering, wiring, printed circuit boards (PCBs), documentation, software quality assessment, etc.

6.11.3 Electromagnetic Compatibility (EMC)

The presence of strong EM fields in electric railway lines and in electric trains leads to significant electromagnetic interference (EMI) which must be taken into account. This is something that does not apply only to onboard electronics, but we have decided to include some general remarks in this chapter.

EMC is related to the capacity of electronic devices to work properly without causing or suffering EMI. There are two types of EM emissions:

1. Radiated EM emissions: They come from unintentional emissions propagated through the air from other devices, both near and far-field emissions, but especially far-field (frequency range is 30 MHz–1 GHz). Typical sources of these emissions are radios, power supplies, electric engines, etc. These devices are commonly found in railway environments. Notice that the frequency band includes several PLMN and T2G bands; so this must be considered when making measurements in the track.
2. Conducted EM emissions: They occur when the emitting device is coupled with another. This can happen via connectors, grounding, shields, etc. These emissions are typically near-field and belong to lower frequencies than radiated emissions (typically, from 150 kHz to 30 MHz).

There are some best practices to design electronic equipment and avoid both radiated and conducted emissions. Some of them are general (like avoiding large circuits, separating from potential sources, etc.) and some others are related to the specific environment.

Finally, the reference standard for EMC in railways (not only rolling stock) is EN 50121, which has several parts, including those for signaling and telecommunications (EN 50121-4 [156]), energy systems (EN 50121-5 [157]), etc.

6.12 Sensor Networks and Others

In some cases, like trains that work in lines at 200 km/h or higher, it is necessary to monitor the condition of some parts of the bogie, such as axle boxes. In other scenarios, this monitoring is triggered by demand, but in all cases, a sensor network must be deployed onboard the train, with accelerometers and other sensors.

The specific aspects of this use case are far from the scope of the book, but not the network technology to provide connectivity to such devices.

There are many different setups and scenarios, but in some cases, accelerometers are CAN devices, whereas in some others, they are analog devices connected to an analog-to-digital converter.

A technology which is frequently used is EtherCAT (Ethernet for Control Automation Technology). It is standardized by IEC 61158 [158] (as type 12 devices) and is much used in automation. An optimization of IEEE 802.3 Ethernet standard to make data frames more oriented to real-time applications.

6.12.1 Summary of the Chapter

In this chapter, we have explained several technologies for the train communication network (TCN), which are the infrastructure for two different types of networks: consist and train backbone networks. This comes from the operational need to couple and decouple trains on a normal basis, which means that the network may need to rearrange upon these changes (backbone) or not (consist network). Notation and precise wording is more needed than ever here!

Regarding TCN for consists, we introduce several technologies: RS-485, CAN, MVB, and ECN. Only MVB and CAN are railway specific, being the other two adaptations to railways of technologies already defined in other sectors. For the train backbone, we have two specific technologies: WTB and ETB.

On the other hand, there are some specific aspects to be addressed when dealing with onboard electronics: EMC and the requirements for all electronic equipment onboard trains (compiled in the reference standard EN 50155). EMC, however, applies for wayside systems as well and can be a major issue due to the high electric currents and fields present in several railway locations.

Further Reading

Unlike others, in this chapter, the references for further reading are the standards that play the leading role:

- All parts of the IEC 61375 for the different TCNs.
- All parts of the IEC 61580 standard, for multimedia applications onboard trains.
- Finally, EN 50155 is essential to understand the onboard scenario and the implications it has on electronics.
- Finally, the European Technical Specifications for Interoperability (TSI) has some points for onboard networks, so its impact on interoperability is also noticeable.

In Chap. 8, we provide more information about the future of TCN standards, with the projects that have prototyped them.

Exercises

1. What is a consist?
2. What is an open train and a closed train?
3. What is the difference between a consist network and a backbone network?
4. Why are EMC aspects relevant in railways?
5. Which TCN technologies are Ethernet based?
6. What is the main use of the EtherCAT onboard trains?
7. What is OMTS? What are the differences between OMTS and TCMS?
8. Explain what the train inauguration is.
9. Does the EN 50155 standard consider vibrations and EMC for onboard electronic equipment?
10. What are the differences between MVB, EMD, and ESD?
11. What can happen if an ECN switch stops working?
12. What can happen if an ECN switch is powered off? Does the onboard network continue working?

Operation and Maintenance of Communications Systems in Railways

7

This final chapter is focused on an essential aspect of railways, which receives a lot of resources and attention in real-world scenarios but is, paradoxically, largely ignored in the literature in general. Therefore, here we provide some general ideas about the maintenance of many of the systems we have described in the other chapters of the book.

7.1 Railway Maintenance: General Concepts and Terminology

As we learned in Chap. 1, there are many types of maintenance (corrective maintenance, preventive maintenance, predictive maintenance, legal maintenance, and condition-based maintenance).

Corrective maintenance: Fixes only what ceases to work properly. This is the normal way to work when the consequences of this malfunction are not potentially dangerous for safety. Indeed, it is the simplest way to work but can have a notable reputational impact if there are many breakdowns.

Preventive maintenance: Performs periodical checks to see that everything is fine. If not, it replaces the damaged parts. This is costly because most of the time these periodic checks will be useless, but a notable workforce must be assigned to get these tasks done. Moreover, the availability of the railway assets can go down due to this (for example, trains are stopped for preventive maintenance once per month or every two months, which means that extra trains are needed to provide the committed service).

Predictive maintenance: Tries to determine in advance when an asset will stop working and replaces or fixes it just before. This type of maintenance is optimal (both guarantees safety and minimizes costs with no service disruptions) but requires an accurate model for the assets to maintain which can predict the condition of the asset. Many times, this is impossible to achieve due to the great complexity of the railway systems (for example, predicting when an axle box will fail is very complex, as many different factors converge there). However, in some scenarios where the behavior of the assets is predictable (because it is linear with time or depends on one or two variables only), this strategy is worth following. A good example of this is rail switches: monitoring the electrical current in the engines can allow us to predict the condition of the engine and estimate when it will fail.

Legal maintenance: This is the maintenance which must be done by legal mandate. In railways, there are many regulations and applicable laws, and this maintenance must fulfill all the requirements that come from them.

Condition-based maintenance (CBM): This type of maintenance is based on continuously sensing the assets to maintain to react in case it is needed. No models to predict its condition are needed, because it is assumed that it can be directly measured every time. Current technology (as of 2023) is ready to make CBM work but requires large expenditures to place sensors, connect them to the network, transmit the data, collect them, etc.

Indeed, there are many more, like reliability centered maintenance (RCM) or risk-based maintenance (RBM) which are not exclusionary to the others. Indeed, they all belong to the same process, focused on providing the best service to passengers. For example, an operator can set periodic checks of some subsystems, every month or every 2 months, functionally testing that everything is OK. Passenger Information System (PIS) displays, rail switches, CCTV cameras or recorders, etc., can fail unnoticeably, so perhaps having a strategy focused on detecting hidden failures could work. On the other hand, there are other subsystems that, when they fail, everybody notices: like train axles, the track itself, train or platform doors, etc. Furthermore, some failures can lead to accidents, so they must be prevented. The historic approach here has been to do both legal and preventive maintenance: this involves replacing some components at a certain time, even if they are working cor-

Supplementary Information The online version contains supplementary material available at https://doi.org/10.1007/978-3-031-82247-6_7.

J. Moreno García-Loygorri, *Telecommunication Systems for Modern Railways*, Textbooks in Telecommunication Engineering,
https://doi.org/10.1007/978-3-031-82247-6_7

rectly. For example, rubber parts of train couplers must be replaced every 8 years, because the manufacturer says so, that is, the manufacturer does not assume any responsibility for older faulty rubbers.

Regarding communications devices and networks, in general terms, it is not possible to implement CBM or predictive strategies for maintenance. The degradation of these systems is more subtle than mechanical or electrical ones. Therefore, although corrective and preventive maintenance still apply, other strategies must be set. We'll learn about this in the next sections.

7.2 Objectives and Organization of Maintenance: Onboard Systems and Wayside Systems

Typically, maintenance costs take a large share of both operators and infrastructure managers, so its organization must be carefully planned. Typically, it is divided into three levels:

Level 1: Maintenance staff is distributed across the railway network, as close as possible to the assets, that is, deployed in maintenance yards to be near the trains, or in stations or other places to be near the potentially failing asset. These people are focused on solving the issue very fast, but not on repairing the faulty component or system. This system is delivered to level 2.

Level 2: Deployed centrally, their work is to configure and repair replacement parts to allow level 1 staff to have what they need. Tasks on this level include formatting disks, soldering parts, replacing PCB, etc. Things that, if they were to be done onboard the train or on the tracks, would be much less efficient.

Level 3: Sometimes, level 2 does not have enough information, skills, or the required tools to repair something. In this case, the component or system must be forwarded to a third party, perhaps the manufacturer or a specialized company, to have it fixed.

All these levels can be organized in many ways (in-house, outsourced, etc.), but this approach emphasizes the maintainability and availability of the complete system.

It is also very common to separate the maintenance into areas or specializations, like power electronics, networks, RF, etc. The objective here is to have the staff trained and focused on what they are best at. Typically, in those railway companies which integrate both the train operator and the infrastructure manager, the maintenance departments for both trains and wayside are separated.

Another relevant issue is the way to compute the quality of the service (i.e., RAMS parameters). In wayside systems, the focus is on working hours, and in onboard systems, it is on kilometers run by the train. For example, the main parameter associated with reliability in wayside systems is mean time between failures (MTBF), but in rolling stock, it is mean kilometers between failures (MKBF).

$$MKBF = \frac{\text{Total km}}{\text{Total failures}}$$

Sometimes, MKBF is expressed in millions of km or hundred thousand, like this:

$$MKBF = \frac{\text{Total km}}{\text{Total failures} \cdot 10^5}$$

$$MKBF = \frac{\text{Total km}}{\text{Total failures} \cdot 10^6}$$

Either way, the information provided by the MKBF is the same.

However, this parameter does not consider the impact of the failures (for example, it does not discriminate severe failures, which have a strong impact on the service or in other systems).

To do this, failures are usually classified into levels of severity. Moreover, the failures of a system are usually represented as a tree, addressing the individual contribution of the components and subsystems (it makes no sense to replace a complete radio unit if the problem is a faulty power supply, for example).

Another relevant issue regarding communication systems and MKBF is that, unlike mechanical or pneumatic systems, the main driver for degradation is time, not kilometers. So, there is always pressure to account for reliability for electronic systems (that is, not only communications) using MTBF and not MKBF. The use of MKBF comes from historic reasons, the early days of railways, where no electronic systems were onboard.

In rolling stock, availability is usually accounted for one to two times per day. This includes accounting available trains only at peak hours, whatever they are (perhaps 7.30 a.m. or maybe 6.00 p.m.), but this depends heavily on the type of service. For example, some companies only give service at off-peak hours, which means that the hour of interest will not be 7.30 or 9.00.

On the wayside, this may change or not. In some cases, only availability at peak hour is accounted for, whereas in some others, it is done the whole day.

Finally, maintainability is usually accounted for as the time needed to fix the issue once the maintenance staff is on site. There may be differences across operators and infrastructure managers (IMs), but the general idea is the same for all of them, either wayside or onboard, etc.

7.3 Tools for Maintenance of Communications Systems: SNMP

Once we know the basics, let's see what can be done to maintain communications systems. The idea is to have a centralized platform, capable of acquiring status data from several devices in the field (onboard trains, on the wayside, etc.), store them in a database, and graphically represent them to allow maintenance staff to make the corresponding decisions.

This centralized maintenance platform can be for maintenance only or, perhaps, for operational tasks also. The idea is to skip costly preventive tasks and to react fast to incidents in the network. However, all this requires that devices send periodic status messages, acknowledging their condition. This can be done by several means, using proprietary protocols or standards, etc., but almost every solution has in common one thing: the Simple Network Management Protocol (SNMP).

It is both a protocol with some functions and a network management system ((NMS), which is the platform we mentioned before). The reader may notice the word "network" in both SNMP and NMS, but this system is not limited to just networking hardware. All IP-based devices can benefit from this technology.

The current version of this protocol is v3, which includes improved security and authentication features.

Regarding the functionality, SNMP is not only for monitoring but also for remote configuration of devices and management of the system. However, in some cases, operators or IMs restrict what can be done remotely, and therefore, over-the-air (OTA) upgrades are banned.

SNMP is an application layer protocol, which runs on top of the Transmission Control Protocol (TCP) / Internet Protocol (IP) stack. More precisely, all SNMP messages use User Datagram Protocol (UDP) packets.

The way SNMP works is the following:

- An SNMP manager requests information (queries) from an SNMP agent.
- The agent responds to the query.

Therefore, it is a manager-agent protocol. SNMP v3 is a significant change in some aspects of previous versions, as it includes security aspects into the protocol. Security is addressed by introducing encryption and a strong authentication mechanism. The first two versions (v1 and v2), having no cyphered data, allowed attackers to eavesdrop on the SNMP messages.

Indeed, SNMP v3 has also some vulnerabilities but is much stronger than v1 and v2. Indeed, the only security measure in older versions was a password, which made older versions of SNMP a nightmare for cybersecurity staff.

Therefore, SNMP v3 is a very helpful tool for the maintenance and management of communications systems in railways, both onboard and wayside. The requirements are very simple: IP connectivity and some room to run an SNMP agent on the device. So, what happens with non-IP systems? What can be done in legacy communications systems which lack an IP network?

7.3.1 Other Tools for Maintenance

In case the typical manager-agent scheme is not possible, there are other ways to deal with the monitoring of communications systems. Examples of this are RS-485, ProfiBus, Multifunction Vehicle Bus (MVB), and some others. Here, a polling scheme can be implemented from the bus master (on any other centralized entity) to query the status of the other devices, sometimes using programmable logic controller (PLC) and sometimes using other mechanisms.

This is typical onboard trains, where the train driver must notice the status of several onboard systems, and from stations, with all their equipment (lifts, intercoms, etc.). However, the railway trend is to have IP networks everywhere and to implement SNMP-based platforms. Moreover, these platforms tend to aggregate several systems from different vendors to centralize the maintenance.

Finally, all devices and subsystems require a maintenance guide, with details about the maintenance procedure, fault-finding steps, instructions for corrective maintenance, and testing.

The general idea is to make maintenance something routine, to be done on a fixed and predictable time basis, with all tools within reach of the technician. As discussed in other parts of this book: technology, people, and processes are the three main ingredients of a successful service.

Moreover, maintenance tasks of communications systems and components must be integrated in the general maintenance plan of the assets, like the train fleets or the wayside systems of a railway line. Having a good knowledge of the tasks to be done allows for proper resource allocation and smooth maintenance cycle of all railway assets.

Autoprovisioning

Autoprovisioning is perhaps the best friend of maintenance staff. This function allows electronic equipment to configure itself once it is connected to the network or powered up. This way, not only several resources but also human errors are saved.

Implementing a successful autoprovisioning strategy requires an engineering effort to allow the reliable configuration (and download of updated or new configuration details) in a safe and efficient manner.

7.3.2 Obsolescence and End-of-Life Estimation

In this section, we describe briefly two relevant aspects of the life cycle and maintenance of communications systems in railways.

When a system is commissioned, the most visible part of its life begins. When it stops working, it must be fixed. In this cycle of being working, under reparation, etc., electronic equipment can stay for a long period of time, but the problem comes when the manufacturer stops providing support for this device (or even for a software product). In this case, there are many risks to be tackled, like cybersecurity aspects: without support, there may be no security updates, so the system is in peril. On the other hand, some failures may not be fixed as the manufacturer does not provide any related support. The limited availability of some parts may have a strong impact as well. This one is related to the second one, but it is not the same, as in this one, the manufacturer does not repair anything, and it is only required to provide electronic parts. Finally, in some cases, new software is not able to work with old hardware.

All of them are relevant reasons to have a suitable management plan for obsolescence. We have just addressed the technical obsolescence, but there are two more types: the knowledge obsolescence, which is related to the knowledge of the staff, training, etc., and the functional obsolescence, which is about systems or devices which are less functional or desirable compared with the new ones (like having a 10-year-old mobile phone compared with a brand new smartphone). All of them must be managed with suitable procedures, like the International Railway Industry Standard (IRIS) [159] which addresses more things than obsolescence but is a good certification for suppliers and manufacturers to deal with it.

Typically, manufacturers provide an estimation of the end of life for their products, that is, a guess of how long their product will be able to work as it was conceived (without many failures, etc.). To do this, many different approaches can be followed, like pre-statistical modeling, more complex system models (i.e., reliability models), or accelerated tests to emulate a longer period (see EN 50155 for more information about these tests). Anyway, IMs and operators must have a plan to avoid service disruptions or unnecessary pain due to end-of-life issues of the electronic equipment.

7.4 Summary of the Chapter

In this chapter, we have learned about the different types of maintenance, and more importantly, how to remotely monitor communications systems: using an SNMP-based centralized platform.

On the other hand, we have introduced some useful ideas about railway maintenance for both wayside and onboard equipment.

This chapter is very brief, so I only provide 3 exercises:

1. We have a fleet of trains formed by 10 consists, each one of them formed by 3 cars. In this month, they have made 50 000 km each. Our maintenance staff detected and fixed 20 failures in the onboard communications systems of the entire fleet of 10 trains. Please compute the MKBF. Remember to take into account the total km of the fleet in "km · car" and provide the MKBF per 1 million km. You can assume 30 days per month. Solution: MKBF = 50000x3x10/(20x1e6)= 0.075

2. Now let's compute the availability. Assume that one day we had 3 complete trains stopped in the yard (this is, unavailable) because two faulty GSM-R cab radios and a broken antenna connector, respectively. Compute the monthly availability of the GSM-R system of the complete fleet for this month (assume 30 days per month). Solution: 30 days x 10 trains = 100 %; 1 day x 3 trains = 1 % unavailable; Availabilty = 100 – 1 = 99%

3. An MPLS network experiences 10 failures in one year. Please compute the MTBF. Solution: 365 days / 10 failures = 36.5 days. MTBF is usually given in hours, so 36.5 x 24 = 876 h.

8 Future Technologies and Communications Systems in Railways

8.1 Structure of the Chapter

Following the same scheme as the book itself, this chapter provides a vision about the future developments for the railway sector. We have done our best to write this chapter on a neutral way, that is, not giving our opinion about a particular technology, but summarizing the most promising developments of the mid-term future of railways. Therefore, this chapter is not an insight into the future (which would have been not very formal and, more importantly, not very useful as a textbook), but a concise description of some research projects that have been developed in recent years (2013–2023) and will probably have an important role in the future of railways.

We have also addressed some general trends in the communications sector, that is, not only things related to railways but to the common world, like cloud-edge computing, 5G, and some others that will presumably be very relevant in the future.

The most relevant research program that has had the railway sector is the EU-funded Shift2Rail. Most of the details that follow come from there, and we encourage readers to know more about it, its philosophy, and, specially, its outcomes. This initiative ended in 2022 but had a successor: the Europe's Rail Joint Undertaking. Given that it had just started, by the date of writing this book, there were almost no relevant results, so this chapter does not cover them.

Surely, this chapter will be outdated in the mid-term, but this is very good news for the railway sector and, more particularly, for communications engineers and everyone involved in this challenging and amazing sector.

Let's start with some future solutions for railways in general, that is, the topic we covered in Chap. 1. The other sections will follow the same order as in former chapters of this book.

Supplementary Information The online version contains supplementary material available at https://doi.org/10.1007/978-3-031-82247-6_8.

8.2 Future Developments for Railways

In the first section, we highlight those technologies that impact railways in general, that is, not communications systems. The very first to mention is the construction of train car body shells using composite materials instead of the typical aluminum manufacture. This is something that railways has adopted following the example of aircraft manufacturers, like Boeing with the 787 Dreamliner, whose wings and fuselage are largely made of polymers and other materials.

Apart from the obvious impact on certification and the uncertainties on the life cycle costs, this kind of train will require a paradigm shift in both the inspection duties and the sensors needed to monitor the condition of these structures.

Regarding traction systems, the trend is to use power electronics based on silicon carbide (SiC). Diodes based on this material can be lighter, smaller, have less power losses, are more efficient on the dissipation of higher temperatures, and can switch faster than its insulated-gate bipolar transistor (IGBT) equivalents, which are the most popular technology as of 2023.

Noise is another important research field in railways, as many people are affected by it. Typically, this field is addressed as noise, vibration, and harshness (NVH) to include other related disciplines that could share a common background and methodologies when studied. In particular, the current focus for railway noise is to reduce it below a certain level for daytime and nighttime, but this requirement has a strong impact on many fields. For example, track and infrastructure design, to prevent transmission of vibrations and noise to nearby buildings; train wheelset design, or AQ2more frequent wheel reprofiling (which has a strong impact on costs), Heating, Ventilation, and Air Conditioning (HVAC) and other onboard systems (as they are an important source of noise), use of cast iron brakes, among many others. However, the focus is not only to develop more silent systems but also to work on the prediction of the noise levels before these new components are integrated either in the wayside or in the train. Finally, not only exterior noise but

J. Moreno García-Loygorri, *Telecommunication Systems for Modern Railways*, Textbooks in Telecommunication Engineering,
https://doi.org/10.1007/978-3-031-82247-6_8

also interior noise is under the microscope (that is, the noise levels perceived by the passengers onboard the train). To sum up, it is a very important business that impacts all fields defined in railways.

In the next section, we briefly explain future trends and developments for signaling systems.

8.3 Future Signaling Solutions

8.3.1 The Future of ETCS

As we have seen in Chap. 2, European Rail Traffic Management System (ERTMS) is the most common technology for nonurban trains. We also have explained levels 1 and 2, but not level 3. This level was specified at the same time as the two others, but its deployment has not followed the same story of success. Indeed, it has not been adopted anywhere. European Train Control System (ETCS) level 3 allows the removal of track circuits as they are no longer required to safely determine the location of the train. This implies a relevant CAPEX and OPEX saving for infrastructure managers (OPEX savings come from the maintenance tasks no longer needed). However, as we said, level 3 remains undeployed. But, why? Perhaps the motives are a translation of costs toward operators, which are now responsible for a more complex system onboard the train, including the safe detection of the train integrity (that is, to detect that no cars or wagons are accidentally decoupled.[1]).

This lack of acceptance of ERTMS level 3 brings to the table another important issue, which is the heavy costs of this technology that make it inconvenient for low-density lines. Therefore, a more active trend for future ERTMS are those deployments intended for lines like these.

For example, regional ERTMS [160] has been explored to determine its feasibility. It requires the low-density line to be divided into dark zones, where only one train is allowed at a time. Global System for Mobile Communications-Railway (GSM-R) coverage is punctual, and track circuits are not required, both implying a relevant reduction of costs.

Another cost reduction approach involves the safe train location function. That is, it investigates the replacement of wayside balises (and perhaps many track circuits as well) with satellite positioning systems. The SATLOC project, finished some years ago, is a good example of this.

On the other side, the removal of GSM-R can remove complexity and costs. This is the idea behind the 3InSAT Project, which also skips balises and track circuits for positioning, replacing T2G and positioning with satellite technologies.

However, apart from the technical issues to be solved, the main impediment for the deployment of these alternative solutions is certification. This is even more relevant if interoperability is a major requirement (as we know it is, due to TSI). Certification is a long process, which involves many different actors (industry, operators, IM, etc.), may have legal implications, impact safety, and requires large testing and validation workforce, among many other challenges. Moreover, railways have a strong commitment to safety (as airliners and other transportation modes), which narrows the margin for incentives to push new technologies. To be more specific, SIL4 functions over satellite systems can be a remarkable feat, as the European LOCOLOC project highlighted.

[1]The irony comes from the fact that the train integrity problem is much of a nineteenth century issue for a system born in the 2000s.

8.3.2 ATO for Nonurban Trains HST

As we explained in Chap. 2, ETCS implements an automatic train protection (ATP) function, but not an automatic train operation (ATO) one. ATO has relevant benefits in terms of train regularity, energy efficiency, and some others that make it interesting for operators. Notice that an ATO function for ETCS must address interoperability, because this is what ETCS is about.

This research field has been explored in parallel in Europe, Japan, and some other places. For example, in Japan, JR Central has implemented an ATO system for the famous Tokaido Shinkansen, but it is not interoperable.

Regarding Europe, the Shift2Rail initiative has launched some projects to develop and test this technology. In particular, X2Rail-1 project aimed at ATO over ETCS, whereas one of its project successors, X2Rail-4, has the development of GoA4 over ETCS as one of its objectives. As we explained in Chap. 2, it's more complex GoA4 operation than GoA2.

8.3.3 Remote Driving

When we say "driverless" or "unattended," it should say "mostly unattended" or "mostly driverless," because in some scenarios or maneuvers, a driver is required. This is like the "last mile" in delivery. However, there are some urban rail networks that have their depots fully automated.

The TC-Rail project [161] studied the feasibility of a remote driving system for trains, based on a high-capacity train-to-ground system. It was required that the T2G had enough capacity to transport the data of both a video camera and an infrared camera that allowed the remote driver to see what the train was actually "sensing." Even a demonstrator was built, using a French locomotive, and a remote driving cabin, connected by an Long-Term Evolution Frequency-Division Duplex (LTE-FDD) system by Nokia working at 2.6 GHz.

Another demonstrator, based on a Ku-band satellite, was built and tested within the same TC-Rail project.

Technology Readiness Levels (TRL)

In research world, it is very frequent to categorize the maturity of a technology, depending on where it has been tested. It is not the same to have something working for just a moment in the lab, to manufacture a complete product and test it for 1 year on a train that makes 1 million km. This categorization of the readiness of a technology is called TRL, was first proposed by the NASA, and is formed by nine levels from 1 to 9:

TRL 1—basic principles observed.
TRL 2—technology concept formulated.
TRL 3—experimental proof of concept.
TRL 4—technology validated in lab.
TRL 5—technology validated in relevant environment (industrially relevant.
environment in the case of key enabling technologies),
TRL 6—technology demonstrated in relevant environment (industrially relevant environment in the case of key enabling technologies).
TRL 7—system prototype demonstration in operational environment.
TRL 8—system complete and qualified.
TRL 9—actual system proven in operational environment (competitive manufacturing in the case of key enabling technologies or in space).

This list is not the original one by the NASA (which is very aerospace-oriented) but a more general one based on that one provided by the EU for research projects.

Even though this approach is not free of criticism, undoubtedly, it has helped to clarify many discussions about the maturity of technology developments. Sometimes, it is hard to determine adequately if something is in TRL 6 or 7 or any other two consecutive levels, but in our opinion, it is a simple approach with great advantages.

8.3.4 Evolution of Urban Signaling: Train-to-Train CBTC

In Chap. 2, we explained the actual Communications-based train control (CBTC) system, which relies on T2G communications. But the next evolution of this technology could be based on direct train-to-train communications, which could replace the T2G link in some scenarios, therefore decreasing the end-to-end latency. This concept has been studied by some railway companies like Alstom and in academia as well [162], but is not yet implemented anywhere.

In current T2G CBTC systems, the critical data is double: first, trains estimate and sends to the ground systems its location and speed. Actually, also acceleration data and route information are usually delivered in CBTC systems as well, but for the sake of the model simplicity, we can skip them both. Second, ground systems (its name depends on the vendor) compute the movement authorities for each train and send them to the trains.

The most relevant limitation of this approach is the end-to-end latency. If the computing of the movement authorities could be safely done onboard trains and the location/speed data could be sent very fast to other trains, theoretically, we could have closer trains than in T2G CBTC. This is the basis of CBTC-T2T (train to train), as depicted in Fig. 8.1. Given that trains are not always at short distances to communicate directly, the T2T link would be complementary to the T2G, only to be used when it is available.

However, there are many issues to solve, such as:

- How to make the decision to switch from T2G to T2T communications. Because it will only compensate if the T2G link would be established for a certain time.
- Many cybersecurity and safety issues, as well as certification of the solution. It has also the drawback that it will not probably allow the removal of the costly T2G infrastructure, so the business case is challenging as well.
- It is a big step from the classical safety approach based on interlockings, because some movements on a CBTC-T2T do not fall under the interlocking domain. However, these movements can be banned and leave out of the interlocking only those who do not imply any signal, rail switch, etc.

For the communications point of view, CBTC-T2T requires a capable, reliable, and ultralow latency system, able to deal with direct communications as well. Finally, it has many things in common with the topic of the next subsection: virtual coupling.

8.3.5 The Next Step in Train Control: Virtual Coupling

Virtual coupling can be easily introduced as a generalization of the CBTC-T2T concept, and we have just explained in the previous section. The key idea here is to have two or more than two trains, but only one movement authority, because they behave as if they were only one train. This is, only one train but not mechanically coupled: virtually coupled by a safe wireless communications link.

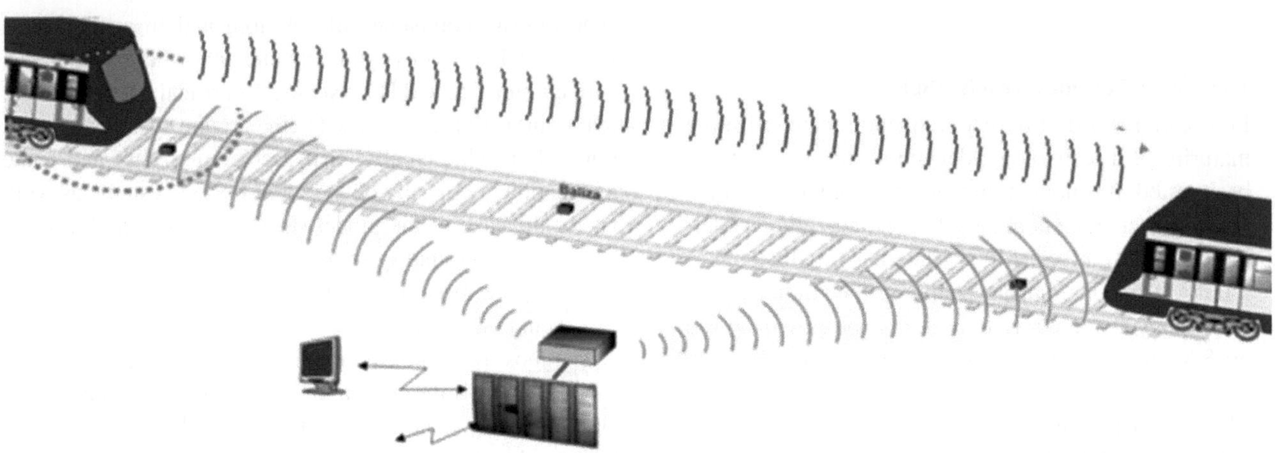

Fig. 8.1 CBTC T2T (red) and classic CBTC T2I (green) [163]

Some authors call it "ETCS level 4," but as it has no relationship with this standard, we prefer not to use this terminology. However, it is very good to see virtual coupling as an evolution of the moving block paradigm.

This is a futuristic system that was conceived by the end of the twentieth century [164][2] that has the potential to revolutionize railways, especially, freight trains. A problem that freight trains have now is related to the capacity of the line. Trains must circulate at distances larger than the braking distance, which drastically limits the number of trains in the line. With a virtual coupling system, freight trains could run closer to the braking distance, as they act as a single train (indeed, they share the same traction and brake commands, with very short latencies between units). Please notice that virtual coupling is not limited to freight trains, so it is applicable in high-speed trains and even in urban scenarios.

Consider that in "wired trains" (that is, mechanically coupled), not all train cars or wagons receive the brake/traction commands at the very same time. There are delays due to wires, processing, etc., but those systems are designed to have very limited delays and to be resilient to them.

This technology has been studied by many research groups and also by the railway industry. For example, X2Rail-3 [165] and MovingRail [166, 167] projects continued some of the work done in previous projects like Roll2Rail [168], looking for related experiences in the automotive industry and checking the feasibility of existing radio communications systems.

As in CBTC-V2V, there are still many engineering problems to be solved, such as cybersecurity and safety issues, as well as certification of the solution, having a very reliable and ultralow latency radio communications link. Moreover, in case the radio system fails (or is disrupted), the virtually coupled train will stop. Therefore, a fallback function must be defined to restart the journey. Perhaps, the regular T2G system could do the job, but on a very degraded performance. Finally, a relevant aspect still missing is the definition, testing, and validation of the essential safe coupling and decoupling of trains.

There was even a demo of the virtual coupling concept using real trains in a controlled scenario in 2018. For an extended summary of this promising technology, see [169], and for specific details on control techniques, see [170].

8.4 Future Noncritical Solutions and Other Developments for Railway Services

The noncritical services ecosystem is in constant evolution, with more cameras, video calls, more immersive PIS, or, on the other side, mobile apps that go further in the bring your own device (BYOD) paradigm, where some high-speed operators do not have onboard screens, so their infotainment platforms rely entirely on apps that run on passengers' mobile phones and WiFi connectivity onboard the train. Usually, these developments for noncritical services come from trends in the IT world, not limited to railway needs. Therefore, in this section, we highlight some of them.

8.4.1 Cloud-Edge Computing

As we have more capable communication systems with very low latencies, the border between cloud and edge is largely removed, so we say "cloud continuum" instead. Therefore,

[2] Or, more precisely, this is the oldest reference to virtual coupling we've been able to research.

the typical separation based on which one provides a service is not applicable. This redefinition started some years ago with the recentralization of services for stations, where instead of providing them in the edge and interfacing with the Operations Control Center (OCC) and other centralized systems, devices at stations were simplified on behalf of more centralized computing. This paradigm enabled the massive digitalization of railway assets, as it enabled the centralized storage of large volumes of data from the network. However, trains remained largely isolated because the T2G network was not capable enough. Recently, with the adoption of LTE and even more, with 5G, trains can adopt this approach as well.

PIS systems, CCTV, and intercoms benefit from this paradigm, as edge devices can be simplified to perform more functions on a centralized basis. Also, due to the capabilities of the network and the cost reduction of electronics, some functions can be performed on the edge (like video processing in CCTV cameras, for example), or to orchestrate this cloud-edge resource allocation dynamically, based on the available resources. This is something which is already being done in other sectors, and perhaps will happen in railways too.

8.4.2 IT-OT Convergence

Bridging information technologies (IT) and operational technologies (OT) is, perhaps, the most general trend that may engulf all the others. IT systems are designed for data processing, storage, and processing, meanwhile OT systems are for controlling real-world assets and machinery (sometimes with strict time constraints). IT systems are data based; OT systems is about the physical world, and data is its secondary product (for example, as data logs or sniffing data from a train Multifunction Vehicle Bus (MVB) bus).

This IT/OT convergence brings up many challenges, for example, in cybersecurity. As we explained in Chap. 6, many railway systems in both trains and wayside rely on Power Line Communication (PLC) and other electronic systems, sometimes not connected to any IP network and subject to different life cycles than IT systems. For example, management of security patches differs largely between OT and IT. Typically, this is very common in Information technology (IT) systems, something deeply rooted in their culture, but not in Operational Technology (OT).

8.4.3 CCTV

The trend is to have much more cameras both wayside and onboard and rely on automated analytics to trigger alarms in the OCC, for example, if somebody stays onboard a train that will head to the depot or somebody enters an unauthorized area in a station. These analytics can be either AI based or not, as AI models are getting powerful every day in video processing. There are many AI-related use cases that could be beneficial for railways like detecting smoke or fire, abandoned objects, audio analysis (to detect certain situations, like fights or graffities), and so on.

8.4.4 Voice Communications

Voice communications tend to include video as well to provide a better user experience. It is still not very extended, but perhaps in the future, we'll see video communications between passengers in stations or onboard trains and the OCC. Maintenance staff can benefit largely from video calls to get support from their colleagues, therefore improving their efficiency.

8.5 Future Network Technologies for Railways: SD-WAN

In Chap. 4, we introduced MPLS as the state-of-the-art technology for networks, not only in railways. However, there is already another technology available in the market which is seen as the next step in network evolution: software-defined wide area network (SD-WAN).

The key concept for SD-WAN is that it is software defined. That is, SD-WAN is based on software-defined networks (SDN), which is a technology based on a separation between the network software and the hardware. Here, the control plane is completely software based. It differs from the classical approach of having specific network devices (i.e., routers, switches) for specific network functions (routing, switching), as in SDN we find general-purpose devices running specific software to provide these functions. Consequently, the network can be virtualized, and the hardware can be commercial-off-the-shelf (COTS). It is highly centralized, therefore reducing the management and operation needs. Both SDN and network function virtualization (NFV) are two essential pillars of 5G. See T2G section in this chapter for more details. Moreover, it is more flexible and customizable (resources can be assigned and de-assigned dynamically, based on real-time requirements) and is safer (as cybersecurity can be managed on a centralized way, with better visibility of the threats). Another important benefit is the dependence reduction from vendor platforms for the network management.

There are many types of SDN, such as:

- API-based SDN: Control plane uses an application programming interface (API) to interact with network elements.

- Open SDN: The control plane relies on open protocols such as OpenFlow. These protocols can also be seen as API based but are not proprietary technology. Therefore, we can consider Open SDN as a particular case for API-based SDN.
- Hybrid SDN: Both SDN and classic network control protocols coexist in the same network, often as a backup, but sometimes only a part of the network is SDN based.

Perhaps the most important difference between MPLS and SD-WAN is that MPLS relies on engineered (that is, static and predefined) traffic and networking, whereas SD-WAN is based on dynamic traffic profiles and network requirements. On the other hand, MPLS is for private networks, and SD-WAN can route traffic through public networks (like Internet) using tunnels. The approach is different, but SD-WAN looks more flexible and, perhaps, under some circumstances, less expensive. On the other side, MPLS has fulfilled the strict requirements for railway networks in the last 20 years, providing a large sense of control for infrastructure managers.

Finally, it is important to notice that neither SD-WAN nor MPLS are network technologies for railways, but general-purpose ones instead. This fact implies a trend in railways, which relies on general standards more and more every time, rather than relying only on railway-specific solutions (that is, not reinventing the wheel). This does not mean that a large network integration and configuration must be done when implementing solutions from other industries and sectors in railways.

8.6 Future T2G Technologies and Related Developments

8.6.1 Train-to-Ground Future: FRMCS

From Chap. 5, we know that GSM-R is obsolete for several reasons. Indeed, now we know that 2035 is the key date where this equipment will not be supported from vendors anymore. The answer from the rail industry is Future Railway Mobile Communications System (FRMCS), which addresses two important requirements: first, the obsolescence of GSM-R mentioned before, that is, supporting ETCS and voice communications, and second, providing connectivity for the railway digitalization, which means all the services highlighted in Chap. 3: digital train, intercoms, CCTV, etc.

To achieve its requirements by the required deadline (2035), a detailed plan has been envisaged, and several working groups have been formed (functional, frequency affairs, onboard, another group to interface with the 3GPP). Indeed, the high-level requirements (which are the only available right now) have been included in an annex of the 2023 edition of the Technical Specifications for Interoperability (TSI).

Some of the essential building blocks of FRMCS have been researched in the European Shift2Rail initiative under the umbrella of several projects, like X2Rail-1, X2Rail-3, and EmulRadio4Rail, where many key agents of the rail industry were consortium members.

The development of FRMCS has been envisaged in three steps:

- FRMCS V1: The initial specification with many inputs for the TSI and other detailed specifications. This step has been already achieved.
- FRMCS V2: This step involves the development of a real-world demonstrator, which will be carried out in the 5GRAIL project [171]. In this stage, the industry has a leading role.
- FRMCS V3: This final stage is about market readiness, with the results of demonstrators and a clearer mind about cross-border procedures, more mature system-level, and functional requirement specifications (SRS and FRS). The objective is to have the earliest rollouts by 2025, which is very ambitious.

Learning from GSM-R mistakes, from the very beginning, FRMCS' development aimed 3GPP recognition, that is, the standardization body for both LTE and 5G. As the UIC acknowledges, this recognition from the 3GPP and the inclusion of railway-related functions as 3GPP use cases protects the investment from both operators and infrastructure managers in FRMCS. Moreover, the market will be larger and more competitive, avoiding "niche effects" that made GSM-R an expensive technology. Therefore, FRMCS use cases are included in 5G releases, starting with R16.

Another major issue is the migration strategies from GSM-R. Both wayside and onboard equipment must be migrated, and there are many different alternatives for this, with different approaches for the coexistence of legacy GSM-R equipment and brand new FRMCS devices. The key element onboard the train for this migration is the so-called TOBA Box (Telecom On-Board Architecture) which acts as an onboard hub, with all the communications systems onboard the train connected to it.

The 5GRail project, which started in 2020 and finished in 2023, was focused on prototyping FRMCS devices, among other objectives, like the development of the very first TOBA box prototypes, the development of the first 5G FRMCS–compliant network devices, interfacing with other systems, like ETCS, ATO, TCMS, and, last but not least, real-world testing.

Finally, a critical point in FRMCS development and standardization is the spectrum available for railway use. Not being able to have allocated more spectrum than for GSM-R is identified as a major risk for FRMCS (that is, 4 + 4 MHz in the 900 MHz band). However, ten more MHz in the 1900 MHz band has been allocated recently for railway use. As FRMCS is conceived as a flexible communications system, perhaps, it could use public bands from mobile operators as well, but this could be controversial.

8.6.2 WIFI 7

WiFi 7 is the market name for IEEE 802.11be, which will be released in May 2024. It is defined in all WiFi bands: 2.4, 5, and 6 GHz and represents another step further in terms of channel width (up to 320 MHz), modulation index (an impressive 4096-QAM, which would not be suitable for almost every scenario), and multi-link operation, to enforce seamless handovers (that is, being connected to an AP after starting the roaming procedure to connect to another AP). On the other hand, it is backward compatible with 802.11 ac and WiFi 6, which means that legacy devices could connect to a WiFi 7 AP, but under degraded conditions.

Time will tell if this new standard plays a major role in railways, as IEEE 802.11 ac does.

8.6.3 WiFi in Mmwave

WiFi is also defined for mmwave, more particularly in the 60 GHz band. This band enables high-capacity communications as very wide bandwidths can be used. IEEE 802.11ad [172] is the standard and works in the 57–66[3] GHz band (depending on the country or region, as frequency-band allocations may change). There are four channels defined on this band, each one of them 2.16 GHz wide.

IEEE 802.11ay is the follow-up standard for 60 GHz and defines other channel widths.

For railways, the typical application for mmwave WiFi is high-capacity data offload and download. For example, when a train stops at a terminal station or a depot, it gets connected to an mmwave AP and can transfer large files, like video recordings to be stored in the wayside (Uplink (UL) only), large train communication network (TCN) logs (UL only), and multimedia and infotainment files (Downlink (DL) only).

On the other hand, a system like this could be used for high-capacity communications between cars in the same consist, to form the TCN backbone of the train. This is the approach of the Shift2Rail Safe4Train3 project, which tested this technology in this environment.

[3] 57–71 GHz in the USA and Europe; 57–64 GHz in China. For all the other cases, 57–66 GHz is the most common band allocated for this purpose.

Another application, less mature but potentially more disruptive, is to use mmwave links for direct communications between trains (i.e., T2T communications). This link benefits from the high stability and the high capacity of IEEE 802.11 in mmwave. In Fig. 8.2, we show some latency results from some tests in two subway trains. We can see that the latency is around 1 ms, which is remarkable if we consider that both trains were in motion. Indeed, the delay is always under 1 ms, which is very relevant as well.

8.6.4 Satellite Communications

Including satellite communications as a future communications system is a little tricky because railways have used satellite communications (satcoms) for a long time. One of the first examples for this was the Thalys trains in Europe, which used the Hispasat satellite to provide onboard Internet connectivity. And many more have followed its steps in recent years. However, we include satcoms here because the rail industry has a growing confidence on this technology which probably make it more extended soon. Therefore, in this section, we highlight some specific features of satcoms and discuss some related issues, to allow readers to understand challenges, advantages, and drawbacks.

The first thing to consider is the type of orbit, as it could be:

Low Earth orbit (LEO), typically between 200 and 2000 km away, which means delays of up to 10 ms, is very suitable for many types of communication services, but they travel very fast (25,000 km/h), which means high Doppler shifts and also frequent handovers, as they stay visible from a fixed Earth surface location about only 15 min. Examples of LEO satellite constellations are Starlink and Iridium.

Medium Earth orbit (MEO) satellites orbit our home planet at distances between 6000 and 25,000 km, take 5–15 h to complete an orbit, which implies 2–4 h of coverage for a certain location. Orbits are close to circles; the delay is 80–100 ms, and we need much less satellites to provide global coverage than in LEO. Typical examples of MEO satellites are GNSS constellations, like GPS, GLONASS, Galileo, or Beidou (that is, not for communications).

Geostationary Earth orbit (GEO): Only three satellites are needed to provide planetary coverage, as they remain static over a certain point of the Earth surface (GEO satellites orbit the Earth at a speed of 11,000 km/h, but there is no Doppler as the relative speed to the Earth ground is 0 km/h). The main drawback here is the delay, which is around 240 ms as they orbit at 35786 km. A summary of all three orbits is shown in Fig. 8.3.

To use satellites for T2G communications, we must consider several issues, such as:

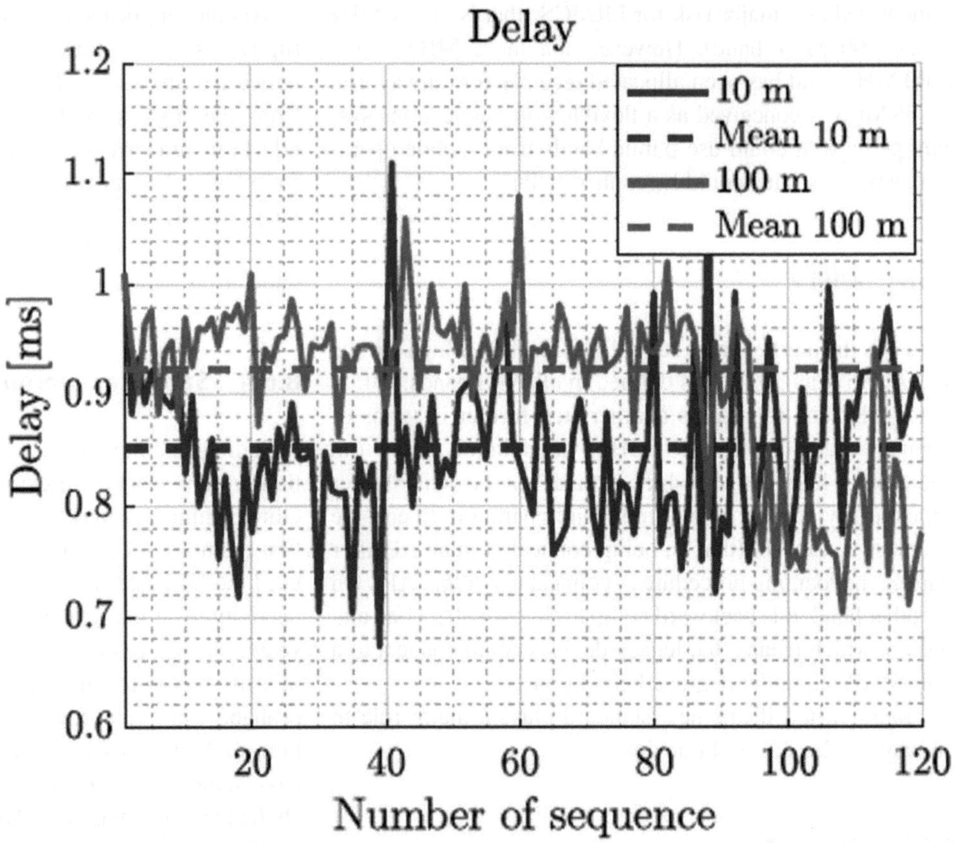

Fig. 8.2 Delay of the mmwave T2T tests at distances of 10 m and 100 m between trains [162]

- Accounting Doppler, especially in LEO orbits.
- Considering onboard antennas, which are typically large, because an automatic pointing system is needed. Therefore, some gauge issues may arise, even impacting the aerodynamic profile of the train.
- Latency can be high in some cases.
- Tunnels and other special scenarios are not suitable for satcoms. There are solutions to overcome this limitation, like gapfillers, but have many limitations.
- Current technology is not able to provide very high throughputs (typically, 0.5 Mbps; recently, 10 Mbps), but this will probably improve soon.
- As of today, satcoms do not meet EIRENE requirements, not to say FRMCS, specially regarding voice and latency.

8.6.5 Future Antenna Developments for Railways

The rollout of new T2G systems will be related to the development of new antennas. Larger multiple-input multiple-output (MIMO) arrays and the combination of several antennas into a single one will still be a trend in railways. Moreover, the replacement of mechanically steered antennas for satellite use by flat electronically scanned arrays can be beneficial in terms of aerodynamics and maintenance.

8.6.6 Mobile Communications: 5G and 6G

5G is the fifth-generation mobile communications; it is in line with LTE, and its key requirements are:

- Ultralow end-to-end latencies: 1 ms.
- Data rate: 1 Gbps.
- Density of connected devices: 1 million per km^2.
- High availability: 99.99%.

Another essential aspect of 5G is flexibility: several different bandwidths (up to 400 MHz), different subcarrier spacings (from 15 kHz to 240 kHz), Time Division Duplexing (TDD) and FDD, and many bands, not all devices on 5G, are required to support all defined bandwidths (unlike LTE), and 5G supports adaptive bandwidth, which allows devices access to a low-bandwidth, low-power configuration when appropriate. Even more important, 5G is defined in two different frequency ranges called FR1 and FR2. FR1 is for sub-6 GHz bands. This is fully consistent with previous generations of mobile. See Chap. 5 for more information.

FR2 is for mmwave bands. An absolute greenfield technology for mobile, millimeter-wave communications or, in short, mmwave communications are those bands where the wavelength is 1–10 mm. In terms of frequencies, formally, it is usually defined as 30–300 GHz, but in practice, the

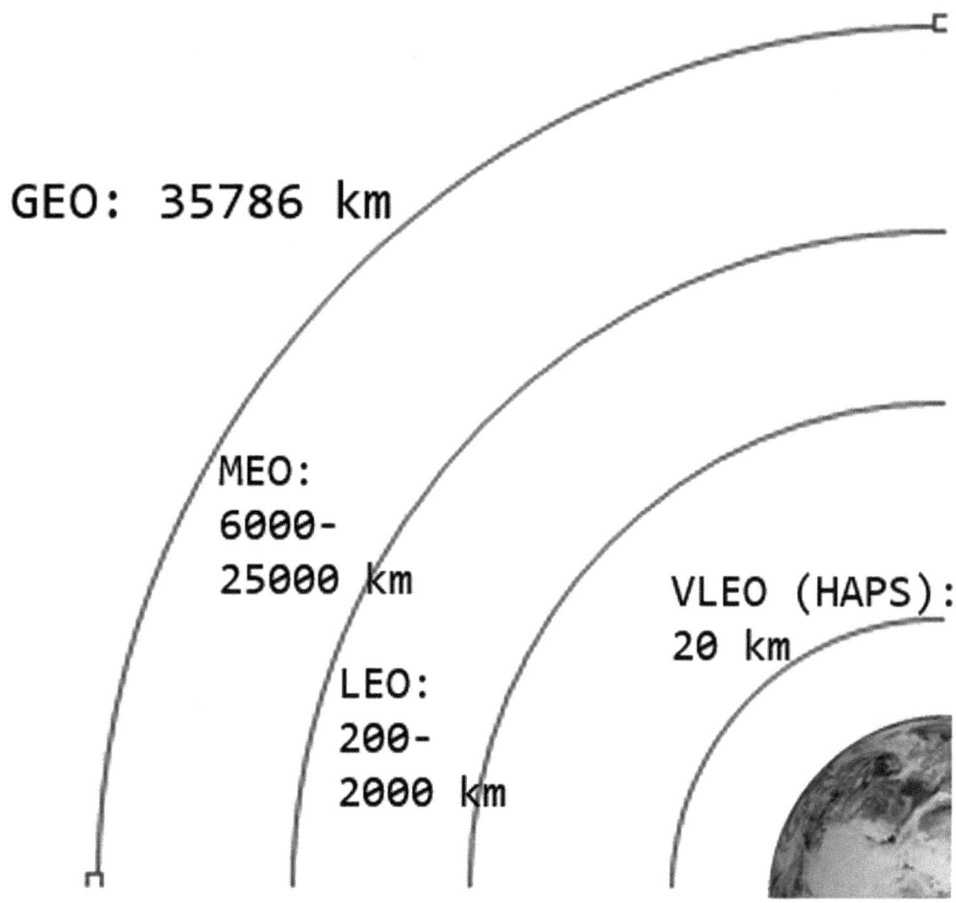

Fig. 8.3 Types of satellites depending on the orbit

24-GHz band is included there as well. In these frequency bands, the propagation is slightly different from FR1, because even small objects can block the signal and attenuate it largely. Moreover, there are some bands where some gases available in the atmosphere like O_2 and H_2O have peaks of absorption, which can attenuate the signal as well. Even the leaves of a tree have a significant impact on path loss. Finally, there are large losses due to free-space attenuation because the carrier frequency is now high. Therefore, a solution for all these losses is needed, and beamforming is the answer. It is an essential technology in 5G to compensate for extra path loss and all the others.

Regarding its attractiveness for operators, efficiency plays a key role. Measured both in W/bps and bps/Hz, that is, power efficiency and spectrum efficiency, 5G has better figures than LTE and its predecessor technologies.

There are some usage cases where 5G is defined: eMBB, MMTC, URLLC, and V2X. Let's discuss them.

Enhanced mobile broadband (eMBB): For very high traffic volumes, in mobility (medium to high). Same idea as LTE but higher data rates, for both business and consumer markets. Data download and upload. eMBB is the first use case that 5G addresses, but not the most relevant for the long term. Is the most mature today, in non-standalone (NSA) rollouts. NSA rollouts are those where the core is still LTE, but the radio access network is 5G (or a mix between LTE and 5G). Standalone (SA) 5G networks are those where both the RAN and the core are 5G. The applications are many, such as augmented reality, 8 K video, connectivity in ultra-dense environments, mobile-cloud computing, and connectivity for HST and urban trains.

Massive machine-type communications (MMTC): Connectivity for a large number of IoT devices (low-power, low-traffic profiles and batteries must last for a long time). Devices are assumed to be cheap but able to communicate even if the received signal is weak. In IoT devices, the individual data rate could be low, but the aggregated for all devices is high. In MMTC, another challenge is to avoid (mobile) signaling traffic overload the network (reconnections). MMTC applications for railways are large: IoT networks in stations and trains, train telemetry, fleet management, etc.

Ultra-reliable low-latency communications (URLLC): Communications that need very low latencies (<1 ms) and also ultrahigh reliability. LTE latency is suitable for VoIP calls, but not for real-time applications, like those needed for remote control of machinery (like trains) or virtual coupling between trains.

Vehicle-to-everything (V2X): Communications related to vehicles; this is not only for automotive but also for trains, aircrafts, ships, etc. In railways, it has not played a relevant role yet (the only relevant one is T2G so far). This use case has been assigned spectrum on an unlicensed band and, more recently, another one (ITS band, 5.9 GHz). For the side link, that is, the direct communication between devices without need for a network, a new interface is defined: PC5.

All these usage cases mean a complex mix of requirements to be fulfilled. Actually, there is no need to meet all requirements at the same time. The architecture of 5G is similar to LTE and admits both standalone and non-standalone deployments.

There are three key enabling technologies for 5G: NFV, SDN, and slicing.

Network function virtualization (NFV): If networks are formed by specific hardware which provides specific network functions (i.e., switches, routers, etc.), this leads to hardware-software coupling, high prices, migration problems, etc. In NFV, network elements run on a COTS Hardware (HW) (generic and cheaper), with a different software for every different network function, which implies easier to maintain SW.

Software-defined networking (SDN): In previous mobile generations, the control plane is very complex. Separation between control plane and user plane allows centralization of control plane. It is complementary to NFV.

Network slicing: A slice is a virtual and isolated subnetwork. Slices are implemented using NFV and SDN, and slices could be set for different use cases (eMBB, URLCC, MMTC, V2X), different users, and even enables mobile virtual network operators (MVNO). This approach provides a remarkable flexibility to operators, allowing them to optimize each slice for the traffic they carry. For example, mobile operators using 5G with this technology could service railway operations with an agreed quality of service (QoS).

8.6.7 6G

Although its definition and requirements are not ready yet, 6G will be a reality in some years. Without current knowledge of the discussion on 6G, let's see some technologies that could be helpful for the railway sector: reconfigurable intelligent surfaces (RIS) and joint communications and sensing (JCAS).

Reconfigurable intelligent surfaces (RIS) is a technology which aims to dynamically control the received signal, for example, can transmit a received signal toward a certain direction. The European Telecommunications Standards Institute (ETSI) says that RIS turns "wireless environment into a service." For example, RIS can be placed to overcome the extra attenuation of mmwave transmissions due to penetration into buildings.

Joint communications and sensing (JCAS): This is to exploit radio waves already available from current communication systems for sensing purposes. Like RIS, it is a topic which is receiving a lot of attention and funding, as it looks very promising.

8.7 Future TCN and Other Onboard Technologies

Another relevant research field is the future TCN. In Chap. 6, we explained the different technologies for TCN and LAN, and the trend is clear: replace train lines with industrial, sometimes proprietary buses, and replace these buses with Ethernet-based networks (ECN and Ethernet train backbone (ETB), for the consist and the train network, respectively). As we explained in Chap. 6, ETB is still a bottleneck in terms of performance, so perhaps wireless networks can save the day. Another limitation we have is that providing high-safety functions (i.e., SIL 4) replacing all train lines with data networks is very difficult [173]. This is for several reasons, among them the reliability and availability of the link, plus the required end-to-end latencies.

Consequently, today, we have two relevant trends regarding TCN: include wirelessly connected devices in the TCN and, a more deterministic wired communications, to achieve higher safety levels and integrate many TCMS functions in data networks.

A wireless TCN would have some obvious advantages, like less weight,[4] and less wires and connectors, which means lower costs. The main requirement here is to provide the same safety levels and performance than a wired network. On the other hand, the wireless TCN may coexist with wired segments of the same TCN, all of them with the same safety requirements than classical ECN and ECB networks (see Chap. 6 for more details). This is the approach for a complete research line, started by the Roll2Rail project [168] in 2015 and continued by its successors Connecta-1, 2, and 3, and Safe4Rail-1, 2, and 3, which finished in 2023, providing many interesting results as well as a complete set of inputs to the upcoming versions of several IEC 61375 standards.

Another relevant requirement comes from the fact of the dynamic coupling/uncoupling of consists to form trains. This implies to address train-integrity and train inauguration functions. Both are relevant in terms of safety, because the consist sequence can't be safely determined at inauguration time in some degraded modes (for example, if an unpowered consist is at one of the two train ends) [168]. This is called the "dead consist" problem. Apart from the mentioned before,

[4]As in automotive, cabling weight may seem not relevant but can be huge. More than 10 years ago, some metro trains had 1 km of wire for every linear metro of the train.

the problem here is to guarantee the backbone continuity even with a dead consist in between.

Consequently, the determination of the train length must be done by other means (for example, radio frequency identification (RFID) devices, which must be integrated in the TCN). These are just some examples of the complexity of some TCMS functions, with impact safety that must be addressed every time a technology upgrade is proposed for standardization.

Therefore, the proposal from this research line implies two new types of networks for the TCN: wireless train backbone (WTBN), analogous to the ETB and, wireless consist network (WLCN), analogous to the ECN.

Many different technologies have been considered for the WTBN, but it seems that 3GPP is the best to fulfill the requirements. Regarding the WLCN, there may be room for other technologies, like IEEE 802.11, even at mmwave, but time will tell. Remember that train backbone networks may change every day (as we couple and decouple trains) but consist networks are much more stable than that.

Another technology that has been explored in these projects is time-sensitive networks (TSN), which is based on several IEEE 802.1 standards, such as IEEE 802.1Qbv, Qav, Qbu, etc., allows a more deterministic TCN, resembling them to MVB and WTB. Notice that, by definition, wireless networks are more random in nature than wired networks, so TSN can help to fix this perhaps for access to media and for resource management. The aim of these projects is to check if TSN profiles can help to control latency, prioritize traffic, and synchronize.

Apart from the network technologies, other topics for future TCMS and TCN were addressed, such as functional distribution framework (FDF), which aims to solve two main issues: the abundance of CPUs onboard trains and the coexistence of both safety critical and non-safety critical functions in the same network. The former implies low reliability and more complexity, rather than a more centralized approach, while the latter leads to recertification even if the new software version is a very small change. So, FDF is a framework to allow cost-efficient development of applications, upgrades of already existing ones, etc. The idea comes from a similar one in the automotive industry called AutoSAR. It happens from time to time, but it is not a bad idea to look around and try to copy good ideas from other industries.

Finally, it is impossible to roll out wireless networks without talking about cybersecurity, for obvious reasons. Even in wired networks, it is a major topic, not to say in wireless.

By the date of writing this book (2024), the status of all the aforementioned TCN new developments was tested in some degree (some of them with real trains in operation) and proposed for new versions of the IEC 61375 standards.

8.7.1 Digital Automatic Coupling

Digital automatic coupling (DAC) is to automatically couple freight trains mechanically, electronically, that is, both the power supply and data link and pneumatically.

The main objective is to decrease the needed time to couple and decouple trains manually. Notice that the current standard is the "buffer and chain" coupler, which is costly preparation time.

There are other benefits from this technology, such as allowing longer trains, which increases the capacity of the line, reducing the needed maintenance, provisioning of the safe-train integrity determination function, which is a key element for the development of ETCS L3, and finally, higher safety for railwaymen.

This has been researched in the DAC4EU project, funded by the Shift2Rail program. The main objective of this initiative is to replace old couplers like the "buffer and chain" (see Fig. 8.4).

8.8 The Future of Maintenance in Railways

The future of maintenance in railways appears to be based on two pillars: more GUI for systems' administration, maintenance, and operation, and an extensive use of big data to develop predictive models of the railway assets, that is, both non-IA and IA-based algorithms.

Therefore, as data acquisition and extraction techniques are already mature enough, it is the turn of the algorithms to provide input for more efficient maintenance plans, based on the real condition of the component rather than in general statistics. Consequently, overmaintenance, which is a consequence of both the culture of safety and the logistic limitations, will be drastically reduced.

Summary of the Chapter

In this chapter, we went over the details of some future developments related to railway communications systems. The structure of this chapter is the same as for the book: railways, signaling, noncritical services, network technologies, T2G, and onboard.

We highlight the impact of public-private partnerships like the European Shift2Rail program in the removal of bottlenecks in the development of some railway fields.

Of course, it is very hard to make predictions, specially about the future. But there are some trends that are shaping railways right now: the convergence toward standards (therefore, avoiding niche and ad hoc solutions), IT/OT convergence, FRMCS, 5G rollout (which addresses some railway-related use cases), and, perhaps above all, interoperability.

Fig. 8.4 Buffer and chain coupler [174]

Finally, we encourage readers to be up to date reading papers, deliverables, white papers, and newsletters focused on railway research. The future looks bright for railways!

Further Reading

To be up to date about research projects in railways, we recommend three ways: first, browse websites of large research programs like Shift2Rail or Europe's Rail, which have links to specific projects deliverables, brochures, and papers.

Second, read research papers. This method is only recommended for researchers, PhD students, etc., as it is the most technical and least divulgative. There are many ways to do this, but perhaps IEEE Xplore, Google Scholar, ResearchGate, and some others are the most popular ways to start with.

Finally, browse the websites of the main railway agents, which are very focused on research: CAF, Thales, Hitachi, Alstom, Siemens, Schunk, Faiveley, and so on.

Acronyms

3GPP	3G Partnership Project
AC	Alternate Current
ACF	Auto Correlation Function
AFD	Average Fading Duration
AI	Artificial Intelligence
AMPS	Advanced Mobile Phone System
APC	Automatic People Counters
API	Application Programming Interface
APN	Access Point Name
AREMA	American Railway Engineering and Maintenance-of-Way Association
ASFA	Anuncio de Señales y Frenado Automático
ATACS	Advanced Train Administration and Communication System
ATC	Automatic Train Control
ATO	Automatic Train Operation
ATP	Automatic Train Protection
ATS	Automatic Train Supervision
AuC	Authentication Centre
BAN	Body Area Network
BER	Bit Error Rate
BGP	Border Gateway Protocol
BS	Base Station
BSC	Base Station Controller
BSS	Base Station Subsystem
BTM	Balise Transmission Module
BTS	Base Transceiver Station
BYOD	Bring Your Own Device
CAD	Computer-Aided Design
CAN	Controller Area Network
CAPEX	Capital Expenditure
CARS	China Academy of Railway Sciences
CBM	Condition-Based Maintenance
CBTC	Communications-Based Train Control
CCN	CANopen Consist Network
CCTV	Closed-Circuit Television
CDF	Cumulative Distribution Function
CDMA	Code Division Multiple Access
CE	Conformité Européenne
CEN	European Committee for Standardization
CENELEC	European Committee for Electrotechnical Standardization
CEPT	European Conference of Postal and Telecommunications Administrations
CIR	Channel Impulse Response
CNIS	China National Institute of Standardization
COS	Customer-Oriented Service
COTS	Commercial Off-the-Shelf
CP	Cyclic Prefix
CPU	Central Processing Unit
CSS	Chirp Spread Spectrum
CSV	Comma-Separated Values
CTC	Centralized Traffic Control
CTCS	Chinese Train Control System
CTF	Channel Transfer Function
DAC	Digital-to-Analog Converter
DC	Direct Current
DECT	Digital Enhanced Cordless Telecommunications
DFS	Dynamic Frequency Selection
DFT	Discrete Fourier Transform
DHCP	Dynamic Host Configuration Protocol
DL	Down Link
DMI	Driver-Machine Interface
DNS	Domain Name System
DPI	Deep Packet Inspection
DS-ATC	Digital Communication & Control for Shinkansen-Automatic Train Control
DSSS	Direct Sequence Spread Spectrum
DTG	Distance-to-Go
DTO	Driverless Train Operation
Ebicab	European Balise-Based Cab Signalling
ECN	Ethernet Consist Network
ED	End Device
EH	Hybrid Electromagnetic
EHF	Extremely High Frequency
EIR	Equipment Identity Register
EIRENE	European Integrated Railway Radio Enhanced Network
EIRP	Effective Isotropic Radiated Power

J. Moreno García-Loygorri, *Telecommunication Systems for Modern Railways*, Textbooks in Telecommunication Engineering,
https://doi.org/10.1007/978-3-031-82247-6

EM	Electromagnetic
eMBB	Enhanced Mobile Broadband
EMC	Electromagnetic Compatibility
EMD	Electromagnetic Disturbance
EMI	Electromagnetic Interference
EN	European Norm
eNB	Evolved Node B
ENISA	European Network and Information Security Agency
EPC	Evolved Packet Core
ERA	European Union Agency for Railways
ERTMS	European Rail Traffic Management System
ESD	Electrical Short Distance
ETB	Ethernet Train Backbone
ETCS	European Train Control System
EtherCAT	Ethernet for Control Automation Technology
ETSI	European Telecommunications Standards Institute
E-UTRAN	Evolved Universal Terrestrial Radio Access Network
EVC	European Vital Computer
FCC	Federal Communications Commission
FDD	Frequency Division Duplex
FDF	Functional Distribution Framework
FGC	Ferrocarrils de la Generalitat de Catalunya
FHSS	Frequency-Hopping Spread Spectrum
FIR	Finite Impulse Response
FLD	Flange Lubrication Detectors
FOD	Foreign Object Debris
FRMCS	Future Railway Mobile Communication System
FRS	Functional Requirement Specification
FWA	Fixed Wireless Access
GCR	Group Call Register
GEO	Geostationary Earth Orbit
GLONASS	Global Navigation Satellite System
GMSK	Gaussian Minimum Shift Keying
GNSS	Global Navigation Satellite System
GoA	Grade of Automation
GPS	Global Positioning System
GSM	Global System for Mobile Communications
GSM-R	Global System for Mobile Communications-Railway
GUI	Graphical User Interface
H.264	Advanced Video Coding
HD	High Definition
HF	High Frequency
HLR	Home Location Register
HPBW	Half Power Beam Width
HSL	High-Speed Line
HSR	High-Speed Rail
HST	High-Speed Train
HTTP	Hypertext Transfer Protocol
HVAC	Heating, Ventilation, and Air Conditioning
ICI	Intercarrier Interference
IDFT	Inverse Discrete Fourier Transform
IDS	Intrusion Detection System
IEC	International Electrotechnical Commission
IEEE	Institute of Electrical and Electronics Engineers
IGBT	Insulated Gate Bipolar Transistor
IM	Infrastructure Manager
IoT	Internet of Things
IP	Internet Protocol
IPS	Intrusion Prevention System
IRSE	Institution of Railway Signal Engineers
ISI	Inter-Symbol Interference
ISM	Industrial, Scientific, and Medical
ISO	International Organization for Standardization
IT	Information Technology
ITS	Intelligent Transportation Systems
ITU	International Telecommunication Union
JCAS	Joint Communications and Sensing
JPEG	Joint Photographic Experts Group
JSON	JavaScript Object Notation
KPI	Key Performance Indicator
LAN	Local Area Network
LCC	Life Cycle Cost
LCR	Level Crossing Rate
LCX	Leaky Coaxial Cable
LDP	Label Distribution Protocol
LED	Light-Emitting Diode
LEO	Low Earth Orbit
LER	label Edge Routers
LEU	Lineside Electronic Unit
LF	Low Frequency
LIFO	Last In, First Out
LORA	Long Range
LORAWAN	Long Range Wide Area Network
LOS	Line of Sight
LPWAN	Low-Power Wide Area Network
LSP	Labeled Static Paths
LSR	Label Switch Router
LTE	Long-Term Evolution
LTE-U	Long-Term Evolution in Unlicensed spectrum
LTV	Linear Time Variant
LZB	Linienzugbeeinflussung
MAC	Media Access Control
MAGLEV	Magnetic Levitation
MCS	Modulation and Coding Scheme
MEO	Medium Earth Orbit
MF	Medium Frequency
MIMO	Multiple-Input Multiple-Output
MISO	Multiple-Input Single-Output
MKBF	Mean Kilometers Between Failures

ML	Machine Learning
MME	Mobility Management Entity
mMTC	Massive Machine Type Communications
MNO	Mobile Network Operator
MORANE	Mobile Radio for Railway Networks in Europe
MPC	Model Predictive Control
MPEG	Moving Picture Experts Group
MPLS	Multiprotocol Label Switching
MS	Mobile Station
MTBF	Mean Time Between Failures
MTTR	Mean Time to Repair
MU-MIMO	Multi-User Multiple-Input Multiple-Output
MVB	Multifunction Vehicle Bus
NAC	Network Access Control
NASA	National Aeronautics and Space Administration
NFV	Network Function Virtualization
NIS	Network Information Systems
NLOS	Non-Line-of-Sight
NMS	Network Management System
NMT	Nordic Mobile Telephone
NSA	Non-standalone
NTP	Network Time Protocol
NVH	Noise, Vibration, and Harshness
NVR	Network Video Recorder
O&M	Operations and Maintenance
OCC	Operations Control Center
OFDM	Orthogonal Frequency Division Multiplexing
OFDMA	Orthogonal Frequency Division Multiple Access
OGF	Optical Glass Fiber
OHL	Overhead Line
OMSS	Operation and Management Subsystem
OMTS	Onboard Multimedia and Telematic Services
ONVIF	Open Network Video Interface Forum
OOS	Operator Oriented Services
OPEX	Operational Expenditure
OSI	Open Systems Interconnection
OT	Operational Technology
OTA	Over the Air
PABX	Private Automatic Branch Exchange
PAPIS	Passenger Announcement and Public Information System
PAPR	Peak-to-Average Power Ratio
PAS	Passenger Announcement Systems
PC	Personal Computer
PCR	Puesto Central de Radio
PDF	Probability Density Function
PDP	Power Delay Profile
PFR	Puesto Fijo de Radio
P-GW	Packet Data Network Gateway
PHY	Physical Layer

PIS	Passenger Information System
PLC	Programmable Logic Controller
PLMN	Public Land Mobile Network
PMR	Private Mobile Radio
PMR	Puestos Moviles de Radio
POE	Power over Ethernet
POTS	Plain Old Telephone Service
PSD	Platform Screen Doors
PSIM	Physical Security Information Management
PSO	Public Service Obligation
PSTN	Public Switched Telephone Network
PTC	Positive Train Control
PTT	Push-to-Talk
QAM	Quadrature Amplitude Modulation
QOS	Quality of Service
QPSK	Quadrature Phase Shift Keying
RAC	Radio Admission Control
RAMS	Reliability, Availability, Maintainability, and Safety
RAN	Radio Access Network
RB	Resource Block
RBC	Radio Block Center
RBM	Risk-Based Maintenance
RCM	Reliability Centered Maintenance
RF	Radio Frequency
RFID	Radio Frequency Identification
RIS	Reconfigurable Intelligent Surface
RMS	Root Mean Square
ROSCO	Rolling Stock Operating Company
RRM	Radio Resource Management
RRU	Remote Radio Unit
RS	Recommended Standard
RSSB	Rail Safety and Standards Board
RSVP	Resource Reservation Protocol
RTLS	Real-Time Location System
SA	Standalone
SCADA	Supervisory Control and Data Acquisition
SDH	Synchronous Digital Hierarchy
SDN	Software-Defined Networking
SD-WAN	Software-Defined Wide Area Network
S-GW	Serving Gateway
SHA	Secure Hash Algorithm
SHF	Super High Frequency
SiC	Silicon Carbide
SIL	Safety Integrity Level
SIMO	Single-Input Multiple-Output
SINR	Signal-to-Interference-Plus-Noise Ratio
SIP	Session Initiation Protocol
SISO	Single-Input Single-Output
SL	Side Link
SLA	Service Level Agreement
SLL	Secure Sockets Layer
SMS	Short Message Service

SNMP	Simple Network Management Protocol
SNR	Signal-to-Noise Ratio
SRS	System Requirements Specification
SSL	Secure Sockets Layer
STM	Synchronous Transport Module
T2G	Train-to-Ground
T2T	Train-to-Train
TACS	Total Access Communication System
TBS	Transmission-Based Signaling
TC	Track Circuit
TCMS	Train Control and Management System
TCN	Train Communication Network
TCP	Transmission Control Protocol
TDD	Time Division Duplex
TDL	Tapped-Delay Line
TDMA	Time Division Multiple Access
TE	Transverse Electric
TETRA	Terrestrial Trunked Radio
TGV	Train à Grande Vitesse
THR	Tolerable Hazard Rate
TLS	Transport Layer Security
TM	Transverse Magnetic
TRAU	Transcoder and Rate Adaptor Unit
TRDP	Train Real-Time Data Protocol
TRL	Technology Readiness Level
TS	Technical Specification
TSN	Time-Sensitive Networking
TTL	Time to Live
TWT	Target Wake Time
UDP	User Datagram Protocol
UE	User Equipment
UHF	Ultrahigh Frequency
UI	User Interface
UIC	International Union of Railways
UITP	International Association of Public Transport
UK	United Kingdom
UL	Uplink
ULLC	Ultralow Latency Communications
UMTS	Universal Mobile Telecommunications System
UNIFE	Union of the European Railway Industries
UNII	Unlicensed National Information Infrastructure
USA	United States of America
UTO	Unattended Train Operation
V2X	Vehicle-to-Everything
VHD	Very High Density
VHF	Very High Frequency
VLAN	Virtual Local Area Network
VLEO	Very Low Earth Orbit
VLF	Very Low Frequency
VLR	Visitor Location Register
VMS	Video Management System
VOIP	Voice over Internet Protocol
VPLS	Virtual Private LAN Service
VPN	Virtual Private Network
VSWR	Voltage Standing Wave Ratio
VTC	Virtual Track Circuit
WAN	Wide Area Network
WGB	Workgroup Bridge
WIFI	Wireless Fidelity
WLAN	Wireless Local Area Network
WLC	Wireless LAN Controller
WLCN	Wireless Consist Network
WSSUS	Wide-Sense Stationary Uncorrelated Scattering
WTB	Wire Train Bus
WTBN	Wireless Train Backbone
XML	Extensible Markup Language

Solutions' Manual

Exercises: Chapter 1

1. Choose any railway line near your home and find its main parameters, such as length, number of train services, maximum speed, minimum speed, average speed, number of stations, average/maximum/minimum distance between stations, and voltage (if any).

   - Madrid-Barcelona, UIC track gauge, double track high-speed line.
   - Length: 620.9 km.
   - Maximum speed: The line is designed for speeds of up to 350 km/h, but trains usually do not run faster than 300 km/h.
   - Average speed: The shortest journey time is 2:30 hours; the average speed is approximately 250 km/h.
   - Number of train services: 20 trains.
   - Electrification: 25 kV AC 50 Hz.
   - The line had +4 million passengers.
   - The line has the following stations: Madrid-Atocha, Guadalajara, Calatayud, Zaragoza, and Camp de Tarragona and Barcelona.

2. On a metro line, discuss the impact of long stops (larger than 1 min) in the headway of the line. What about only one stop with more than 3 min?

   Every railway line is as fast as the slowest train it has. If a train is significantly slower than the others or stops for a long time at the stations across the route, this will affect all the other vehicles. However, this depends on the headway between trains: it is not the same to have 60 second headways or to have 15 min.
3. Tramways usually do not have an onboard signaling system. Discuss why they do not have such a system and what should be done in order to let them have an efficient one.

   Tramway's platform is usually shared with cars, buses, etc., that is, all the other traffic in the street. This means that the safe movement of the train does not rely on the signaling system as other railway vehicles do.

   In order to have a useful signaling system in railways, a possible solution is to allocate them on a dedicated infrastructure (that is, not shared with other vehicles). The problem here are the extra costs, and the benefits are higher regularity and shorter trip times.
4. Power calculations. Consider any of the HST from Table 1 and compare it with your favorite automobile. Calculate the ratio between power and number of passengers.

   Typically, on a normal automobile, 100–200 hp (1 hp = 735.5 W) and one to five passengers. This means 14.7–147.1 kw/passenger.

   Looking at Table 1, we have 8800 kw for many high-speed trains and around 400 passengers. The ratio here is 22 kw/passenger, and certainly, traveling at a much faster speed.
5. Why railway crossings are usually not present in HSL?

   For safety reasons. Indeed, in those lines with a high interval, it is even less recommended.
6. We have explained how a single-track stretch implements the track circuit detection technology. But what about a line stretch with more than two ends (like a junction)? Draw a line with a junction and place the track circuit to detect any train on that stretch.

   This stretch has three ends. Place only one transmitter on one end and two receivers in the other two ends. The key idea is that all ends have either a transmitter or a receiver, but only one transmitter can work on a single track circuit.
7. Track circuits can be used to detect some broken rails, but not of all types. Taking into account what we have learned from this technology, draw a sketch of a broken rail not detectable by track circuits.

   One which does not break the electric circuit and allows the electric current to flow.
8. The system under study was not working for 2 hours last year. Please calculate the availability of this system.

   We need to calculate the percentage of time the system is not available per year. The availability of this system will be 100% minus the time it is not available.

J. Moreno García-Loygorri, *Telecommunication Systems for Modern Railways*, Textbooks in Telecommunication Engineering,
https://doi.org/10.1007/978-3-031-82247-6

$$A = \left(1 - \frac{2}{24 * 365}\right) \cdot 100 = 99.98\%$$

Exercises: Chapter 2

1. Which is the maximum hazard rate allowable for a SIL4 system?

   The tolerable hazard rate (THR) defined by IEC 61508 for a SIL4 function is typically estimated to be between 10^{-9} and 10^{-8} per hour.

2. Which is the maximum hazard rate allowable for a SIL4 system per month?

   As the THR is typically defined for an hour interval, if we increase the basis up to a month (24 hours/day × 30 days/month = 720 hours/month), typically, the THR will go up three orders of magnitude. However, this approach is not correct. The monthly THR must provide a figure for the probability of not having any safety issues on a complete month. Therefore, this probability is:

$$\mathrm{THR_m} = 1 - \left(1 - \mathrm{THR_h}\right)^n$$

   n is the number of hours per month, THR_h is the THR computed on an hourly basis, and THR_m is the same but on a monthly basis. If we compute THRm, we find that:

$$0.7 \bullet 10^{-6} < \mathrm{THR_m} < 0.7 \bullet 10^{-5}$$

3. To which GoA level belongs a system where the driver needs to do everything but setting the maximum speed?

   GoA1.

4. To which GoA level belongs a metro system with no human staff on board?

   GoA4.

5. What are the differences between ETCS L2 and L1?

   The main difference between them is that ETCS L2 is a continuous train protection system, whereas L1 is only punctual. This difference comes from the availability of GSM-R in the tracks.

6. Consider a train able to run up to 300 km/h and to brake at 1 m/s2. Consider that the train braking curve follows the uniformly accelerated motion (UAM) equations. Calculate the time and distance needed to fully stop the train in the worst-case scenario. Now, recalculate both time and distance if the error in speed determination is 5%.

   Time needed to fully stop: 83.33 seconds. Distance needed to fully stop: 3.472km.

   If the error in speed determination is 5%, in the worst-case scenario, the speed could be as high as 315 km/h. It takes 87.5 s to stop the train and 3.844 km. Notice the extra distance to be protected in this scenario, and the UAM model is not accurate in this scenario.

7. Read sections for ETCS L1, L2, and L3. Suggest a potential L4 and sketch the architecture. L4 should be an improvement from L2 on at least one indicator.

   An improvement is the virtual coupling between trains, which is described in Chap. 8.

8. It is required to implement an onboard PIS based on CBTC data. Please describe the information that these two systems must exchange and how often, to have the system working.

   Actually, rather than exchange information, the PIS system must receive information about the location of the train. Depending on the accuracy requirements, this information must be received more or less often. The minimum data to be included in this protocol is:

   - Timestamp
   - ID of the train
   - Location

   The details of the "location" field has to be agreed beforehand, as there may be ambiguities (the location is for the first car, the last car, etc.).

   There are many other optional fields and some other improvements, but these three are the essential ones.

9. There are punctual and continuous train protection systems. Explain the implications of having a punctual train protecting many different points instead of a continuous one.

   A punctual train protection system only protects trains punctually, which means no protection at many other places. On the other hand, the costs are remarkably shorter in the punctual one.

10. Why 100% safety is not achievable? What would be needed to reach that figure?

    Due to the high complexity and the large degrees of freedom, a priori consideration of all the potential events and situations is impossible and will probably increase the costs a lot. On the other hand, testing of all these situations would take a lot of time, even if they are not very likely to happen, so the time needed to put in service the system will be very large as well. Same follows for upgrades or updates of the system.

11. Consider a safety-critical system with very bad maintainability numbers. Does this have an impact on the safety of the system?

    Bad maintainability means high repair times. This does not impact the safety of the system, but can stress the operational aspects of the railway service.

Exercises: Chapter 3

1. If a CCTV camera is used to assist better passengers in ticket machines, to which type of service does it belong?

   Customer-oriented services (COS), as it is focused on assisting them and does not have impact on either safety or the operation itself of the railway network.

2. If a Closed-Circuit Television (CCTV) camera is used to determine the condition of the pantograph-catenary contact, to which type of service does it belong?

   Operator-oriented services (OOS), as it is focused on improving either the maintenance or the operation and does not have impact on either safety or is a service oriented to passengers.

3. Describe the difference between latency and jitter. Why is this important in voice communications?

   Latency is the delay in transmission, while jitter is the variation in the delay. Both are critical for smooth voice communications, because higher figures for both make voice communications annoying and not useful.

4. Explain the difference between public cloud and "private" (on-premise) cloud.

   Public cloud is managed and hosted by third-party providers (like Amazon, Google, etc.), accessible over the Internet. On-premise, this is, private clouds, within the enterprise own network.

   Public cloud is more flexible, but users have less control over it, while private clouds are designed and maintained for a specific purpose, but the time needed to have in service plus the expenditure costs are larger.

5. What are the differences between PIS and an intercom system?

   Intercom is for the communications between passengers and train driver (or, maybe, OCC). PIS is for announcements and providing unidirectional information to passengers.

6. What does functional addressing mean?

   Instead of static addressing, there are several forms of dynamic addressing, like functional. For example, trains receive an operational number (which may change every day or even within a day). If the OCC wants to start a call or a transmission toward a train using only this train number, it is functional addressing.

7. What does regional addressing mean?

   Regional addressing is a form of dynamic and functional addressing, but taking into account the location of one side of the communications session. For example, depending on where you start a 112 call, you may be routed to a different endpoint. Same in railways: the OCC may want to communicate to all trains that are closer to arrive to a specific station, to alert them about something.

8. What is the most challenging parameter on a gaming system?

   Latency is the most important parameter, as it impacts directly real-time responsiveness.

9. We have a switch with 12 ports, 8 of them PoE. The maximum power consumption is 103 W, but the maximum PoE power is 60 W. How many CCTV cameras can be powered over the Ethernet with this switch? Consider that the camera is IEEE 802.3af/802.3at type 1 class 3.

   From the PoE standards: IEEE 802.3af (type 1) class 3 devices can draw up to 12.95W.

   Power available per port: 60W/8 ports = 7.5W per port. This is less than class 3 needs.

   Number of cameras that can be powered:

   60W (total PoE power)/12.95W (power per class 3 device) = 4.63

   This switch can power four CCTV cameras that are IEEE 802.3af/802.3at type 1 class 3 compliant.

10. Consider a large station near your home. Try to estimate the number of potential IP end devices (escalators, elevators, CCTV cameras) in the station, and group them into types of services. How many IP addresses are needed?

    At least one IP address for each end device to be connected to the network. In large stations, this figure can be some thousands.

11. Consider a train the reader usually rides. Try to estimate the number of potential IP end devices (CCTV cameras, speakers, intercoms) onboard, and group them into types of services. How many IP addresses are needed?

    At least one IP address for each end device to be connected to the network. In modern trains, this figure can be several hundreds of end devices.

12. We have a sensing system with accelerometers on each train bogie. The sampling frequency is 2 kHz , the number of bits per sample is 12, and the number of accelerometers is 3 per bogie. If the train has 16 bogies, provide an estimation for the storage system needed to save 10 days of measurements. Consider that the raw data can be compressed up to a 25% of its initial volume, but a redundancy system 1:1 is needed as well. The file formatting overhead and the metadata inclusion imply an extra 30% volume from the raw data.

    $2000 \cdot 12 \cdot 3 \cdot 16 \cdot 10 \cdot 24 \cdot 60 \cdot 60 \cdot 0.75 \cdot 2 \cdot 1.3/8 =$ 226 GBytes. A 256 GB disk will do.

13. For the scenario described in Exercise 12, estimate the size of storage system in case we want 30 days of measurements.

    $2000 \cdot 12 \cdot 3 \cdot 16 \cdot 30 \cdot 24 \cdot 60 \cdot 60 \cdot 0.75 \cdot 2 \cdot 1.3/8 =$ 677.8 GBytes. A 1 TB disk will do.

14. We have a digital train app onboard that retrieves all MVB data from the train. We estimate that there are around 2000 variables, meaning a (more or less) constant data rate of 25 kbps. If we want to send all the data to the cloud, please estimate the monthly 4G/5G data plan needed to avoid service disruption.

    25000 · 60 · 60 · 24 · 30/8 = 7.54 GBytes/month. Notice that some other factors may arise as data compression algorithms (lowering the previous figure), header insertion in communications protocols, or retransmissions (increasing compared to the previous figure).

Exercises: Chapter 4

1. If we want to provide bidirectional connectivity to an end device using fibers, how many of them we need to segregate from the main cable? Why?
2. Two per end device, in order to establish a bidirectional communication.
3. Which is the minimum circuit that can be established on an SDH network?

   The minimum circuit in an SDH network is "STM-1," which is defined for 155.52 Mbps. However, within an STM-1 circuit, lower-level signals can be defined, E-1 (2048 kbps) being the shortest one.
4. What is an MPLS label?

   An MPLS label is an identifier assigned to a particular packet by an MPLS router. The label is used to make forwarding decisions for this packet faster than the typical methods.
5. What is label stacking in MPLS?

   Label stacking in MPLS is assigning more than one label to a packet. These labels form a label stack.
6. How are GSM-R BTS connected to the network? Explain which technology is used, to which domain are connected, and which link is used.

   The access for BTS is carried out by an STM-1 link, but the transport network that connects them all is typically a double STM-16 ring.
7. Explain the differences between OT and IT systems and why this has an impact on cybersecurity.

   Operational technology (OT) controls physical processes (like railway signals), while information technology (IT) typically deals with data.

   Regarding cybersecurity, OT puts reliability on top, while IT is more focused on confidentiality, integrity, etc.
8. Which type of MPLS router is in charge of adding the label to the packet?

   Label edge routers.

Exercises: Chapter 5

1. What is EM spectrum?

   It is a range of frequencies. Typically, in communications, it can range from some KHz (10^3) up to THz (10^{12}).
2. What is the radiation pattern of an antenna?

   It is a graph (strictly speaking, a 3D graph which is more often depicted by two of its orthogonal cuts) which informs on how an antenna radiates/receives in the angular domain.
3. In which bands does GSM-R work?

   876–880 MHz for the uplink and 921–925 MHz for the downlink. Some countries have allocated additional bands or even moved their GSM-R systems into another bands, but these bands are the most common ones.
4. How many different channel bandwidths are defined in IEEE 802.11?

   20, 40, 80, 160, and 320 MHz. However, as the capacity increases each time, we double the channel bandwidth and noise increases more, so the efficiency (capacity/bandwidth) goes down.
5. What is the Doppler effect?

   Doppler shift or Doppler effect is the frequency displacement caused by the relative speed between transmitter and the receiver.
6. Calculate the maximum Doppler shift of a signal transmitted at 2.4 GHz from a train running at 300 km/h. Is this higher or lower than the subcarrier spacing in IEEE 802.11 or LTE?

   The maximum Doppler shift in this case is: $\frac{f_c v}{c} = \frac{2.4 \cdot 10^9 \cdot 300/3.6}{3 \cdot 10^8} = 666.6\,\text{Hz}$ which is significantly less than the subcarrier spacing of both LTE and IEEE 802.11. However, offsets like this can have impact on the overall performance of the T2G system.
7. What can happen if an onboard antenna is out of the gauge envelope defined for a train?

   It will probably crash with something in the infrastructure.
8. What is the directivity of an antenna?

   Directivity is one of the most relevant parameters of an antenna. It provides information on how "focused" is the radiation pattern of an antenna on its best direction. Technically speaking, directivity is the ratio between the intensity in the best direction between all of the others.
9. What is the difference between a Ricean and a Rayleigh fading?

   The main difference between these two is that Ricean typically shows a dominant multipath component over all the others. This component is generally associated to a LOS condition, whereas Rayleigh, with no dominant component, is associated to NLOS. However, this is based on general assumptions, and the best way to distinguish them both is to calculate the Ricean factor "K."

10. What is LOS?

    LOS stands for line of sight, which means that both ends of a radio communications link have nothing in between obstructing the imaginary line that connects them both.

11. Do you think that the prevalence of LOS or NLOS conditions has an impact on the antennas to be placed onboard the train?

    In general terms, NLOS is prevalent (at least in T2G systems, which are the most relevant so far). Indeed, this has a strong impact on the antenna radiation pattern, as the power may arrive from several different angles of arrival.

12. Explain the differences between MIMO, SISO, SIMO, and MISO.

    MIMO is a system with several antenna elements on both the transmitter and the receiver ends. Multiple-input, multiple-output.

    SIMO is a system with several antenna elements only on the receiver end and one antenna element on the transmitter. Single-input, multiple-output.

    MISO is a system with several antenna elements only on the transmitter end and one antenna element on the receiver. Multiple-input, single-output.

    SISO is a system with one antenna element only on both the transmitter and the receiver ends. Single-input, single-output.

Exercises: Chapter 6

1. What is a consist?

   According to the standard, a consist is a "single vehicle or a group of vehicles which are not separated during normal operation, and which contains no, one, or several consist networks." Consists are also called train sets.

2. What is an open train and a closed train?
   - Open train: "Train composed of one or a set of consists, where the configuration may change during operation, as, for instance, locomotive hauled international UIC trains."
   - Closed train: "Train composed of one or a set of consists, where the composition does not change during normal operation, for instance metro, suburban train, or high-speed train units."

3. What is the difference between a consist network and a backbone network?
   - The consist network is a "communication network interconnecting communication devices in one consist."
   - The train backbone is a "bus connecting the vehicles of a train and which conforms to the TCN protocols."

4. Why are EMC aspects relevant in railways?

   Because of the presence of large electromagnetic fields due to electric engines onboard trains, overhead lines, and other systems.

5. Which TCN technologies are Ethernet based?

   Ethernet consist network (ECN) and Ethernet train backbone (ETB).

6. What is the main use of the EtherCAT onboard trains?

   Sensor networks.

7. What is OMTS? What are the differences between OMTS and TCMS?

   OMTS stands for "Onboard Multimedia and Telematic Services." They comprise some services which are less critical than TCMS but more demanding of data rate, like CCTV, PIS, and some others.

8. Explain what the train inauguration is.

   It is a TCMS function that discovers the train topology. It is launched in case of a change in the train composition or each time the train switches on. It has a great importance for the train operation.

9. Does the EN 50155 standard consider vibrations and EMC for onboard electronic equipment?

   Yes. Indeed, EN 50155 points to two different standards: EN 50121 for CEM and IEC 61373 for vibrations.

10. What are the differences between MVB-EMD and MVB-ESD?

    EMD has galvanic isolation and allows larger distances than ESD, which is not isolated.

11. What can happen if an ECN switch stops working?

    All end devices connected to it will stop communicating with other EDs in the TCN network. An exception for this are all the ED that implement dual homing and are connected to another switch. If the network topology is a ring or other redundant topology, the network will heal itself, enabling alternative routes rapidly that will avoid further problems.

12. What can happen if an ECN switch is powered off? Does the onboard network continue working?

    Same as when a switch stops working.

References

1. C. Wolmar, Blood, Iron and Gold: How the Railways Transformed the World, 2009
2. C. Wolmar, The Great Railway Revolution: The Epic Story of the American Railroad, 2012
3. M.S.C.C.T.M. Simon Iwnicki, *Handbook of Rail Vehicle Dynamics* (CRC Press, Boca Raton, 2020)
4. E. Hand, Arsenic heats up iron superconductors. Nature **452** (2008)
5. A. G. Álvarez, Manual de ferrocarriles. El sistema ferroviario español (2022)
6. Iowa DOT, [En línea]. Available: https://iowadot.gov/iowarail/Highway-RailRoad-Crossings/Federal-Aid-Crossing-Safety-Program-SEC-130
7. L. Charlier, V150: Power-packed train proves AGV technology in record sprint, 2007. [En línea]. Available: https://web.archive.org/web/20120227102245/http://www.railwaygazette.com/news/single-view/view/v150-power-packed-train-proves-agv-technology-in-record-sprint.html
8. A. G. Álvarez, Automatic track gauge changeover for Trains in Spain. Fundación de Ferrocarriles Españoles (2010)
9. L. L. Travelled, Artist, *Southbound Bakerloo Line Train Arrives at Kensal Green.* [Art]
10. SnippyHolloW, Artist, *IMGP1989.* [Art]
11. M. Berbineau et al., EmulRadio4Rail. Deliverable D1.1: Application layer requirements for communications systems in railway environments (stream a) (2019)
12. International Standardisation, The European Rail Association's vision, UNIFE , [En línea]. Available: https://www.unife.org/wp-content/uploads/2021/05/INTERNATIONAL-STANDARDISATION-THE-EUROPEAN-RAIL-ASSOCIATIONS-VISION.pdf
13. M. F. London, Artist, *Farringdon Elizabeth line platforms.* [Art]
14. Chelseagirl, Artist, *File:Overhead lines, Puidoux.jpg.* [Art]
15. J. Griffiths, Artist, *Standard class 5 on the Southern Region BR.* [Art]
16. Regulation of Railways Act 1889, [En línea]. Available: https://www.legislation.gov.uk/ukpga/Vict/52–53/57 (1889)
17. F. J. González, Señalización ferroviaria: del guardagujas a la operación sin, 2008
18. C.N. Pyrgidis, *Railway Transportation Systems* (CRC Press, 2016)
19. H.J. Pragnell, *Early Britich Railway Tunnels: The Implications for Planners, Landowners and Passengers Between 1830 and 1870* (University of York, 2016)
20. C. Wolmar, *Cathedrals of Steam: How London's Great Stations Were Built – and How They Transformed the City* (Atlantic Books, 2021)
21. T.R. Quinn, Train stations: iconography, wayfinding, and the evolution of a type. Univ. Arkansas Undergrad. Res. J. **9**(1), 68–77 (2008)
22. F. A. T. W. Anthony Dewar, Station design guidance. design manual NR/GN/CIV/100/02, Network Rail , 2021. [En línea]. Available: https://www.networkrail.co.uk/wp-content/uploads/2021/06/NR_GN_CIV_100_02_Station-Design.pdf
23. Railway Stations, Boosting the City, UIC Passenger Department, November 2019. [En línea]. Available: https://uic.org/IMG/pdf/2019_nextstation_railway_stations_boosting_the_city.pdf
24. UIC, High-Speed Rail: fast track to sustainable mobility, [En línea]. Available: https://uic.org/IMG/pdf/high_speed_brochure.pdf
25. J.-C. Raoul, How high-speed trains make tracks. Sci. Am., 100–105 (1997)
26. G. de Rus Mendoza (ed.), *Economic Analysis of High Speed Rail in Europe* (BBVA, 2009)
27. D. H. R. T. Lewis Pickett, The future of rail, 2022. [En línea]. Available: https://researchbriefings.files.parliament.uk/documents/CBP-8961/CBP-8961.pdf
28. A.N. John Welsby, The privatisation of Britain's railways: an inside view. JTEP **33**(1), 55–76 (1999)
29. K. Akita, Y. Hasegawa, History of COMTRAC: development of the innovative traffic-control system for Shinkansen. IEEE Ann. Hist. Comput. **38**(2), 11–21 (2015)
30. T. Ishihara, Railroads: green light for the bullets: advances in vehicles, communications, and signals keep plans for the Shinkansen line on schedule. IEEE Spectr., 71–74 (1977)
31. K. Asano, JR East high-speed rolling stock development. JR EAST Tech. Rev. **36**, 1–6 (2017)
32. IEEE 1474.1-2004, *IEEE Standard for Communications-Based Train Control (CBTC) Performance and Functional Requirements* (IEEE, 2004)
33. B. Karlsen, Railway applications – the specification and demonstration of reliability, availability, maintainability and safety (RAMS) – part 1: generic RAMS process (2024)
34. B. Karlsen, Railway applications – the specification and demonstration of reliability, availability, maintainability and safety (RAMS) – part 2: systems approach to safety (2024)
35. B. Karlsen, railway applications – the specification and demonstration of reliability, availability, maintainability and safety (RAMS) – part 1: generic RAMS process (2012)
36. EN50129, C. E. N. E. L. E. C, Railway applications – communication, signalling and processing systems – safety related electronic systems for signalling (2003)
37. Structure and content of the European Identification Number, 013SST1139, [En línea]. Available: https://www.era.europa.eu/system/files/2022-11/technical_doc_ein_codification_en.pdf
38. IEEE P1474.3, Recommended practice for communications-based train control (CBTC) system design and functional allocations, I (2018)

J. Moreno García-Loygorri, *Telecommunication Systems for Modern Railways*, Textbooks in Telecommunication Engineering,
https://doi.org/10.1007/978-3-031-82247-6

39. F.J. González, *Señalización ferroviaria: del guardagujas a la operación sin conductor* (Madrid, 2008)
40. EN 62990, Railway applications – urban guided transport management and com-mand/control systems
41. IEEE P1474.2, *Standard for User Interface Requirements in Communications-Based Train Control (CBTC) Systems* (IEEE, 2018)
42. IEEE P1474.4, *Recommended Practice for Functional Testing of a Communications-Based Train Control (CBTC) System* (IEEE, 2023)
43. Profibus – Profinet, [En línea]. Available: http://www.profibus.org/
44. CCS TSI, Appendix A – mandatory specifications (ETCS B4 R1, RMR: GSM-R B1 MR1 + FRMCS B0, ATO B1 R1), [En línea]. Available: https://www.era.europa.eu/era-folder/1-ccs-tsi-appendix-mandatory-specifications-etcs-b4-r1-rmr-gsm-r-b1-mr1-frmcs-b0-ato-b1
45. System Requirements Specification, Version 16.1.0, [En línea]. Available: https://www.era.europa.eu/system/files/2023-09/index033_-_EIRENE_SRS_v1610.pdf
46. Functional Requirements Specification, Version 8.1.0, [En línea]. Available: https://www.era.europa.eu/system/files/2023-09/index032_-_EIRENE_FRS_v810.pdf
47. M. Ghazel, Formalizing a subset of ERTMS/ETCS specifications for verification purposes. Transp. Res. Part C Emerg. Technol., 60–75 (2014)
48. Nooberella, Artist, *Blank map Europe with Artsakh.* [Art]
49. ETCS Implementation Handbook, UIC, Paris, France (2008)
50. En línea., Available: https://www.era.europa.eu/content/braking-curves#:~:text=The%20ETCS%20braking%20curves%20algorithm,the%20ETCS%20on%2Dboard%20equipment
51. J. S. Stover, Bombardier Cityflow 650: System Overview, 2008
52. T.Y.K.F.S.O.S.L. Kenta Nakashima, New CBTC system for smart operation. Hitachi Rev. **67**(7), 66–70 (2018)
53. B. Ning, T. Tang, K. Qiu, C. Gao, Q. Wang, CTCS – Chinese train control system. WIT Trans. Built Envir. **74** (2004)
54. ATACS: The Japanese Level 3, [En línea]. Available: https://www.railengineer.co.uk/atacs-the-japanese-level-3/
55. En línea., Available: https://www.transportation.gov/tags/ptc
56. En línea., Available: https://www.era.europa.eu/system/files/2023-01/sos1_index001_-_era_ertms_003204_v500.pdf
57. En línea., Available: https://uic.org/IMG/pdf/srs-16.0.0_uic_951-0.0.2_final.pdf
58. ETCS Implementation Handbook, UIC, Paris, France (2008)
59. En línea., Available: https://www.ertms.net/wp-content/uploads/2021/07/ERTMS_Factsheet_8_UNISIG.pdf
60. 802.3at-2009, IEEE Standard for Information technology (2009)
61. 802.3af-2003, *IEEE Standard for Information Technology – Telecommunications and Information Exchange Between Systems* (IEEE, 2003)
62. I.802.3bt-2018, *EEE Standard for Ethernet Amendment 2: Physical Layer and Management Parameters for Power Over Ethernet Over 4 Pairs* (IEEE, 2018)
63. EN 50155:2018, Railway applications – rolling stock – electronic equipment (2018)
64. EN 62262:2002, Degrees of protection provided by enclosures for electrical equipment against external mechanical impacts (IK Code) (2002)
65. EN 60529:2018, Degrees of protection provided by enclosures (IP Code) (2018)
66. P. Read, M.-P. Meyer, Restoration of motion picture film, in *Conservation and Museology*, (2000), pp. 24–26
67. EN 62676-1-1:2015, Ideo surveillance systems for use in security applications – part 1–1: system requirements – general
68. EN 12299:2010, Railway applications – ride comfort for passengers – measurement and evaluation (2010)
69. EN 14363:2017, Railway applications – testing and simulation for the acceptance of running characteristics of railway vehicles – running behaviour and stationary tests (2017)
70. P. N. Abad, Artist, *Tren laboratorio A330 Séneca. Antequera.* [Art]
71. V. R. H. José María Muñiz Aza, Las telecomunicaciones y el ferrocarril, *Cuadernos técnicos.* Fundacion de Ferrocarriles Españoles, vol. Junio, pp. 3–36 (2011)
72. EN 16334-1:2015, Railway applications – passenger alarm system – part 1: system requirements for mainline rail
73. EN 16334-2:2020, Railway applications – passenger alarm system – part 2: system requirements for urban rail
74. IEC 60268-16:2020, Sound system equipment – part 16: objective rating of speech intelligibility by speech transmission index
75. EN 62267:2011, Railway applications – automated urban guided transport (AUGT) – safety requirements
76. EN 50849:2017, Sound systems for emergency purposes
77. EN 61131-1:2004, Programmable controllers – part 1: general information
78. EN 61131-3:2013, Programmable controllers – part 3: programming languages
79. EN 61131-5:2001, Programmable controllers – part 5: communications
80. Bosch. Technical documentation for video analytics, [En línea]. Available: https://www.boschsecurity.com/xc/en/solutions/video-systems/video-analytics/technical-documentation-for-video-analytics/
81. V.A.M. Moreno, Design and implementation of a telecommunications network high-speed rail. Revista Vía Libre Técnica, 49–70 (2012)
82. M.T.R.N. Dimitra Liveri, *Railway Cybersecurity: Security Measures in the Railway Transport Sector* (ENISA, 2020)
83. TS 50701, Railway applications – cybersecurity, european committee for elec-trotechnical standardization technical specification, standard CLC/TS 50701, CENELEC (2020)
84. D.M.M.H.J.I. Simone Soderi, Cybersecurity considerations for communication-based train control. IEEE Access, 92312–92321 (2023)
85. EN 50126-1, Railway applications – the specification and demonstration of reliability, availability, maintainability and safety (RAMS) – part 1: generic RAMS process
86. IEC 62443, Industrial communication networks
87. J. Braband, A Survey of Cybersecurity in Railway Signaling, [En línea]. Available: https://www.discotec.org/2019/slidestoshare/Braband_DiscoRail2019.pdf
88. ISO/IEC 27001:2022, Information security, cybersecurity and privacy protection – information security management systems – requirements
89. S. Soderi, D. Masti, Y.Z. Lun, Railway cyber-security in the era of interconnected systems: a survey. IEEE Trans. Intell. Transp. Syst. **24**(7), 6764–6779 (2023)
90. C. Squatriglia, Polish Teen Hacks His City's Trams, Chaos Ensues, Wired.com, 11 April 2008. [En línea]. Available: https://www.wired.com/2008/01/polish-teen-hac/
91. H. Huifeng, Passenger Wi-fi freezes third Shenzhen Metro train in a week, scmp.com, 09 November 2012. [En línea]. Available: https://www.scmp.com/news/china/article/1078165/passenger-wi-fi-freezes-third-shenzhen-metro-train-week
92. N. Levanti, UK railways had four major cyber-attacks in one year, Infrastructure intelligence, 20 July 2016. [En línea]. Available: http://www.infrastructure-intelligence.com/article/jul-2016/uk-railways-had-4-major-cyberattacks-one-year
93. S. Gibbs, Ransomware attack on San Francisco public transit gives everyone a free ride, TheGuardian, 28 November 2016. [En línea]. Available: https://www.theguardian.com/technology/2016/nov/28/passengers-free-ride-san-francisco-muni-ransomeware

94. A. Greenberg, The cheap radio hack that disrupted Poland's railway system, Wired.com, 27 August 2023. [En línea]. Available: https://www.wired.com/story/poland-train-radio-stop-attack/
95. M. S. Curtis, A. Siller Jr, SONET/SDH: A Sourcebook of Synchronous Networking (1996)
96. L.D. Ghein, *MPLS Fundamentals* (Cisco Press, 2007)
97. Railway Cybersecurity Training, [En línea]. Available: https://www.railway-cybersecurity.com/Railway_cybersecurity.html
98. ENISA, Videos & presentations – ENISA-ERA conference: cybersecurity in railways, ERA, 2021. [En línea]. Available: https://www.enisa.europa.eu/events/ENISA-ERA_Conference/Agenda_ENISA-ERA_Conference
99. 4SECURAIL Project, [En línea]. Available: http://www.4securail.eu/
100. A.F. Molisch, *Wireless Communications* (Wiley, 2011)
101. J. Parsons, *The Mobile Radio Propagation Channel* (Wiley, 2000)
102. T.K. Sarkar, R. Mailloux, A.A. Oliner, M. Salazar-Palma, D.L. Sengupta, *History of Wireless* (Wiley-IEEE Press, 2006)
103. mnchilemom, Artist, *IMG_2023-B Very Large Array.* [Art]
104. F. C. Comission, Radio Spectrum Allocation, [En línea]. Available: https://www.fcc.gov/engineering-technology/policy-and-rules-division/general/radio-spectrum-allocation
105. X. Zou, W. Bai, W. Chen, P. Li, B. Lu, G. Yu, W. Pan, B. Luo, L. Yan, L. Shao, Microwave photonics for featured applications in high-speed railways: communications, detection, and sensing. J. Lightwave Technol., 4337–4346 (2018)
106. K. Smith, Europe boosts spectrum allocation for Intelligent Transport Systems, https://www.railjournal.com/, 19 October 2020. [En línea]. Available: https://www.railjournal.com/telecoms/europe-boosts-spectrum-allocation-for-intelligent-transport-systems/
107. E. T. 1. 5. V1.1.1, Urban Rail ITS and Road ITS applications in the 5.9 GHz band. Investigations and use of spectrum., ETSI, 2019
108. P. Bello, Characterization of randomly time-variant linear channels. IEEE Trans. Commun. Syst. **11**(4), 360–393 (1963)
109. S. Salous, *Radio Propagation Measurement and Channel Modelling* (Wiley, 2013)
110. Wireless World Initiative New Radio, part II. Winner II
111. ETSI TR 138 900 V14.2.0. LTE; 5G, Study on channel model for frequency spectrum above 6 GHz (3GPP TR 38.900 version 14.2.0 Release 14)
112. M. Hata, Empirical formula for propagation loss in land mobile radio services. IEEE Trans. Veh. Technol., 317–325 (1980)
113. EmulRadio4Rail Project, D1.3 Characterization of the railway environment: channel models & general characteristics
114. H. Qiu, J.M. Garcia-Loygorri, K. Guan, D. He, Z. Xu, B. Ai, M. Berbineau, Emulation of radio technologies for railways: a tapped-delay-line channel model for tunnels. IEEE Access, 1512–1523 (2020)
115. M. Berbineau, R. Behaegel, J.M. Garcia-Loygorri, R. Torrego, R. D'Errico, A. Sabra, Y. Yan, J. Soler, Channel models for performance evaluation of wireless systems in railway environments. IEEE Access **9**, 45903–45918 (2021)
116. C.A. Balanis, *Antenna Theory: Analysis and Design*, 3rd edn. (Wiley, 2005)
117. T.A. Milligan, *Modern Antenna Design* (IEEE Press, 2005)
118. D. Chizhik, G.J. Foschini, M.J. Gans, R.A. Valenzuela, Keyholes, correlations and capacities of multielement transmit and receive antennas. IEEE Trans. Wirel. Commun. **1**(2), 361–368 (2002)
119. J. Moreno, L. de Haro, C. Rodriguez, L. Cuéllar, J.M. Riera, Keyhole estimation of a MIMO-OFDM train-to-wayside communication system on subway tunnels. IEEE Antenn. Wirel. Propag. Lett. **14**, 88–91 (2015)
120. L. Zheng, D. Tse, Diversity and multiplexing: a fundamental tradeoff in multiple-antenna channels. IEEE Trans. Inf. Theory **49**(5), 1073–1096 (2003)
121. C.E. Shannon, A mathematical theory of communication. Bell Syst. Tech. J. **27**, 623–656 (1948)
122. A. Goldsmith, S.A. Jafar, N. Jindal, S. Vishwanath, Capacity limits of MIMO channels. IEEE J Sel Areas Commun **21**(5), 684–702 (2003)
123. G.Z. Muñoz, Experiencias en la construcción de redes GSM-R en las líneas de alta velocidad de ADIF. BIT **157**, 53–55 (2006)
124. EN 301515 V2.3.0, Global System for Mobile communication (GSM); Requirements for GSM operation on railways
125. FINAL REPORT TR 1038, MORANE "MObile radio for RAilway Networks in Europe", TELEMATICS APPLICATIONS PROGRAMME (1999)
126. Functional Requirements Specification. Version 8.1.0
127. System Requirements Specification. Version 16.1.0
128. M.-B.P. Michel Mouly, *The GSM System for Mobile Communications* (Telecom Publishing, 1992)
129. U. E.-R. O. Group y G.-R. I. Group, FFFS for Voice and Data Services Functionality at borders between GSM-R networks, Available: https://uic.org/IMG/pdf/gsmr8300-2.0.pdf (2014)
130. I. D. M. Ariza, Analisis, diseño e implementacion de una red de comunicaciones moviles GSM-R en un tramo ferroviario de red convencional (2020)
131. What is Work Group Bridge? [En línea]. Available: https://www.kareemccie.com/2018/05/what-is-work-group-bridge.html
132. Aruba VHD VRD Theory Guide, [En línea]. Available: https://howwirelessworks.com/wp-content/uploads/Aruba_VHD_VRD_Theory_Guide.pdf
133. Aruba VHD VRD Engineering Configuration Guide, [En línea]. Available: https://higherlogicdownload.s3.amazonaws.com/HPE/MigratedAttachments/7BC8710B-BC01-4229-A170-41F8F5A5E6F8-4-Aruba_VHD_VRD_Engineering_Configuration_Guide.pdf
134. Technical Committee (TC) Terrestrial Trunked Radio and Critical Communications Evolution (TCCE), [En línea]. Available: https://www.etsi.org/committee/1416-tcce
135. J. M. M. Espinosa, TETRA: servicios de radio digital para el transporte, *International Rail Forum* (2010)
136. S.P.J.S.E. Dahlman, *4G: LTE/LTE-Advanced for Mobile Broadband* (Academic Press, 2013)
137. J.D.O. Sanchez, J.I. Alonso, A two-hop MIMO relay architecture using LTE and millimeter wave bands in high-speed trains. IEEE Trans. Veh. Technol. **68**(3), 2052–2065 (2019)
138. Y. Chen, K. Niu, Z. Wang, Adaptive handover algorithm for LTE-R system in high-speed railway scenario. IEEE Access **9**, 59540–59547 (2021)
139. Tren-tierra radiotelefonía ferroviaria, [En línea]. Available: https://www.trenvista.net/a-fondo/tren-tierra-radiotelefonia-ferroviaria/
140. EN IEC 61375-2-6:2018, Electronic railway equipment – train communication network (TCN) – part 2–6: on-board to ground communication, Copyright © 2012 IEC Geneva, Switzerland. www.iec.ch
141. Cisco competitors analysis, [En línea]. Available: https://business-chronicler.com/competitors/cisco-competitors-analysis/
142. D5: FINAL REPORT Coexistence of GSM-R with other Communication Systems ERA 2015 04 2 SC, ERA (2016)
143. EN 61375-1:2012, Electronic railway equipment – train communication network (TCN) – part 1: general architecture., Copyright © 2012 IEC Geneva, Switzerland. www.iec.ch
144. EN 61375-2-1:2012, Electronic railway equipment – train communication network (TCN) – part 2–1: Wire Train Bus (WTB)., Copyright © 2012 IEC Geneva, Switzerland. www.iec.ch
145. EN 61375-3-3:2012, Electronic railway equipment – train communication network (TCN) – part 3–3: CANopen Consist Network (CCN)., Copyright © 2012 IEC Geneva, Switzerland. www.iec.ch

146. IEC 61375-3-1:2012, Electronic railway equipment – train communication network (TCN) – part 3–1: Multifunction Vehicle Bus (MVB), Copyright © 2012 IEC Geneva, Switzerland. www.iec.ch
147. EN 61375-3-4:2014, Electronic railway equipment – train communication network (TCN) – part 3–4: Ethernet Consist Network (ECN)., Copyright © 2012 IEC Geneva, Switzerland. www.iec.ch
148. IEC 61375-2-5:2014, Electronic railway equipment – train communication network (TCN) – part 2–5: ethernet train backbone., Copyright © 2012 IEC Geneva, Switzerland. www.iec.ch
149. RS-485 cable specification guide [En línea]. Available: https://www.analog.com/en/technical-articles/rs485-cable-specification-guide%2D%2Dmaxim-integrated.html
150. ISO 11898-1:2015, Road vehicles. Controller area network (CAN). Part 1: data link layer and physical signalling
151. J. C. A. B. Han, Research on MVB conformance test system, *MATEC Web of Conferences* (2022)
152. ISO/IEC 13239:2002, Information technology. Telecommunications and information exchange between systems. High-level data link control (HDLC) procedures
153. EN 50155:2021, Railway applications – rolling stock – electronic equipment
154. EN 61373:2011, Railway applications – rolling stock equipment – shock and vibration tests, Copyright © 2012 IEC Geneva, Switzerland. www.iec.ch
155. EN 50121-3-1:2017, Railway applications – electromagnetic compatibility – part 3–1: rolling stock – train and complete vehicle
156. EN 50121-4:2017, Railway applications – electromagnetic compatibility – part 4: emission and immunity of the signalling and telecommunications apparatus
157. EN 50121-5:2018, Railway applications – electromagnetic compatibility – part 5: emission and immunity of fixed power supply installations and apparatus
158. EN IEC 61158-1:2019, Industrial communication networks – fieldbus specifications – part 1: overview and guidance for the IEC 61158 and IEC 61784 series
159. International Railway Industry Standard (IRIS), [En línea]. Available: https://www.iris-rail.org/
160. ERTMS Regional – Status and Future Enhancements, *UIC ERTMS World Congress* (2014)
161. Y. Alsaba, M. Berbineau, I. Dayoub, E. Masson, G.M. Adell, E. Robert, 5g for remote driving of trains, in *International Workshop on Communication Technologies for Vehicles*, (Springer International Publishing, Cham, 2020), pp. 137–147
162. C.G. Ruiz, J.M. García-Loygorri, B.M. Velázquez, Train-to-train connectivity for safety-critical use cases, in *International Workshop on Communication Technologies for Vehicles*, (Springer International Publishing, Cham, 2021), pp. 71–82
163. J. Moreno, J.M. Riera, L. De Haro, C. Rodriguez, A survey on the future railway radio communications services: challenges and opportunities. IEEE Commun. Mag., 62–68 (2015)
164. U. Bock, J.U. Varchmin, *Enhancement of the Occupancy of Railroads Using Virtually Coupled Train Formations* (World Congress on Railway Research, 1999)
165. S.A. Canesi, X2Rail-3 (GA No 826141), D6.1 – virtual train coupling system concept and application conditions (2020)
166. MOVINGRAIL (GA No 826347), D3.1 – virtual coupling communication solutions analysis (2018)
167. M. Schenker, S. Stickel, H. Dittus, S. Canesi, S.D. Iovino, V. Riquier, F. Parrilla Ayuso, J. Goikoetxea, Virtually coupled train sets – a comprehensive analysis. WCRR, 1–6 (2022)
168. J.M. García-Loygorri, J. Goikoetxea, E. Echeverría, A. Arriola, I. Val, S. Sand, P. Unterhuber, F. Del Rio, The wireless train communication network: Roll2Rail vision. IEEE Veh. Technol. Mag. **13**(3), 135–143 (2018)
169. J. Felez, M.A. Vaquero-Serrano, Virtual coupling in railways: a comprehensive review. Mach. Des., 1–37 (2023)
170. Q. Wu, X. Ge, Q.L. Han, Y. Liu, Railway virtual coupling: a survey of emerging control techniques. IEEE Trans. Intell. Veh. **8**(5), 3239–3255 (2023)
171. V. Nikolopoulou, D. Mandoc, S. Gogos, 5GRAIL paves the way to the future railway mobile communication system introduction, in *2022 IEEE Future Networks World Forum (FNWF)*, p. 2022
172. 802.11ad-2012, IEEE standard for information technology – telecommunications and information exchange between systems – local and metropolitan area networks – specific requirements-part 11: Wireless LAN Medium Access Control (MAC) and Physical Layer (PHY)
173. J. Goikoetxea, Shift2Rail CONNECTA: the next generation of the train control and monitoring system, in *Proceedings of 7th Transport Research Arena TRA 2018, April 16–19, 2018, Vienna, Austria*
174. M. Kozlenko, 2020-08-01 Buffers and chain coupling on train carriage at at New South Wales Rail Transport Museum 4 [Art]
175. The Cambridge Dictionary, [En línea]. Available: https://dictionary.cambridge.org/. [Último acceso: 2023]
176. En línea., Available: https://es.wikipedia.org/wiki/TGV_Postal#/media/Archivo:TGV_la_poste_3.jpg
177. Shift2Rail – Europe's Rail, [En línea]. Available: https://projects.shift2rail.org/
178. About UIC, [En línea]. Available: https://uic.org/about/about-uic/
179. CEN-CENELEC, [En línea]. Available: https://www.cencenelec.eu/
180. En línea. Available: https://en.wikipedia.org/wiki/File:Track_gauge.svg
181. A. Đ. HRT, [En línea] (2019). Available: https://en.wikipedia.org/wiki/File:Level_crossing_in_Hanoi.jpg
182. T. Page, [En línea] (2019). Available: https://en.wikipedia.org/wiki/File:Crossrail_platform_at_Farringdon.jpg
183. Jailbird, [En línea]. Available: https://en.wikipedia.org/wiki/File:Feste_Fahrbahn_FFB%C3%B6gl.jpg
184. BesigedB, [En línea] (2005). Available: https://commons.wikimedia.org/wiki/File:Rails.and.ballast.bb.jpg
185. ChelseaGirl, [En línea]. Available: https://en.wikipedia.org/wiki/File:Overhead_lines,_Puidoux.jpg
186. Tennen-Gas, [En línea]. Available: https://en.wikipedia.org/wiki/File:Overhead_lines_JR_West_001.JPG
187. Rail privatisation: top 30 failures in 30 years, [En línea] (2023). Available: https://weownit.org.uk/blog/rail-privatisation-top-30-failures-30-years
188. UIC Webiste, [En línea]. Available: https://uic.org/
189. Wikipedia: Railway speed record, [En línea]. Available: https://en.wikipedia.org/wiki/Railway_speed_record
190. O. Wallach, Visualizing the Fastest Trains in the World, Visual Capitalist, [En línea]. Available: https://www.visualcapitalist.com/visualizing-the-fastest-trains-in-the-world/
191. Wikipedia: List of track gauges, [En línea]. Available: https://en.wikipedia.org/wiki/List_of_track_gauges
192. How Much Do You Know About Railway Track Gauge? Railroad Rails, [En línea]. Available: https://railroadrails.com/knowlege/railway-track-gauge/
193. En línea., Available: https://en.wikipedia.org/wiki/Maglev#/media/File:A_maglev_train_coming_out,_Pudong_International_Airport,_Shanghai.jpg
194. En línea., Available: https://en.wikipedia.org/wiki/Loading_gauge#/media/File:Why_London_Underground_is_nicknamed_The_Tube.jpg
195. En línea., Available: https://en.wikipedia.org/wiki/High-speed_rail#/media/File:Viaduc_de_la_Haute-Colme_01_09.jpg

196. Pantoine, Wikimedia, [En línea]. Available: https://en.wikipedia.org/wiki/Train_wheel#/media/File:Rollingstock_axle.jpg
197. En línea., Available: https://lb.m.wikipedia.org/wiki/Fichier:Fst_ICE_3MF.jpg
198. En línea., Available: https://es.wikipedia.org/wiki/Anuncio_de_Se%C3%B1ales_y_Frenado_Autom%C3%A1tico#/media/Archivo:ASFA_-_Baliza.jpg
199. En línea., Available: https://www.railway-technology.com/projects/european-rail-traffic-management-system-ertms/?cf-view
200. ERTMS.net, [En línea]. Available: ertms.net
201. Alstom Solutions, [En línea]. Available: https://www.alstom.com/solutions/signalling/alstom-cbtc-range-world-leading-high-capacity-signalling
202. En línea., Available: https://www.ertms.net/wp-content/uploads/2021/06/24.-ERTMS-in-China.pdf
203. En línea., Available: https://www.hitachi.com/rev/archive/2017/r2017_02/11/index.html
204. En línea., Available: https://www.ertms.net/facts-figures/deployment-statistics
205. En línea., Available: https://www.ertms.net/newsroom/ertms-news/
206. ERTMS History, [En línea]. Available: https://www.ertms.net/wp-content/uploads/2021/06/8.-ERTMS-History.pdf
207. En línea., Available: https://www.railway-technology.com/projects/european-rail-traffic-management-system-ertms/
208. Study on feasibility of SATCOM for railway communication, [En línea] (2016). Available: https://www.era.europa.eu/system/files/2022-11/Study%20on%20feasibility%20of%20satcom%20for%20railway%20applications%20by%20INDRA_ALG_0.pdf
209. En línea., Available: https://www.railwaypro.com/wp/ertms-the-eu-large-scale-project
210. ERTMS para dummies, En línea. Available: https://www.geotren.es/blog/ertms-para-dummies/
211. EN 12299:2010, Railway applications – ride comfort for passengers – measurement and evaluation
212. General Data Protection Regulation, [En línea]. Available: https://gdpr-info.eu/
213. Relational database, [En línea]. Available: https://en.wikipedia.org/wiki/Relational_database
214. En línea. Available: https://es.wikipedia.org/wiki/Serie_330_de_ADIF#/media/Archivo:Albacete_(10965996115).jpg
215. En línea., Available: https://en.wikipedia.org/wiki/File:DFI_D%C3%BCsseldorf_1.jpg
216. Bosch. Lens calculator, [En línea]. Available: https://www.boschsecurity.com/LensCalculator/html/lens-calculator.html
217. W.H. Hay, The beginnings of telephone dispatching. Railroad History **130**, 55–60 (1974)
218. En línea., Available: https://en.wikipedia.org/wiki/Optical_fiber#/media/File:Optical-fibre.svg
219. Wikipedant, [En línea]. Available: https://en.wikipedia.org/wiki/File:USA.NM.VeryLargeArray.02.jpg
220. Huber-Suhner Sencity Antenna, [En línea]. Available: https://www.hubersuhner.com/en/shop/product/antennas/antennas/vehicles/84027011/sencity-rail-excel-antenna
221. En línea., Available: https://www.engeniustech.com/span/go-guide-channel-transmit-power-wi-fi-networks-2/
222. En línea., Available: https://help.keenetic.com/hc/en-us/articles/360012060379-Available-channels-on-the-5-GHz-Wireless-network
223. Workgroup Bridge Mode, Cisco, [En línea]. Available: https://www.cisco.com/c/en/us/td/docs/routers/access/wireless/software/guide/RolesWGB.html
224. En línea., Available: https://study-ccnp.com/wi-fi-device-location-tracking-in-wireless-networks
225. LTE Network Architecture, tutorialspoint.com, [En línea]. Available: https://www.tutorialspoint.com/lte/lte_network_architecture.htm
226. En línea., Available: https://es.wikipedia.org/wiki/Acceso_m%C3%BAltiple_por_divisi%C3%B3n_de_frecuencias_ortogonales#/media/Archivo:OFDMA_subcarriers.png
227. L. Alliance. [En línea]. Available: https://lora-alliance.org/
228. LoRa architecture, [En línea]. Available: https://www.3glteinfo.com/lora/lora-architecture/
229. The Future of 5 G and LoRaWAN ® Friends or Foes?, [En línea]. Available: https://blog.semtech.com/the-future-of-5g-and-lorawan-friends-or-foes
230. Semtech: LoRa, [En línea]. Available: https://semtech com/LoRa
231. LoRa alliance, Case study paper, [En línea]. Available: https://lora-alliance.org/wp-content/uploads/2021/03/GS1-LoRa-Alliance-Case-Study-Paper-1.pdf
232. En línea., Available: https://eprints.lse.ac.uk/118234/1/Myers_spectrum_auctions_published.pdf
233. En línea., Available: https://www.fcc.gov/auctions
234. En línea., Available: https://www.cramton.umd.edu/papers2000-2004/01hte-spectrum-auctions.pdf
235. En línea., Available: https://www.policytracker.com/blog/the-top-five-biggest-spectrum-issues-of-2021/
236. En línea., Available: https://priceonomics.com/the-spectrum-auction-how-economists-saved-the-day/
237. En línea., Available: https://milgrom.people.stanford.edu/wp-content/uploads/1998/01/Game-Theory-and-the-Spectrum-Auctions-1998.pdf
238. M. Sauter, Wireless Moves. Cell Site Power Consumption, [En línea]. Available: https://blog.wirelessmoves.com/2019/08/cell-site-power-consumption.html
239. ETCS and its viability over TETRA, 4 December 2018. [En línea]. Available: https://www.teltronic.es/etcs-y-su-viabilidad-en-tetra/
240. Test prove TETRA as good or better than GSM-R, African Wireless Communications, 09 November 2018. [En línea]. Available: https://www.africanwirelesscomms.com/product-service-details?itemid=2278&post=tests-%E2%80%98prove%E2%80%99-tetra-as-good-or-better-than-gsm-rtests-%E2%80%98prove%E2%80%99-tetra-as-good-or-better-than-gsm-r-488219
241. Wikipedia: List of WLAN Channels, [En línea]. Available: https://en.wikipedia.org/wiki/List_of_WLAN_channels
242. Wikipedia: CAN bus, [En línea]. Available: https://en.wikipedia.org/wiki/CAN_bus
243. En línea., Available: https://en.wikipedia.org/wiki/IEC_metric_screw_sized_connectors#/media/File:M12A5.webp
244. 802.1Q-2014 – Bridges and Bridged Networks, [En línea]. Available: https://www.ieee802.org/1/pages/802.1Q-2014.html
245. Shift2Rail, [En línea]. Available: https://projects.shift2rail.org/
246. Europe's Rail Joint Undertaking, [En línea]. Available: https://rail-research.europa.eu/
247. ETCS Implementation Handbook, UIC, Paris (2008)
248. Satloc: a high-tech saviour for low-density lines, International Railway Journal, [En línea]. Available: http://www.railjournal.com/index.php/telecoms/satloc-a-hightech-saviour-for-low-density-lines.html
249. 3INSAT – Train Integrated Safety Satellite System, [En línea]. Available: https://artesapps.esa.int/projects/3insat
250. LOCOLOC Project, [En línea]. Available: http://esamultimedia.esa.int/docs/NavigationProjects/Locoloc_final_presentn_dec04%20.pdf
251. En línea., Available: https://www.railjournal.com/technology/jr-central-trials-ato-at-goa2-on-tokaido-shinkansen/
252. K. Smith, Remote driving clears first hurdle. Int. Railw. J. (2020)

253. En línea., Available: https://ec.europa.eu/research/participants/data/ref/h2020/wp/2014_2015/annexes/h2020-wp1415-annex-g-trl_en.pdf
254. En línea., Available: https://twitter.com/goikotrains/status/1038396656507670528
255. UIC, [En línea]. Available: https://uic.org/IMG/pdf/brochure_frmcs_v2_web.pdf
256. En línea., Available: https://en.wikipedia.org/wiki/Buffers_and_chain_coupler#/media/File:Chain_coupler_detail.jpg
257. SNMPv3, [En línea]. Available: https://www.rfc-editor.org/info/std62
258. B. S. Mark Curran, Basics of Fiber Optics
259. M. G.-J. Veci, Estudio y comparativa de señalización en un tramo de línea de alta velocidad, 2015
260. IEC 61375-1, Electronic railway equipment – train communication network (TCN) – part 1: general architecture, Copyright © 2012 IEC Geneva, Switzerland. www.iec.ch

Index

J. Moreno García-Loygorri, *Telecommunication Systems for Modern Railways*, Textbooks in Telecommunication Engineering,
https://doi.org/10.1007/978-3-031-82247-6

MIX
Papier aus verantwortungsvollen Quellen
Paper from responsible sources
FSC® C105338

If you have any concerns about our products,
you can contact us on
ProductSafety@springernature.com

In case Publisher is established outside the EU,
the EU authorized representative is:
Springer Nature Customer Service Center GmbH
Europaplatz 3, 69115 Heidelberg, Germany

Printed by Libri Plureos GmbH
in Hamburg, Germany